AF292978

Springer

London
Berlin
Heidelberg
New York
Barcelona
Budapest
HongKong
Milan
Paris
Santa Clara
Singapore
Tokyo

Maria Marinaro and Roberto Tagliaferri (Eds)

NEURAL NETS
WIRN VIETRI-96

Proceedings of the 8th Italian Workshop on Neural
Nets, Vietri sul Mare, Salerno, Italy, 23–25 May 1996

International Institute for Advanced Scientific Studies
"E.R. Caianiello" (IIASS)
Istituto Italiano per gli Studi Filosofici (Naples)
Dept. of Informatics and Applications, University of Salerno
Dept. of Theoretical Physics and S.M.S.A, University of Salerno
Artificial Intelligence Software – AIS S.p.A.
Dept. of Information Science, University of Milan
IEEE NNC
IRSIP – CNR (Naples)
Società Italiana Reti Neuroniche (SIREN)

Springer

Professor Maria Marinaro
Dipartimento di Fisica Teorica e S.M.S.A.,
Università di Salerno, 84081 Baronissi (SA), Italy
and
IIASS "E.R. Caianiello", Via G. Pellegrino 19, 84019 Vietri sul Mare, (SA),
Italy

Dr Roberto Tagliaferri
Dipartimento di Informatica ed Applicazioni "R.M. Capocelli"
Università di Salerno, 84081 Baronissi (SA), Italy
and
IIASS "E.R. Caianiello", Via G. Pellegrino 19, 84019 Vietri sul Mare, (SA),
Italy

Series Editor

J.G. Taylor, BA, BSc, MA, PhD, FInstP
Director, Centre for Neural Networks,
Department of Mathematics, King's College,
Strand, London WC2R 2LS, UK

ISBN-13: 978-1-4471-1240-2   e-ISBN-13: 978-1-4471-0951-8
DOI: 10.1007/978-1-4471-0951-8

British Library Cataloguing in Publication Data
    Neural nets : Wirn Vietri-96 : proceedings of the 8th Italian Workshop on Neural
    Nets, Vietri Sul Mare, Salerno, Italy, 23-25 May 1996
    1. Neural networks (Computer science)
    I. Marinaro, Mario II. Tagliaferri, Roberto III. Workshop on Neural Networks (1996 :
    Vietri, Italy)
    006.3

Library of Congress Cataloging-in-Publication Data
A Catalog record for this book is available from the Library of Congress

Typesetting: Camera ready by contributors

34/3830-543210 Printed on acid-free paper

PREFACE

This volume contains the Proceedings of the *8th Italian Workshop on Neural Nets WIRN VIETRI-96*, organized by the International Institute for Advanced Scientific Studies "Eduardo R. Caianiello" and by the Società Italiana Reti Neuroniche (SIREN). We would like to thank Professor L.B. Almeida who agreed to deliver an invited lecture which is included in these Proceedings. The spectrum of contributors and participants covers the activity of Italian research in the field and of some European groups. The highly qualified and motivated attendance is proof of the interest with which this annual event has met in the Italian scientific community.

Maria Marinaro
Roberto Tagliaferri

Organizing – Scientific Committee: B. Apolloni *(Univ. Milano)*, A. Bertoni *(Univ. Milano)*, D.D. Caviglia *(Univ. Genova)*, P. Campadelli *(Univ. Milano)*, M. Ceccarelli *(CNR Napoli)*, A. Colla *(ELSAG – Genova)*, M. Frixione *(IIASS)*, C. Furlanello *(IRST – Trento)*, G.M. Guazzo *(IIASS)*, M. Gori *(Univ. Firenze)*, F. Lauria *(Univ. Napoli)*, M. Marinaro *(Univ. Salerno – IIASS)*, F. Masulli *(Univ. Genova)*, P. Morasso *(Univ. Genova)*, G. Orlandi *(Univ. Roma)*, E. Pasero *(Politecnico Torino)*, A. Petrosino *(CNR Napoli)*, M. Protasi *(Univ. Roma II)*, S. Rampone *(IIASS)*, R. Serra *(Gruppo Ferruzzi Ravenna)*, F. Sorbello *(Univ. Palermo)*, R. Stefanelli *(Politecnico Milano)*, R. Tagliaferri *(Univ. Salerno)*, R. Vaccaro *(CNR Napoli)*.

Referees: Apolloni B., Bertoni A., Campadelli P., Cantoni V., Caviglia D., Ceccarelli M., Cesa-Bianchi N., Colla A.M., d'Acierno A., Di Claudio E., Esposito A., Frixione M., Furlanello C., Gori M., Lauria F., Tonella P., Marinaro M., Masulli F., Morasso P., Orlandi G., Palmieri F., Parisi R., Parisini T., Pasero E., Petrosino A., Raiconi G., Rampone S., Serra R., Sorbello F., Tagliaferri R.

The sponsorship and support of:

- International Institute for Advanced Scientific Studies "E.R. Caianiello" (IIASS)
- Istituto Italiano per gli Studi Filosofici (Naples)
- Dept. of Informatics and Applications, University of Salerno
- Dept. of Theoretical Physics and S.M.S.A, University of Salerno
- Artificial Intelligence Software – AIS S.p.A.
- Dept. of Information Science, University of Milan
- IEEE NNC
- IRSIP – CNR (Naples)
- Società Italiana Reti Neuroniche (SIREN)

are gratefully acknowledged

CONTENTS

SECTION 1

INVITED PAPER

A CLASS OF COST FUNCTIONS FOR INDEPENDENCE

LUIS B. ALMEIDA

Instituto Superior Técnico and INESC
Lisbon, Porutgal

and

GONÇALO C. MARQUES
INESC
Lisbon, Porutgal

ABSTRACT

This paper addresses the problem of transforming a set of patterns into new patterns whose components are mutually statistically independent. This is a problem that, depending on the specific setting, is often designated as *factorial coding, independent components analysis, source separation* or *nonlinear principal components analysis*. The reasons for the growing interest on this problem are briefly examined. Some of the most important methods that have been proposed to solve the problem are overviewed. A new class of objective functions for solving the problem is presented. These new objective functions have the advantage of being continuous and differentiable even when they are computed from the empirical distribution of the training data. Some examples of the use of these objective functions are given.

1. Introduction

Within the area of unsupervised learning, a problem that has been receiving increasing attention is the one of transforming a set of patterns into new patterns whose components are mutually statistically independent.

Consider that we are given d-dimensional input data vectors $\mathbf{x} = (x_1, x_2, \cdots, x_d)$ obeying a probability distribution with density $p_\mathbf{x}$. In general, the various components x_i of the data will be statistically interdependent. The problem that we wish to address consists of finding output vectors

$$\mathbf{y} = (y_1, y_2, \cdots, y_{d'}) = f(\mathbf{x}) \tag{1}$$

such that the output components y_i are mutually independent.

4

If $d' = d$ and f is invertible, we are simply recoding the data without any loss of information. If $d' < d$ we are reducing the amount of information present in the data. In the latter case, we usually wish to ensure that the extracted features y_i are the most important ones, in some appropriate sense.

In this paper we will discuss the first situation, $d' = d$, which is usually designated *factorial coding* (FC). However, most of the methods developed for that situation can also be used in the $d' < d$ case. The *"factorial coding"* designation comes from the fact that if the output components are independent, then

$$p_{\mathbf{y}}(\mathbf{y}) = \prod_{i=1}^{d'} p_{y_i}(y_i) \tag{2}$$

i.e., the probability density can be factored into a product of the marginal densities of the output components.

If we assume that the data $\mathbf{x}$ result from a linear combination of independent components, then we can restrict the function f to be linear. In this case, the problem is often designated *blind source separation, independent components analysis* (ICA) or *nonlinear principal components analysis*. The reason for the latter designation comes from the fact that ICA is sometimes performed by introducing nonlinearities into an algorithm that would otherwise perform principal components analysis (PCA).

There are several reasons for the growing interest that FC and ICA have been receiving in recent years:

- They may afford a means to perform *source separation*. Assuming that the observed data $\mathbf{x}$ result from an unknown transformation of independent variables z_i, i.e.

$$\mathbf{x} = g(\mathbf{z}) \tag{3}$$

  where the $\mathbf{z} = (z_1, z_2, \cdots, z_d)$ are unknown *sources* or *causes*, one may ask whether the independent output components y_i that we obtain will coincide with the original z_i. We will discuss this issue ahead.

- Related to this is the fact, verified in practice, that the output components y_i often have a simpler, more intuitive interpretation than the original components x_i.

- If the data are to be stored or transmitted, the y_i are a more efficient representation, because the redundancy due to the statistical interdependence of the components has been eliminated.

- Humans and animals, as well as artificial systems operating in complex environments, often have to estimate probabilities of events. If a factorial representation is found, then according to Eq. 2 only the marginal densities p_{y_i} are needed to estimate these probabilities. Otherwise, the much more complex joint density has to be stored.

- It has been argued that the brains of humans and animals often perform factorial coding operations[1].

In the following subsections we shall discuss in some more detail the linear and nonlinear cases, respectively.

1.1. Linear Case

In this case the functions g (Eq. 3) and f (Eq. 1) are assumed to be linear. The components of the observed data, x_i, are linear mixtures (with unknown weights) of the original sources z_i. If we consider the vectors as column matrices, we can write

$$\mathbf{x} = \mathbf{G}\mathbf{z}$$
$$\mathbf{y} = \mathbf{F}\mathbf{x}$$

where $\mathbf{G}$ and $\mathbf{F}$ are $d \times d$ matrices.

Jutten and Hérault[2] were the first to show that this problem could be solved. They also showed that the solution could lead to the recovery of the original sources. An important reference in this domain is a paper by Comon[3]. In that paper he proved that, in the linear case, the original sources can be recovered, if at most one of them has a Gaussian distribution (all other sources being non-Gaussian). The sources z_i can appear in $\mathbf{y}$ permuted and multiplied by arbitrary scale factors. This should not come as a surprise, since permuting and/or scaling variables doesn't affect their independence.

Principal components analysis is a well known procedure for obtaining uncorrelated features (ranked in order of importance) from correlated data. One may wonder whether ICA and PCA are the same. Figure 1 shows, by means of an example, that they are not. Assuming that the data $\mathbf{x}$ are bidimensional, uniformly distributed within the diamond shape, the principal components are oriented along the horizontal and vertical directions, while the independent components are parallel to the diamond's sides. There is an infinite number of linear transformations that decorrelate the data, PCA being one of them. Among the decorrelating transformations, ICA is the one that not only decorrelates the data, but also makes them independent. Naturally, in order to achieve this, ICA needs to use statistics of order higher than two, since second-order statistics only yield decorrelation.

1.2 Nonlinear Case

In this case the function f in Eq. 1 is allowed to be nonlinear. It is also assumed that the unknown data generating function g in Eq. 3 can be nonlinear. In this case there is an infinite number of solutions to the problem of obtaining independent components y_i. This is illustrated by the example in Fig. 2. Assume that two-dimensional data were distributed according to some continuous distribution within the closed region depicted in the figure. We can arbitrate that the first component, y_1, will be measured along the solid line. Then take an auxiliary component w, linearly measured along the direction orthogonal to y_1 (dashed lines), and choose a nonlinear transformation ϕ such that the

6

second component, defined as $y_2 = \phi(w, y_1)$, will have an arbitrarily chosen distribution (e.g. a uniform distribution between 0 and 1) for all values of y_1. The distribution of y_2 will be independent from y_1, therefore the two coordinates will be independent. Since y_1 was arbitrarily chosen, as was the distribution of y_2, there is an infinite number of solutions to the problem.

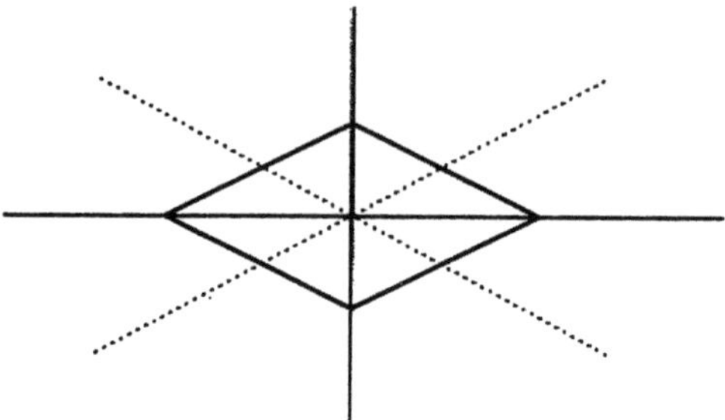

Fig. 1. PCA and ICA are different. Thin solid lines: PCA directions. Dashed lines: ICA directions.

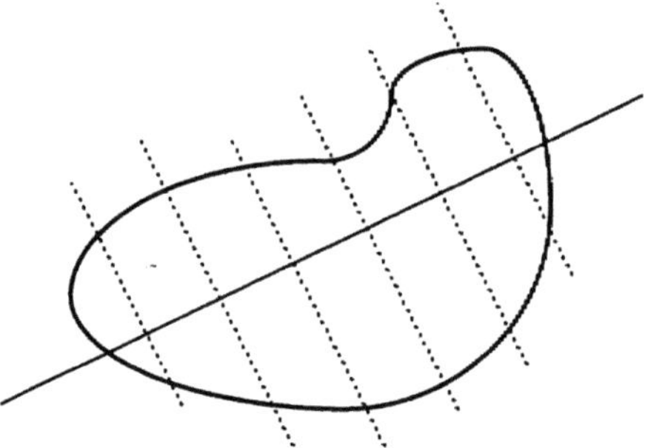

Fig. 2. Illustration of a nonlinear decomposition into independent components. Thin solid line: y_1. Dashed lines: the auxiliary component w.

The existence of an infinite number of solutions means that the problem is ill-posed, and that there is no hope of recovering the original sources z_i unless some extra information is given. This extra information will be problem dependent, but in many cases it may be reasonable to assume that the transformation g was not too complex, and consequently to constrain the solution of the problem by some form of regularisation. In such cases the recovery of the original sources may still be possible, at least in an approximate way.

2. Methods to obtain independence

Several methods to obtain independent components have appeared in the literature. We will briefly review some of them here. In the nonlinear case the methods have to fully enforce independence. On the contrary, in the linear case, the fact that the transformation f is quite restricted allows the use of methods that only partially enforce independence (e.g. by dealing with statistics up to a certain order only). Note that this comment does not imply any qualification of such methods as inferior: some of them are quite adequate for the cases they apply to, and are much simpler than full independence methods.

2.1 Partial Independence Methods

The first method for linearly separating independent components, proposed by Jutten and Hérault[2], is depicted in Fig. 3 for the case of two variables. As shown in the figure, the linear separating network is recurrent. The weights w_{12} and w_{21} are adjusted according to

$$\frac{dw_{ij}}{dt} \propto u(y_i)\, v(y_j)$$

where u and v are two suitably chosen nonlinear functions, e.g. $u(y) = y^3$ and $v(y) = \tanh(y)$. If these two functions were linear, the network would simply perform decorrelation. The nonlinearities bring into play higher order moments, which allow the separation into independent components. It should be noted, however, that the method doesn't always work: the convergence to independent components depends on the statistics of the input data, as well as on the choice of u and v. Several variants of this method have appeared in the literature.

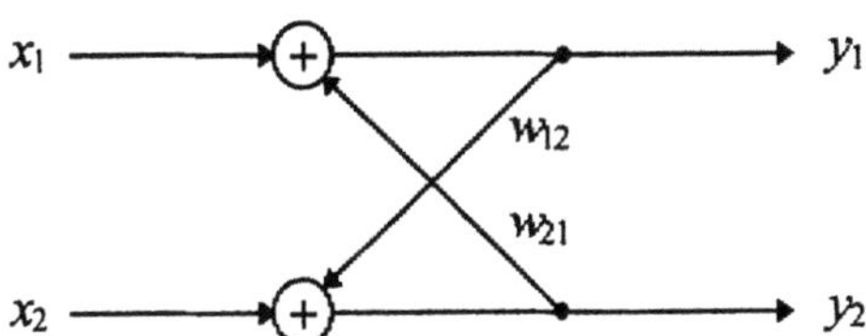

Fig. 3. The separation method of Jutten and Hérault

Another method, proposed by Wang, Karhunen and Oja,[4] involves two successive steps: (1) decorrelate the input data and normalize them, making the variances of all components equal to 1 (this is sometimes called *sphering* the data); (2) obtain the output y through a linear transformation that minimizes $\sum_i k_4(y_i)$, where $k_4(y_i) = \langle y_i^4 \rangle - 3\langle y_i^2 \rangle^2$

is the fourth cumulant, or kurtosis, of y_i; $\langle \cdot \rangle$ denotes expectation. This method is known to work if the sphered components obtained in step (1) all have negative kurtosis. If they all have positive kurtosis, then the sum of cumulants should be maximized, instead of being minimized. If the kurtoses have mixed signs there is no guarantee of separation. In a more recent work[5] an impressive example of separation of nine linearly mixed images was given.

2.2 Full Independence Methods

Almost all of the methods that we know of, which fully enforce the independence of the outputs, are directly or indirectly based on the use of the entropy concept. The (differential) entropy of a continuously valued component y_i is given by

$$H_{y_i} = -\int p_{y_i}(y_i) \log p_{y_i}(y_i)\, dy_i \tag{4}$$

and the total entropy of the output components, considered one by one, is

$$H_t = \sum_i H_{y_i}$$

On the other hand, the joint entropy of the output components is

$$H_y = -\int p_y(\mathbf{y}) \log p_y(\mathbf{y})\, d\mathbf{y} \tag{5}$$

A property of these entropies is that $H_t \geq H_y$, equality being observed only if the y_i are mutually independent.

One way to exploit this property is by trying to minimize H_t, subject to the constraint that H_y remains fixed. This is the basis of the nonlinear methods proposed by Redlich[6] and by Deco and Brauer[7]. In both cases, they enforce $H_y = H_x$, by restricting the architectures of the nonlinear networks that perform separation.

Hinton and Zemel[8] proposed a nonlinear separation method in which the y_i are discrete, the transformation f being non-invertible (implying a reconstruction error in the backward transformation from $\mathbf{y}$ to $\mathbf{x}$). Their method is based on the minimum description length principle of Rissanen[9]. It minimizes the sum of the description lengths of the y_i and of the reconstruction error. This method can be seen to indirectly use the above mentioned property of the entropies. In fact, if we assume that efficient coding is used, the description length of the y_i is approximately given by H_t, and the one of the reconstruction error by $H_x - H_y$. Since H_x is fixed, the method is indirectly minimizing $H_t - H_y$.

Comon[3], as well as Amari, Cichocki and Yang[10], have proposed linear separation methods that try to minimize the Kullback-Leibler divergence between the product of the marginal densities, $\prod_i p_{y_i}$, and the joint density p_y. This divergence is given by

$$D = \int p_{\mathbf{y}}(\mathbf{y}) \log \frac{p_{\mathbf{y}}(\mathbf{y})}{\prod_i p_{y_i}(y_i)} \, dy$$

and can be shown to be equal to the difference between entropies,

$$D = H_t - H_y$$

Therefore, these methods again indirectly exploit the same property of the entropies that we mentioned above.

Another linear separation method that has appeared recently, proposed by Bell and Sejnowski[11], postprocesses each of the y_i individually, resulting in variables $\hat{y}_i$ that have a bounded range (in their examples, the postprocessing is achieved by passing the y_i through sigmoidal nonlinearities). The method then tries to maximize $H_{\hat{y}}$. Since the vector $\hat{\mathbf{y}}$ is constrained to stay within a hypercube, maximizing its entropy tends to make its distribution uniform, which results in independent $\hat{y}_i$. Since each of these only depends on the respective y_i, these will also be independent.

A method that should also be mentioned in this context, because it probably was the first nonlinear factorial coding method to appear, was proposed by Schmidhuber[12]. It does not explicitly use entropy concepts. Instead, it tries to make each component unpredictable from the values of the other components (and thus independent from them). The method can only be applied to binary components.

3. A New Class of Objective Functions for Independence

3.1 Pros and Cons of Entropy-Based Objective Functions

All of the full-independence methods mentioned in Section 2.2 (except for Schmidhuber's) are based, directly or indirectly, on the use of entropies. These methods suffer from one common drawback. As is clear from the definitions of Eq. 4 and 5, the computation of the entropy demands the knowledge of the densities p_{y_i} and/or $p_{\mathbf{y}}$. Since the training data are finite in number, these densities are not known exactly.

A solution to this problem is to estimate the densities. This is the approach taken, for example, by the methods of Comon and of Amari *et al.* They use the Edgeworth or Gram-Charlier series expansions of densities, which both coincide in their first terms,

$$p(y) = \varphi(y)\left[1 + \frac{1}{3!}k_3\, h_3(y) + \frac{1}{4!}k_4\, h_4(y) + \ldots\right]$$

where k_i is the i-th cumulant of p, h_i is the i-th Hermite polynomial, and φ is the Gaussian distribution that is "closest" (in a certain sense) to p. A similar approach is taken by Deco and Brauer, who take series expansions of the Fourier transforms of densities, of the kind

$$P(\omega) = \exp\left[\sum_{n=1}^{\infty} \frac{j^n}{n!} k_n \, \omega^n\right]$$

where j is the imaginary unit. In all of these methods, the cumulants k_i are estimated from the training data, and the series expansions are truncated at an appropriate order. This truncation can be seen as a way to smooth the discrete distribution of the training data, turning it into a continuous distribution. The method of Bell and Sejnowski also implicitly makes a smoothing of the distribution of the data, through limitations in the form of the postprocessing nonlinearity.

Another solution to the estimation problem would appear to be the use of the entropy of the discrete distribution of the training data itself (the so-called empirical distribution). However, this entropy is a discontinuous, piecewise constant function of the values that the y_i take for the various input patterns, and therefore does not lend itself to the usual gradient based methods of minimization.

It should be noted, however, that in spite of this drawback, the methods that are based on the Kullback-Leibler distortion also have an attractive advantage: this distortion is invariant to nonlinear transformations of the outputs y_i, as long as these transformations are invertible. As a special case, these methods are insensitive to a scaling of the outputs.

3.2 A New Class of Objective Functions

The considerations of Section 3.1, regarding the need to estimate, or smooth, the distribution of the data, led us to propose objective functions that could be computed directly from the empirical distribution of the training data[13], without any smoothing or truncation, and that were continuous, differentiable functions of the outputs y_i, to allow the use of gradient based optimization methods. For the sake of simplicity, we shall describe these objective functions for the case of bidimensional data only, and shall make a change of notation, denoting the outputs by a and b respectively:

$$a = y_1$$

$$b = y_2$$

If the two outputs are independent, then

$$\langle a^m b^n \rangle = \langle a^m \rangle \langle b^n \rangle \tag{6}$$

and we can define an objective function as

$$R = \sum_{m,n=0}^{\infty} c_{mn} \left(\langle a^m b^n \rangle - \langle a^m \rangle \langle b^n \rangle\right)^2$$

where the c_{mn} are positive coefficients, chosen so that the series converges. The objective function R is non-negative, and will be zero only if Eq. 6 is true for all pairs (m,n). The computation of R would seem to again require truncation of a series, but we shall now see how to circumvent this difficulty. Developing the square, we obtain

$$R = \sum_{m,n=0}^{\infty} c_{mn} \left(\langle a^m b^n \rangle^2 - 2\langle a^m b^n \rangle \langle a^m \rangle \langle b^n \rangle + \langle a^m \rangle^2 \langle b^n \rangle^2 \right) \tag{7}$$

Assume that we independently draw four patterns, that we identify by subscripts from 1 to 4, and for which we compute the outputs. Then,

$$\langle a^m b^n \rangle^2 = \langle a_1^m b_1^n a_2^m b_2^n \rangle$$

Similar transformations can be made in the other terms of Eq. 7, resulting in

$$R = \sum_{m,n=0}^{\infty} c_{mn} \left(\langle a_1^m b_1^n a_2^m b_2^n \rangle - 2\langle a_1^m b_1^n a_2^m b_3^n \rangle + \langle a_1^m a_2^m b_3^n b_4^n \rangle \right)$$

$$= \left\langle \sum_{m,n=0}^{\infty} c_{mn} \left(a_1^m b_1^n a_2^m b_2^n - 2a_1^m b_1^n a_2^m b_3^n + a_1^m a_2^m b_3^n b_4^n \right) \right\rangle$$

If we choose coefficients $c_{mn} = \alpha_m \alpha_n$, where the α_m are coefficients of the series expansion of some entire function ψ,

$$\psi(x) = \sum_{m=0}^{\infty} \alpha_m x^m$$

then the objective function is given by

$$R = \langle \psi(a_1 a_2) \psi(b_1 b_2) - 2\psi(a_1 a_2) \psi(b_1 b_3) + \psi(a_1 a_2) \psi(b_3 b_4) \rangle$$

For example, with $\alpha_m = 1/m!$ we have $\psi(x) = e^x$ and

$$R = \left\langle e^{a_1 a_2 + b_1 b_2} - 2e^{a_1 a_2 + b_1 b_3} + e^{a_1 a_2 + b_3 b_4} \right\rangle \tag{8}$$

The relative weights of the moments of different orders can be controlled by a suitable choice of the coefficients α_m. For example, for a lower influence of the high order moments, one could choose $\alpha_m = 1/(2m)!$, resulting in

$$\psi(x) = \begin{cases} \cosh\left(\sqrt{x}\right) & \text{if } x \geq 0 \\ \cos\left(\sqrt{|x|}\right) & \text{if } x \leq 0 \end{cases}$$

which, surprisingly, is a non-convex and non-monotonic function.

This new class of objective functions also has its pros and cons. Its most obvious advantage is that it is continuous and differentiable, even if it is directly estimated from the empirical distribution of the training data. A disadvantage is that it changes with the scaling of the outputs a and b. If we scale them down towards zero, R will also tend towards zero. This is an unavoidable consequence of the function's continuity: if the a and b are always zero, they are independent from one another. Therefore, the function's value should be zero. However, this means that an undesirable way to minimize R is simply to scale the output components down towards zero. Consequently, minimization algorithms will probably tend to progressively reduce the range of variation of the output components.

To counteract this tendency, we have added to the objective function a term that tends to normalize the variances of a and b:

$$\hat{R} = R + \lambda\left[\left(\sigma_a^2 - 1\right)^2 + \left(\sigma_b^2 - 1\right)^2\right] \tag{9}$$

where λ controls the relative influence of the normalization term. This is the form of the objective functions that were used in the examples presented in the following section.

Another drawback of the class of objective functions presented here is that, if we wish to perform batch-mode training, we have to use all 4-tuples of patterns (x_1, x_2, x_3, x_4) in the computation of R and of its derivatives. Thus, the amount of processing per iteration will be proportional to N^4, where N is the number of training patterns (this is for two-dimensional data; more generally, the amount of processing will be proportional to N^{2d}, where d is the number of dimensions of the data). Therefore only stochastic (real-time) training will be a viable option, even for moderately sized training sets. In this latter training mode, four patterns are randomly drawn (allowing repetitions) from the training set to form a 4-tuple, and the parameters of the system to be trained are updated after presentation of each 4-tuple.

4. Examples

Fig. 4-a shows two sinusoidal signals with different frequencies. As inputs to the separation algorithm, two linear combinations of these signals were used (Fig. 4-b). Fig. 4-c shows the result of PCA, which was not able to separate the signals. Fig. 4-d shows the result of a linear separation obtained by minimizing an objective function of the form of Eq. 9. Separation is nearly perfect.

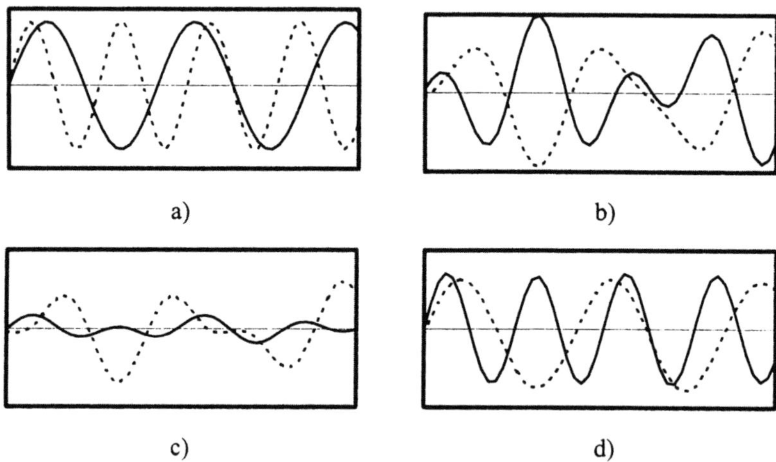

a)   b)

c)   d)

Fig. 4: a) Original signals. b) Mixed signals. c) PCA result. d) Output of the proposed method.

Fig. 5-a shows a set of points, obtained by using a Gaussian distribution in the horizontal direction, and an exponential distribution in the vertical direction. Fig. 5-b shows the data used as input, which were obtained through a linear transformation of the data of Fig. 5-a. The result of PCA is shown in Fig. 5-c. Fig. 5-d shows the result of linear separation, again using an objective function of the form given by Eq. 9. The data became almost completely separated.

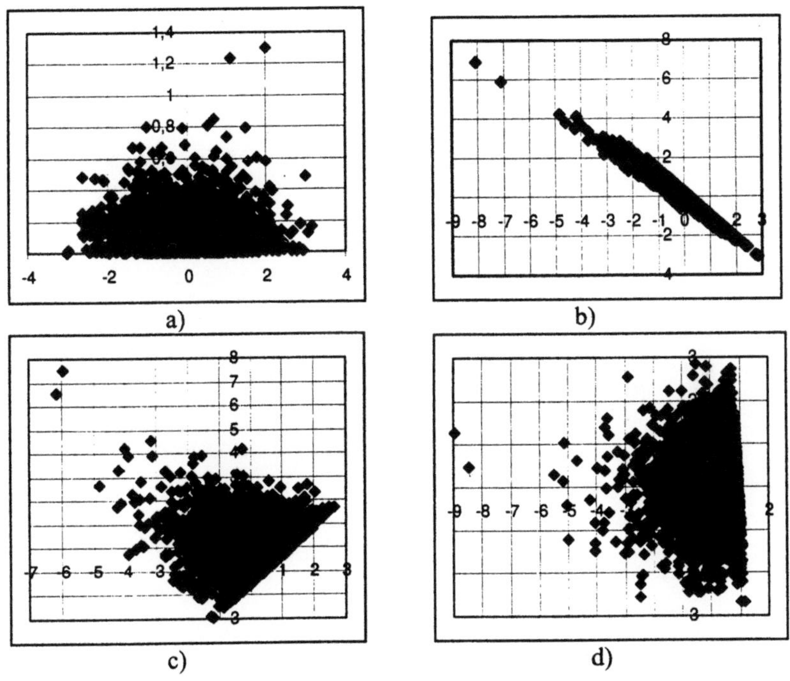

Fig. 5: a) Original data. b) Linear mixture. c) PCA result. d) Output of the proposed method.

Fig. 6-a shows two original images. Two linear combinations (Fig. 6-b) were computed, and used as inputs to the algorithm. Note that the two combinations are very similar to one other, indicating that the two components are highly correlated. Fig. 6-c shows the result of PCA, which again couldn't separate the images. The result of linear separation using the algorithm presented here is shown in Fig. 6-d. Again, separation was almost perfect.

All of the above examples corresponded to linear situations. The last examples, to be presented next, are two very small cases of nonlinear processing. Fig. 7-a shows four points (white diamonds), corresponding to four input patterns in a bidimensional space. To have independent coordinates, these points would have to be arranged in a rectangle.

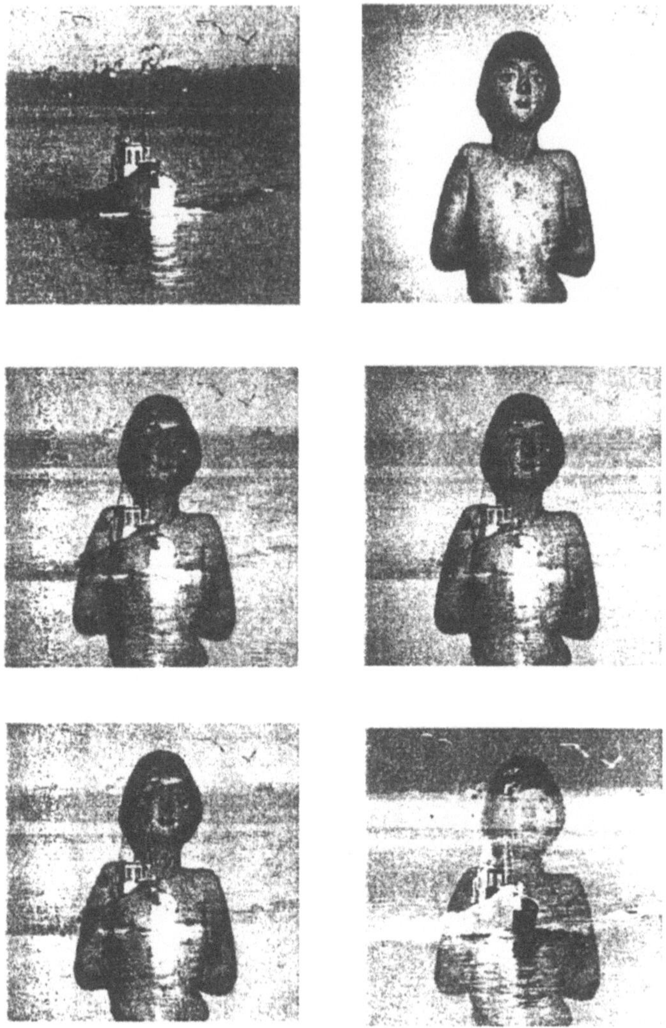

Fig. 6. Separation of images. Top: original images. Middle: mixture images. Bottom: PCA output. (continued in the next page)

Fig. 6 (continued). Output of linear separation.

Since they do not form a parallelogram, there is no linear transformation that will make their coordinates independent. The black diamonds show the result of processing with a nonlinear network of the form shown in Fig. 8, with two hidden units, trained using an objective function of the form of Eq. 9. As can be seen, the patterns were arranged into a rectangle, yielding independent output coordinates.

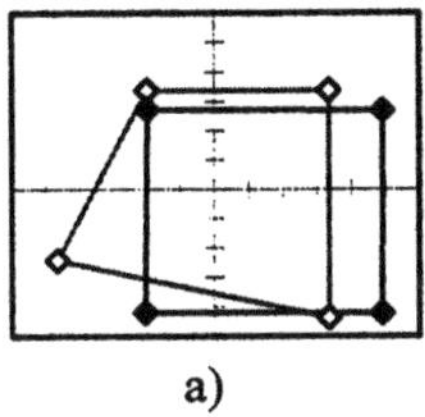

a)

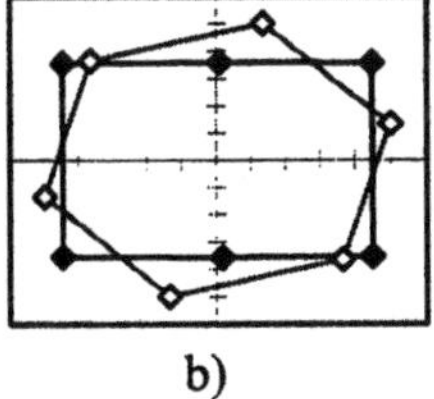

b)

Fig. 7: Nonlinear separation. a) Four point set. b) Six point set.

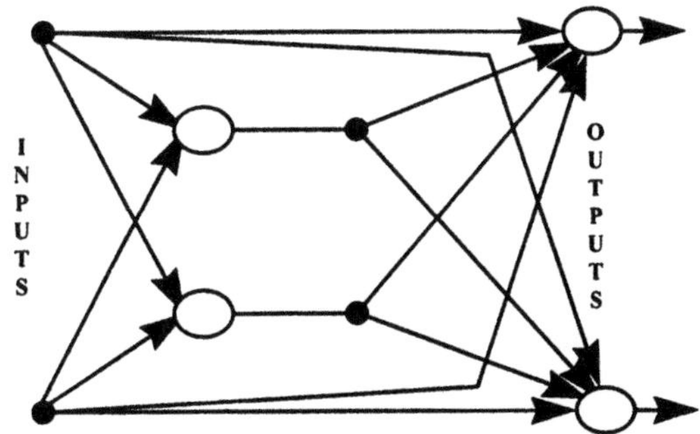

Fig. 8: Non-linear separating network. The output units are linear.

A similar example, now with six points, is shown in Fig. 7-b. The output patterns were arranged by the nonlinear system into a rectangular lattice, again yielding independent output coordinates. The nonlinear network used to process these data was also of the form shown in Fig. 8, but now with three hidden units.

5. Conclusion

We have presented a new class of objective functions for obtaining independent components. These objective functions have some advantages over entropy-based ones, namely the advantage of being continuous and differentiable, even if computed directly from the empirical distribution of the training data. They also have some disadvantages, one being their scale dependency, and another the amount of computation they require if used in batch mode. We have shown some examples of the use of these objective functions, both in linear and nonlinear situations.

6. Acknowledgments

The second author wishes to acknowledge support from a research grant of the PRAXIS XXI program.

References

1. H. B. Barlow, Unsupervised Learning, *Neural Computation*, **1** (1989), pp. 295-311.
2. C. Jutten and J. Hérault, Blind Separation of Sources, Part I: An Adaptive Algorithm Based on a Neuromimetic Architecture, *Signal Processing*, **24** (1991), pp. 1-10.
3. P. Comon, Independent Component Analysis, A New Concept?, *Signal Processing*, **36** (1994), pp. 287-314.
4. L. Wang, J. Karhunen and E. Oja, A Bigradient Optimization Approach for Robust PCA, MCA and Source Separation, in *Proc. ICNN'95*, (Perth, Australia, 1995).
5. J. Karhunen, L. Wang and R. Vigario, Nonlinear PCA Type Approaches for Source Separation and Independent Component Analysis, in *Proc. ICNN'95*, (Perth, Australia, 1995).
6. A. N. Redlich, Supervised Factorial Learning, *Neural Computation*, **5** (1993), pp. 750-766.
7. G. Deco and W. Brauer, Nonlinear Higher-Order Statistical Decorrelation by Volume-Conserving Neural Architectures, *Neural Networks*, **8** (1995), pp. 525-535.
8. G. E. Hinton and R. S. Zemel, Autoencoders, Minimum Description Length and Helmholtz Free Energy, in *Advances in Neural Information Processing Systems 6*,

eds. J. Cowan, G. Tesauro and J. Alspector (Morgan Kaufman, San Mateo, CA, 1994), pp. 3-10.

9. J. Rissanen, *Stochastic Complexity in Statistical Inquiry* (World Scientific Publishing Co., Singapore, 1989).

10. S. Amari, A. Cichocki and H. H. Yang, A New Learning Algorithm for Blind Signal Separation, *Neural Information Processing Systems: Natural and Synthetic* (MIT Press, 1996) in print.

11. A. J. Bell and T. J. Sejnowski, An Information-Maximization Approach to Blind Separation and Blind Deconvolution, *Neural Computation*, **7** (1995), pp. 1129-1159.

12. J. Schmidhuber, Learning Factorial Codes by Predictability Minimization, *Neural Computation*, **4** (1992), pp. 863-879.

13. G. Marques and L. B. Almeida, An Objective Function for Independence, in *Proc. ICNN'96*, (Washington DC, 1996), pp. 453-457, in print.

SECTION 2

REVIEW PAPERS

Virtual Reality and Neural Networks

Nunzio Alberto Borghese
Istituto Neuroscienze e Bioimmagini - CNR
Via Mario Bianco, 9 20131 Milano - Italy
borghese@carla.inb.mi.cnr.it

1 Introduction

Different terms have been applied to what is commonly called, with an oximoric, "Virtual Reality" (VR) or "Artificial Reality" to indicate a world which lies in the computer. At the actual state of the art, it suggests much higher performances than current technology can generally provide [1]. Other terms like "Virtual Worlds", "Virtual Environments" or "Synthetic Environments" seem preferable because they are linguistically conservative, and related to well-established terms like virtual images. VR has a dual nature: it is cultural movement which has its roots in the philosophical thought and a technological nature which derives from computer science advanced technology.

In philosophy, the interest for VR has been linked to the search for what is the essence of reality, the ontology and to what can be known about reality, the gnoseology. This search has permeated philosophical thought in the western world (eastern culture considers the world non existing, e.g. for the Indian philosophy, the world is *maya* which means illusion). The search for the essence of the reality starts in the Ancient Greek, and in particular with Plato (428-348 b.C.), who believed that the real reality lies in a world where everything is perfect and beautiful, the world of the images (ειδα) which is located above the sky above us (υπερ ουρανου). The possibility to get to the real nature of the world was severely questioned by the empirism in the XVII century, and Berkeley (1685-1753) stated that the world could be perceived only with our sensorial systems, too limited to grasp the essence of the reality. In Hegel (1770-1831) extremist view, there was no interest for a world which would be beyond reach: "What is real is rational and what is rational is real". He postulated the existence of an absolute spirit represented by art, religion and philosophy, which can achieve the maximum knowledge of the world summarising the experiences of humans through their evolution; this world being the only one of interest to us. A violent protest against this extreme rationalism was carried out in the sixties by the hippies movement; one of its gurus was Timothy Leary, professor at Harvard, who was interested in exploring the content of the mind in conditions which could not be tested in the physical world but could only be evoked artificially; the obtained mind states go beyond the material world and can be considered virtual although real for the persons who experienced them [2]. In these experiences the essence of the reality could come to life purified from all the myths and the psychological structures created by the humans themselves throughout their history [3]. From the technological view point, the ancestors of VR devices can be tracked down in the sixties to the Head-Mounted display by Philco [4], which was a helmet carrying the monitor of a closed-circuit television system, and in the first synthetic computer-generated 3D display of Ivan Sutherland, that he called the *ultimate display* [5]. From the industrial point of view, the father of VR can be considered Jaron Lanier, founder of VPL Research, the company which in the late 80s engineering and sold the DataGlove along with EyeGlasses for the first VR systems.

Two can be considered the key elements of Virtual Reality technology: immersivity and interactivity. At the actual state of the art, immersivity can be achieved with some approximation while interactivity is beyond reach. *Immersivity* can be achieved when a VR system is able to <u>stimulate the human sensorial systems (vision, hearing, touch, smell and taste) in a coordinated way</u>. *Interactivity* is the ability to modify in real-time the virtual world according to the subject operations. It follows that VR is a transversal discipline which is based on computer graphics, computer vision, sensory physiology and parallel computing.

2 Virtual Reality Systems

A virtual reality system is constituted of four main components: the input systems, the output systems, the world generators and the graphical engine.

2.1 The Input Systems

These systems monitor the location and attitude of the human body or of some of its parts with respect to an external reference system. Moreover they should convey to the VR system the information on the forces which are exerted by the subject <u>on</u> the virtual environment. The input systems can be grouped into *real-time trackers* which output the actual position of the body in motion with a minimum delay and *motion captures* which record and store the entire trajectory of the body to reproduce it later in time, along with other quantities like velocities, accelerations, angles. In the last few years, this distinction has become fuzzy, as under the spur of VR and animation markets, the producers of motion capture systems are proposing new versions which can be used as real-time trackers and producers of real-time trackers are tailoring their products for motion capture. Surveying finger motion is the most challenging task due to the resolution involved (a few millimetres separate the fingers one from the other) and to the complex motion patterns and specific solutions have been proposed which are based on making the subject wearing a glove which carries position sensors. Input systems can also be grouped into optical, magnetic and acoustic according to the adopted methodology [6]. Magnetic trackers were the final output of a research started back in the 1980 in the Machine Group at MIT [7]. In this approach a single or multiple sources solid with the environment, radiate a pulsed magnetic field at different frequencies which is concatenated by small coils attached to the body parts. By measuring the current induced by the concatenated magnetic field, the system reports the location and orientation of the coils with respect to the source. Flock of Birds™ from Ascension Technology and Fastrack™ from Polhemus are the competitors in this technology. They achieve an experimental accuracy ranging from 4cm/3degrees to 10cm/7degrees in position/orientation for operating ranges varying from 60cm to 1.5m depending on the system configuration. They have the advantage to measure the position of a body segment also when it is "hidden" by other segments, but they bear the undesirable drawback to be very sensitive to any metallic object in the field, including the computer monitor; these distort the magnetic field altering the sensor measurements. Moreover, as they require cables for transmitting the current generated in the coils, and the sensors are large, they interfere with the subjects motion in fine manipulation tasks [8]. For these reasons their use is mainly bounded to arm tracking.

Acoustic trackers use high-frequency sounds to triangulate a source within the working area. The source is attached to the body segment and it sends out pings which are picked up by microphones in the environment. These systems rely on the line of sight and suffer from acoustic reflections when surrounded by walls or other acoustically reflective surfaces.

The most ecological approach comes from computer vision. In this approach, the subject is free to move in the environment; the motion is captured by a set of calibrated or uncalibrated TVcameras situated in strategical positions. This approach dates back to the turn of the century when Muybridge in Stanford [9] and Marey in Paris [10] were able to take multiple photographs of the same motion sequence. Muybridge started recording animal motion and his projected images of a horse gallop produced in his audience the illusion "as if the living animal itself was moving" well before the cinema was invented. The method was based on an intense background light (the California sunshine), drop shutters actuated by springs and rubber bands, wet plates and a set of photo cameras equally spaced and oriented with respect to the line of motion. Passing before each camera, the subject broke a thread that triggered the shutter of the camera. The modern counterparts of this system are sophisticated image processors which have basically a two level architecture. The first level filters the image to detect a set of features or primitives. It is usually highly parallel and can be implemented in hardware to get real-time operation [11], [12], [13]. The second level assembles the obtained primitives to reconstruct a 3D representation of the body; this second level requires algorithms which are less parallel and it is usually carried out on-line. One of the main research streams in this vein is Immersive Video. In this approach multiple videos of an event are filmed from different points of view and a full 3D digital video is generated for that event. The viewer can therefore decide to reply the 3D video and to see it from an arbitrary virtual position. In the approach of the group of Jain [14], up to six video cameras are used. A static model of the scene is created and by assembling the data in the image streams from the different TVcameras, a dynamic 3D model of the moving persons (or objects) is created and superimposed to the static 3D model. The obtained result is then rendered from a viewer controlled perspective. To create the dynamic model of the moving person, first the scene is segmented by a subtraction procedure: the static background, as viewed by each TVcamera, is subtracted from the actual image and a set of 2D silhouette of the moving subjects can than be computed [6], [15]. Silhouette extraction is a quite general method of time filtering for feature detection It was first proposed in the framework of a virtual desk by RankXerox at Europark [16] but, as remarked before, the continuos crossings among the fingers and the complex patterns of motion make the fingers very difficult to track when the line of sight is required.

Silhouette tracking is now proposed in the domain of the categorisation of human gestures which involve little finger movements. The second step in the reconstruction of the 3D outline of the subjects moving inside the scene is the assemblage of the silhouettes detected on the different cameras. This is accomplished by back-projecting the silhouettes through the perspective centre of each camera forming a set of three dimensional cones. The 3D volume occupied by each subject is computed as the parallelepiped resulting from the intersection of these cones. In general, the computer vision approach has severe accuracy limitations both in the 2D feature detection and in the assemblage of the features into a single 3D structure. In mathematical terms the problem is ill-posed in the sense that the solution is very sensitive to the noise on the data. A regularised solution would improve the stability of the solution [17], [18]. On the 2D features detection side, a regularized solution can be achieved by filtering multiple measurements over time [19] or by using optimised optical flow detectors [20], [21], [22], [23]. The 3D reconstruction can be improved, at expense of computational time, by tracking the moving subjects at a greater detail level by encapsulating them into models more sophisticated than parallelepipeds. The models most used are the Superquadrics [24] which are 3D geometrical models whose shape can change locally and globally according to a reduced set of parameters and marching cubes [25] which computes a polyhedral approximation through a bilinear interpolation. If the model is known in advance, the identification of the moving bodies becomes even more stable as the problem of identification becomes a problem of fitting a predefined structure to the data. Mesh structures like Kohonen maps [26], Non Uniform Rational B-Splines [27] or Snakes [28] are not considered in this phase because they require too many parameters to be set. A good compromise between motion freedom and reliability/accuracy is achieved adopting a marker based motion detection. This approach has its roots in biomechanics where the human body is modelled as a skeleton, that is a set of rigid links connected by hinges [29]. Each body segment can be identified by its position and attitude in space (six degrees of freedom) or, under the hypothesis of no rotation around the axis joining the two extremes, by measuring the position of its extremes (five degrees of freedom). In this approach, the features to be detected are markers attached to the skin on the joints of the human body. This greatly simplifies the 2D detection procedure. Moreover, thanks to the computation of the centroid of the surveyed markers the accuracy achieved by these systems is of about 0.1 pixels, of an order of magnitude greater than gloves and magnetic trackers [30], [31]. The markers are detected by a dedicated hardware which performs a simple image thresholding (Vicon[TM], Motion Analysis[TM], MacReflex[TM]) or a real-time hardware cross-correlation (Elite[TM]); the latter approach has the advantage to allow the use of much smaller markers (about 1/10th of diameter) retaining the same accuracy. To get the 3D trajectories of the markers and therefore of the body segments, the ensemble of the markers detected on the different cameras have to be grouped and associated to the corresponding body landmarks [32], [33]. This procedure, called tracking, is carried out adopting multiple criteria [34], [35]. First, the projection of the same marker on two TVcameras should lie on a single plane: the epipolar plane, which is the plane containing the perspective centres of the TVcameras and the two projections. Second, as a first approximation, the human skeleton is rigid and two markers attached at the extremes of a segment tend to keep their 3D distance constant throughout the entire motion. Third, due to the viscoelastic properties of the muscles, human motion is generally smooth and future positions can be reliably estimated starting from their previous ones through digital filtering. The computational requirement of the tracking procedure increases greatly with the number of TVcameras and markers and it does not allow to work in real time. On the other side, some "intelligence" has been attributed to the tracking procedure which can be able to autonomously output an interpretation of the scene itself in terms of a set of rigid links connected by hinges [29]. A different approach in motion capture, is the use of active markers which are LEDs flashed sequentially in time (Selspot™ and Optotrack™). This allows to get sampling frequency up to 1kHz for a few LEDs and to avoid markers tracking. On the other side, in this approach the subjects have to wear power cables and synchronisation wires which limit the freedom of motion and for this reason these systems are not widespread in VR markets.

Gloves are a system on their own, largely diffused to record fingers position because they are insensitive to their complex motion patterns. Their technology somehow parallels the evolution of that of the motion capture systems. The first glove was the Sayre glove, first introduced in the market in the 1976. It mounted flexible tubes, one for each finger, with a light source on one end and a photocell at the other. The amount of voltage recorded by the photocell was proportional to the amount of light that could go through and therefore to the degree of bending of the device [36]. MIT LED glove implemented the active marker technique for motion capture but it suffered from instability and inaccuracies due to scattered light [6]. The Digital Data Entry Glove was an interesting special purpose glove [37]: it contained a numerous set of touch, bending and inertial sensors sewn in precise locations and an electronic circuit could recognise eighty unique combinations of sensors readings and to output 80 ASCII characters. This glove was able to recognise the Single Hand Manual Alphabet for the American Deaf but it was never commercialised.

In 1987, Zimmerman and others developed the DataGlove™. It can monitor the amount of bending of ten finger joints along with the location and orientation of hand in the external space. It was a great improvement with respect to the previous gloves and it gave some success to VPL which commercialised it. This glove is lightweight, unobtrusive to the user and comfortable to wear. Physically it consists of a lightweight Lycra glove fitted with specially treated optical fibres along the fingers. Precalibration allows to transform the amount of detected light at one end to the degree of bending of the fingers. The DataGlove can be completed with a magnetic tracker to record the 3D position and attitude of the hand. The PowerGlove™, produced by Mattel for the entertainment industry, was a meteor in the market. It was a plastic gauntlet with Lycra palm, for which the transducer elements were a resistive-ink strips on the back of the fingers. An acoustic tracker was adopted to transduce the location and orientation of the hand. The resolution and accuracy were very low, but the low price made them the perfect interface for game industry. There were people who reversed engineering its interface to adapt the gloves to PC serial ports. The last product is the CyberGlove from Virtual Technologies. It was developed at Stanford to translate the American Sign Language into spoken English. In this approach, the hand is viewed as a three-dimensional joystick plus a system to give commands to the computer (the sign language). It consists of a custom-made cloth glove with up to twenty-two thin foil strain gauges sewn into the fabric to sense fingers and wrist bending [38]. A small electronic box converts the analog signals into a digital stream that can be read by a computer's standard serial port.

A problem common to all these input positional systems is calibration. This can be viewed as the determination of a two-dimensional (in case of camera based systems) or of a three-dimensional map (in case of ultrasound and magnetic trackers) of the systematic error introduced on the images by the electronics or by the device itself. Although theoretical models do exist [39], [40], very often, distortions are very irregular in shape and local algorithms like neural maps may perform better.

As far as aptic input is concerned, its existence is particularly critical for some motor tasks like grasping and manipulation [31]. Up to now few prototypes have been proposed, and they are quite cumbersome. One of the most interesting is the one developed by the group of Bergamasco in Pisa in the framework of the Glove-like Advanced Interfaces for the Control of Manipulatitative and Exploratory Procedures in Artificial Realities (Glad-in-Art, [41]). It is a glove exoskeleton interface system to manipulate virtual objects whose main feature is the. ability to replicate both the forces which are exerted <u>on</u> the environment to the operator and <u>from</u> the environment to it. It is therefore a device to both input and output forces. The glove structure wraps around the hand and supports kinaesthetic sensors for all the fingers. Flexion-Extension as well as Abduction-Adduction are measured as well. Different approaches come from robotics and telemanipulation where these devices have been developed to provide and receive kinaesthetic information to/from the operator (from/on the environment). Three main approaches can be found in the literature. A first approach is the classical direct-drive manipulandum which has usually a very high mass/torque ratio for structural reasons. Parallel structures allow to decrease the mass of the robot arm [42], [43]. In the magnetically-levitated devices [44], [45], [46], the manipulandum is suspended in the air thanks to the magnetic field allowing the maximum freedom of motion with very low friction and viscosity.

Gaze trackers are even more complicated. These systems which have long been used in psychology are quite invasive and they cannot be worn for long time. They are based on thick contact lenses which have a printed coil on their edge to concatenate a magnetic field pulsed from a antenna fixed to the environment. A less intrusive approach is based on the detection of a laser spot coaxial with a camera. By measuring the shape of the reflections from the frontal portion of the cornea, the actual position of the eye can be recovered. A system composed of two infra-red light emitting diodes, a CCD image sensor unit including optical parts, and a processor unit, has been integrated in the stereoscopic eye-wear by Ohshima et al [47].

2.2 The World Generators

The systems which allow to generate synthetical environments for virtual reality are usually integrated tools which are based on hierarchical object-oriented programming and are constituted of a 3D CAD system, lighting, texture, colour modules, animation capacities and special features (particle effects, blurring, anti-aliasing). These systems are computationally demanding and do not work in real-time. Their final output is a 2D image obtained by rendering a 3D image from a certain vantage point and with a certain perspective. Alias™, Wavefront™ and Softimage™ are the most used systems in the high end, and WordToolkit™ from Sense 8 and Virtual Reality™ from Superescape in the low end (PC based).

New features which are partially included in the high end tools will be likely incorporated in the next generation systems. First, it will be possible to reconstruct smooth surfaces both from synthetic and scanned ensemble of 3D points. The methods will be largely based on B-Splines and on their evolution, the NURBS (Non Uniform Rational B-Splines, [27]). This methodology was born in the world of CAD to offer to the users an intuitive mean to modify the shape of the surfaces: by modifying interactively a set of control points, called nodes, it allows a direct control over the final shape of the curve. A different approach are the Snakes models [28] where the surface results from the interaction of two fictitious forces: an elastic force, associated to each point, which attracts the surface to the points, and a smoothing force which penalises brisk variations in the surface. It can be noticed that this approach is very similar to Kohonen maps [26] in its formulation. The analytical shape of the forces used by the snakes can be quite general and multi-layer perceptron has been recently proposed to improve them [48]. Using Snakes, Terzopoulos and Waters [49] have reproduced facial expressions by tracking the position of a set of linear features. An approach which simplifies the computational burden when simple 3D geometrical shapes are considered, is that of the Superquadrics [24], [50]: the basic shape is captured by an object, ellipsoid-like, properly oriented and scaled in space; a set of local parameters can be added to describe local variations with respect to the original shape. This approach is particularly useful when a search in large databases is required or when a rough description of approximately symmetrical objects (e.g. fingers, arm and forearm) suffices. A quite different approach is the Radial Basis Functions (RBF). It derives from the neural networks field [51] and from the regularization theory [52]. In this approach, the surface is represented as a linear combination of Basis Functions, usually Gaussians but as the choice of the structural parameters, namely, the variance, the location and the number of the Gaussians has revealed to be particularly critical [53], [54], they are not diffused in the VR and computer graphics adomains.

Different reconstruction algorithms can be adopted when patchy surfaces can be accepted: in these approaches, the surface is represented as a set of contiguous planar patches (tessellation). These algorithms are based on the Delauney triangulation or on the Kohonen maps. In the Delaunay triangulation, the surface (volume) is partitioned into triangles (tetrahedrons), such that the circumcircle (circumsphere) of every triangle (tethraedron) does not contain any point of the triangulation [55], [56], [57]. In other words, the tessellation minimizes the sum of the angles of the triangles (tethraedrons). Optimised tessellation algorithms like divide and conquer are of the order O(nlogn) where n is the number of the data points [58], [59] and are quite fast: it takes less than one second to tessellate a surface with 18,000 on a R4400, 250Mhz, SGI workstation. In this approach there is not a predetermined structure but a topology can be created by the adjacency relationships among the triangles. This allows the surface to adhere to complex geometrical shapes. Its drawback is that the tessellated surface (volume) has to be convex. A Kohonen map [26] can be seen as a regular grid which is attracted by the ensemble of the points with a soft-max adaptation rule, that is the meshes of the grid are moved and deformed to adhere to the points such that close points will have meshes oriented in similar directions. Its drawback is a certain rigidity in the structure which may create border effects when reconstructing complex surfaces. A composited approach is that of Singh et al. [60] who represented the different segments of the body by Polymesh rigid Models connected at the joints by 3D deformable models. This procedures reduces the number of parameters that have to be given to completely specify the motion of the entire structure. It should be remarked that while Snakes, Kohonen maps and Superquadrics provide an implicit filtering of the data on which the surfaces are reconstructed, Nurbs and Delauney tessellation do not. The use of patchy surfaces is only a theoretical drawback; in fact, as rendering is done on planar polygons, Nurbs, Snakes and Superquadrics have to be sampled and transformed into a set of polygons to be output on the workstations displays. There is still a gap between the rendering capabilities of these systems and the real appearence of the objects, in particular of objects like the human face. In fact, through it we communicate a host of emotions, thoughts, impressions which go beyond the verbal content [61]; this is based on subtle differences in facial expressions achieved thanks to the fluidity, individuality, and complexity of the face musculature which complicate the attempts to reproduce it [62]. Models which specifically take into account the complex physiology underlying human face, and body in general, are under development. The first system can be considered that of Platt and Badler [63], who simulated the human face using a three layered model which comprehended skin, muscles, and bones. Its modern counterpart are the finite-elements models [64]. In particular, Essa and Pentland [65], used these models to extend the work of Platt and Badler by associating to a 3D wire frame model of a face, properties of elasticity, viscosity and inertia derived from the mechanical properties of bones, muscles and skin [66]. Through this model, a regularised and reliable temporal sequence of muscles activation could be recovered. This system was proposed for the categorisation of facial expression by extending the Facial Action Coding System or FACS model proposed by Ekman and Friesen [67].

The problem is now shifted on the acquition of the 3D position of an ensemble of points belonging to the surface which has to be reconstructed. The solution is provided by what are called the 3D scanners. In particular, at the moment, the Cyberware™ is the only commercially available product. It consists of either two or four scanning instruments mounted on a vertical tower 2 meters high for a scanning volume of 2m x 1.20m (height x diameter). The 3D points on the surface are acquired by sampling Infra Red (IR) light spots, projected on the surface by a laser beam, through a stereo optical system. The texture is captured at the same time by a video camera equipped with an IR filter. During the scanning procedure, the heads carrying the laser IR emitters, the cameras and the detectors are moved from bottom to top in about 20 seconds recording up to 60,000 3D points on the surface. The final product is a 3D textured model. Its drawback lies in its rigid structure which cannot be easily adapted to objects of different size.

A similar approach has been proposed by [68]. A set of ten laser spots are projected on the object surface by a automatic or manual scanning procedure. The latter procedure allows a differential scanning of the surface and more dense sampling can be carried out in particularly critical regions. The laser spots are detected by a set of cameras and their position computed by the Elite system at a rate of 1000 points/sec. The cameras position can be decided on the basis of the object to be acquired and this allows the maximum freedom in the set-up which can be tailored according to the particular application. The acquired data are affected by measurement errors due to human tremor, measurement errors, registration errors, which make the reconstruction of the surface a problem theoretically ill-posed: a surface which would touch all the sampled points may exhibit undesirable ripples. A possible solution is to resort to vector quantization techniques which transform the original set of 3D data into a smaller set of 3D reference vectors. This has a double effect: it reduces the number of 3D points to be processed and it allows to filter out the noise in the data. A procedure particularly suitable to this task is Neural-Gas [69] which is an evolved version of Nearest Neighbour quantizer. It is an iterative algorithm, based on a soft-max adaptation rule, which allows to get close to the optimal solution with some reliability. This technique has been further potentiated by the Hyperbox processing procedure [70]. In this procedure, the volume occupied by the 3D data points is partitioned into a set of parallelepipeds (the hyperboxes). A number of reference vectors, proportional to the number of data points in that box, is generated in random location inside each hyperbox. This preprocessing reduces the number of data points which have to be taken into account when updating the position of the reference vectors because only those data points which belong to neighbour hyperboxes have to be examined. Moreover, this feature becomes extremely useful in reducing the computational time when several images compressed at different rates have to be obtained from the same data. Applying the Delaunay triangulation to the obtained reference vectors, a smooth version of the object is reconstructed. The reduction of the number of polygons allows to render an object at different resolutions. This becomes particularly powerful when it can be associated with gaze directed rendering [47] where different perceptual phenomena are exploited, namely: Kinetic, Central/Peripheral and Fusional Vision. For the Kinetic effect, an object is rendered with a resolution inversely proportional to its velocity; Central/Peripheral effect allows to render and objects far from the direction of gaze with a lower resolution with respect to the object on which the eyes are focused. Moreover, objects which are closer and farther away from the plane on which the eyes are focused have a different parallax and the fusion of the two images, required to get the 3D image, is not perfect; these observations allow to reduce the resolution of objects much closer or further away from the object focused [71]. Adaptive texturing can also be proposed, where the texture definition changes with the viewing conditions.

2.3 The Graphical Engine

It allows to display the virtual world from a particular vantage point which is a function of the data coming from the input system. It operates on the set of primitives which constitutes the virtual world. These are the operations carried out by the graphical engine: geometrical transformations (perspective projection), removing hidden lines (z-buffering) and rendering (colour, shading, texture, anti-aliasing). For higher end graphical systems, like SGI Infinite Reality, the world *visual computing* has been proposed: complex 3D objects can be moved in real time, allowing the users to simulate, and often predict, certain characteristics of the real world (e.g. in fluido-dynamics the modification of the flow of particles due to changes of external or internal conditions can be computed and displayed in real-time). The adopted hardware solution are an extensive parallelization, pipe-lining and caching through a primary and a secondary memory.

The bottleneck in achieving VR real-time environment is *collision detection* [72]. For n polyhedral objects with E and F faces respectively, the intersection check for all the pairs of faces which intersect is of the order $O((n-1)^2 EF)$. The most efficient methods are based on a hierarchical approach which establishes a correspondence between the polyhedral faces and the voxel occupancy [73]. A particularly general approach is that proposed by Smith et al. [74]. The objects are first encapsulated into a parallelepiped bounding box, axes-aligned, whose location and orientation is updated with object movement. For each pair of bounding boxes which intersect, the overlap volume is determined. This is a parallelepiped itself and it represents the region in the total volume where an intersection might occur. Then, the object faces which lie inside the overlap region are identified and inserted into a faces checklist, one for each object. to check intersection. The faces in the two lists are now grouped into pairs and a hierarchical octree search is applied: the working volume is progressively subdivided into eight cubes up to a predefined threshold. The voxels obtained with such procedure which contain two or more faces in the check lists are identified. To simplify the computational burden, only those voxels (or children) which contain at least one face in two checklists are further subdivided. For each obtained voxel, all possible pairs of faces are extracted and inserted in a face pair checklist to be tested for intersection. A different approach to objects collision is based on finding the plane separating two objects which is determined starting from the two closest vertices on the two objects [75]. Some advantage can be obtained when the objects motion is not jerky by taking explicitly it into account and checking only those faces towards the direction of motion [76]. The collision systems are still far from being real-time: the intersection of two spheres of about 4000 faces with the algorithm of Smith et al. [74], requires about 70msec on a R4400 150Mhz SGI workstation; and techniques which are based on a parallelization of the computation are under study [77].

Another open problem in VR, due to the restricted bandwidth, is the animation of subjects over a network, a problem encountered in virtual video-conferencing and virtual parks. These problems are well known in the domain of cartoon animation where very complex models are build. Animators use two basic techniques: "Key frame" and "Stop motion". Stop motion is very expensive as each frame is captured live from an actor or a probe positioned in a proper location or orientation. In the Key frame technique, few frames are captured and the interposed frames are realised by interpolation: the motion realism is left to the animator himself. Moreover, Key frame animation is not practical for long sequences (78). Until few years ago, animation techniques required many hand drawings or photocameras pictures to be done. With the new high-end SW packages, this work is going to be computerised and it becomes clear why the motion capture systems are gaining much interest in the field of animation, and in general of VR. They allow the acquisition of complex human motion which can be used to animate realistic models [79]. Once the motion and the 3D structure of the body have been reconstructed, the final result is obtained by applying colour, texture and special effects. The final 3D images displayed on the screen are shown in authentic colours and textures, with subtleties of light and shadow.

2.4 The Output Systems

Two main techniques are used for 3D visualisation: large screen projectors and stereo displays. When a virtual object is supposed to be seen at a distance approximately less than six meters, it produces on the retina of the two eyes an image which is offset horizontally by an amount which is proportional to the interpupillary distance. A visual 3D effect can therefore be obtained by projecting two computer generated images, offset by the inter pupillary distance, separately in the two eyes. Two are the main parameters which characterise stereo-displays: the amplitude of the field-of-view (fov) and the resolution. The fov is particularly critical for the sense of immersivity, in fact humans fov is very large being 180 degrees horizontally and 150 degrees vertically, with a central fov of about 90 degrees [80]. The highest resolution which can be achieved is that of the high resolution monitors, namely 1280x960. At the present state of the art, a 3D Immersive visualisation of high quality is beyond reach because of technological problems.

The simplest stereo display are the eye-glasses. These are glasses which have an incorporated shutter synchronised with the computer monitor. The images are output on the computer screen at a frequency of 120Hz or 144Hz and are seen alternatively by the two eyes producing a stereo effect due to binocular fusion in the visual cortex. This system, although inexpensive (about 1,000$) can be considered only semi-immersive as the surrounding scene in the room can be seen. With a monitor of 19 inches viewed at about 70 cm, the fov is about 37 degrees horizontally and 28 degrees vertically (with a resolution of about 35 pixels/degree). In order to avoid the sight of the surrounding, head-mounted displays (HMD) have been proposed. These systems put two small displays directly in correspondence to the subject eyes. These displays are usually very small CRT screens which are preferred to LCD for two reasons: first the response of fastest LCD is estimated in 40msec which is too slow for real-time applications (60Hz) and second the LCD are not able to display multiple levels of shading as they have a limited contrast ratio (it does not exceed 20:1 with respect to 100:1 of the CRT, [81]). The drawback of HMDs is that resolution degrades rapidly by increasing the fov. Most of the HMDs have a fov in the range of 50-80 degrees which produce a tunnel like effect, that is the image is seen as it were formed at the end of a tunnel. The highest quality HMD is that of n-Vision which allows a 1280 x 960 resolution with a 50 degrees of fov monocular. This field of view can be increased when the two images are projected on the two CRT displays offset to reproduces stereo disparity. Its price is around 60,000$. At lower price and lower quality, the FS5™ of Virtual Research, declares a resolution of 640 x 480 with a 50 degrees of fov monocular (79 degrees with 50% overlap). This means that there is a central region of the fov where vision is binocular and two lateral regions where vision is monocular from the right and the left eye. It should be remarked that with partially overlapped fov, the fov is increased only horizontally. These HMD are heavy and cumbersome for the subjects. A different approach is that of the Boom™ display

(Binocular Omni-Orientation Monitor, [82]). Its main characteristic is that the HMD is mounted over a counterbalanced articulated arm with six degrees of freedom which makes the HMD light; a set of strain gauges on the articulated arm allows to measure the position and orientation of the HMD with respect to the environment with very high accuracy. Its drawback is that subject motion is limited by the supporting arm. The BOOM HMD comes with different optics which allow to get different fovs, up to 140 degrees with very wide lenses (in this extreme condition, the distortion which appears on the edges are supposed to enhance the wide-field effect by mimicking the properties of natural peripheral vision [47], [71]. The colour images on these HMD systems are produced with the field sequential technique: three images are displayed sequentially on three monochrome tubes, each image corresponding to a primary colour: Red, Green and Blue. Between the tube and the lenses is a colour shutter which changes filtering characteristics depending on which colour is being displayed. The frame rate of this technique is 180Hz which allows an image rate of 60Hz by mixing the three images in the primary colours in an even fashion, drawing each colour physically on top of the other. This technique allows to avoid pixellization and to perceive full-colour continuos interlaced images. A particular system which might be a window on the future are the i-glasses!™ These are a very low cost stereo-displays which have become extremely popular in the games industry. They are a pair of glasses carrying a miniaturised CRT for each lens and have a resolution of 480 x 360 and a fov of 30 degrees; they are very compact and easy to carry, not cumbersome as the high-end helmets.

An alternative approach to 3D visualisation is based on the consideration that binocularity is only one of the mechanisms adopted from low-level vision to get 3D information on the surrounding environment. Monocular cues, like shading [83] and optical flow [84] are equally important and they are considered the main source of 3D information at large distances (approximately above six meters) when the eyes convergence at infinite and the two images on the retina are essentially the same. In this approach, the image is projected with very high resolution over very large, about 180 degrees, emi circular screens (IMAX™ and OMNIMAX™ projections). As soon as the imagemoves, the viewer has the compelling feeling that the scene is steady and that he is flying over the projected scene.

The output systems are completed by the audio devices which are typically stereo-headphones which may be integrated in the head-mounted displays. To get real sounds, models which take into account the bi-anural delays as a function of the sound localisation and the selective frequency attenuation induced by the material of the environment, are adopted. The true stereo-effect spatially synchronised with the scene, although it does not increase apparent realism, it does increase the sense of presence [85].

The VR as a whole is not always perceived as a comfortable environment because of what it is usually called VR or sim (simulators) sickness. This is a sense of nausea which surges some time after beginning the immersion in a VR environment. Although it is considered polygenic in nature, the main cause of motion sickness has been ascribed to a "cue conflict" between the visual and vestibular inputs [86] and it is shared with simulators and large screen projections.

3 Applications

One of the main contractors of VR projects is undoubtedly the Army. Systems for simulation and training both on-site and over the network are under development [87], [88], [89].

In medicine, anatomy, surgery and rehabilitation are interested by the VR development. In anatomy, 3D atlas of the human body have been realised. They allow to navigate inside the human body and to the various 3D organs in real colours and texture. In surgical practice, virtual specimens will replace cadavers with a double advantage. First, the surgeon will be trained to repeat many times the same sequence of actions proposed with absolute accuracy; and second it will allow an extensive practice outside the surgery room ("Thousands of operations to become really proficient, who would like to be in the first hundred of cases?"). These systems are already available for training to particular surgeries which require very high accuracy in the position/orientation of the instruments and of the path inside of the body (eye surgery and endoscopy). In plastic surgery VR systems allow to visualise the effect of the surgery and to interactively decide the best solution for the patients. These systems are the natural evolution of the artificial probes which have been used since the most ancient time: in India, b.C., plastic surgeons used to work on a lief to learn and exercise on how to remodel the nose of a person. At the present status of the art, these systems lack of the capability to reproduce the tactile feedback, and in particular the force feedback, which is experienced in the real surgery. The first models which will take this into account are under development [90]. They are focused on the lower leg surgery and they are based on a multi-layer stratification with finite element design [63], [66]. In the different layers detailed biomechanical models of the muscles, tendons, flesh and skin will be added [91], [92] whose parameters are set to reproduce movements and deformations of the real human bodies. VR is also one of the new frontiers in Rehabilitation: a patient can analyse in real-time his impaired 3D movements with respect to the movement prescribed by the physician. Moreover, to reinforce the confidence on a different modality (e.g. tactile versus vision, vision vs vestibular), one of the sensory modalities can be switched off or manipulated .

In science, engineering, physics and chemistry worlds can be displayed three-dimensionally and modifications introduced by the user can be computed and visualised in real-time.

VR can also be an effective tool to investigate the time evolution and the self-organisation of both Artificial Neural Networks and realistic models of the neural networks in brain areas.

3D visualisation and interactivity can boost the sales by mail: markets like clothing on demand will be available in the next future and all the other mail markets will be boosted as well.

Input and output systems are new much more powerful tools to interface users with a computer.

A domain which is going to be a large market in the next years are the theme parks [93]. These are an extension of the Network Cafè, and they will be a place where people can not only talk through the network, but also meet each other, play and chat together, interact in a virtual pleasant place.

Entertainment (games) market is growing extremely fast with two main contenders: Sega and Nintendo. Their systems are based on the use of simplified polygonal models and simplified texture and shading as their bottleneck is in the rendering speed which does not exceed the 3000 polygon/seconds in the new platforms Ultra64™ by Nintendo and Saturn™ by Sega.

VR is extremely valuable in architecture and engineering design, as well as urban planning and car design. Plans and virtual probes can be seen three-dimensionally and modified interactively according to the customs need, possibly by working on the same project over the network.

A chapter on its own is monument architecture in which the Italian company Info Byte is one of the world leaders. It has produced a set of immersive videos on Italian art towns: Rome, Assisi and Florence. These can be visited both in the actual time and at the time of their maximum splendour.

4 Conclusion

VR is an advanced technology which provides a human-computer interface with a sense of immersion, interactivity, navigation and exploration of worlds generated by the computers. When we are interacting with a computer we are not talking to someone else, a new world is being explored; the computer is not bounded to be only a machine, it becomes an instrument for creativity.

Acknowledgements

I wish to thank Mr. S. Sensolo, U. Di Giovanni and F. Neutro for their enthusiastic cooperation and skilled artwork, and I acknowledge the contributions of Dr R. Savarè, S. Ferrari, G. Baroni and G. Ferrigno for some of the presented material.

References

1. Ellis SR. What Are Virtual Environments? IEEE CG&A 1994; 1:17-22
2. Woolley B. Virtual Worlds. Blackwell Publishers, 1992
3. Barthes R. Contemporary Mythes. Wiley & So, 1970
4. Comeau CP. and Brian J.S. Headsight Television System Provides Remote Surveillance. Electronics 1961; 86-90
5. Sutherland IE. Computer Displays. Scientific American 1970; 222(6):56-81
6. Sturman DJ. and Zelter D. A Survey of Glove-based Input. IEEE CG&A 1994; 30-39
7. Bolt RA. Put-that-there: Voice and Gesture at the graphics Interface. Computer Graphics (Proc. Siggraph) 1980; 14(3): 262-270
8. Burdea G and Coiffet P. Virtual Reality Technology. Wiley & So, 1994
9. Muybridge E. The Human Figure in Motion. Chapman & Hall, London, 1901
10. Marey EJ. The history of Chronophotography, Smithsonian Institution, Washington D.C., 1902
11. Tomasi C and Kanade T. Shape and Motion from Image Streams: under Orthography a Factorization Method. Int. J. Comp. Vision 1991; 9(2):137-154

12. Essa IA, Darrell T and Pentland A. Tracking Facial Motion. Proc. IEEE Workshop on Nonrigid and Articulate Motion, Ausin, Texas, 1994
13. Borghese NA, Di Rienzo M, Ferrigno G and Pedotti A. Elite: a goal-oriented vision system for moving objects detection. Robotica 1990; 9:275-282
14. Moezzi S, Katkere A, Kuramura DY and Jain R. Immersive Video. Proc. IEEE VRAIS96, 1996, pp 17-24
15. Goncalves L, Di Bernardo E, Ursella E and Perona P. Monocular tracking of the human arm in 3D. Proc. Int. Conf. Comp. Vision, Cambridge, MA 1995
16. Wellner P. Interacting with Paper on the DigitalDesk. Comm. ACM 1993; 36(7):87-96
17. Poggio T, Torre V and Koch C. Computational Vision and Regularization theory. Nature 1985; 317:314-319
18. Poggio T and Girosi F. Regularization Algorithms for Learning That Are Equivalent to Multilayer Networks. Science 1990; 247:978-982
19. Soatto S, Frezza R and Perona P. Motion Estimation on the Essential Manifold, Springer-Verlag, Heidelberg, 1994, pp 61-72 (Lecture Notes in Comp. Science, no. 801)
20. Hildreth EC and Koch C. The Analysis of Visual Motion: From Computational Theory to Neuronal Mechanisms, Ann. Rev. Neurosci. 1987; 477:533
21. Shi J and Tomasi C. Good Features to Track, Tech. Rep. Dep. Comp. Science, Cornell University, TR 93-1399, 1993
22. Barron JL, Fleet DJ and Beauchemin SS. Performance of Optical Flow Techniques, Int. J. Comp. Vision, 1994; 43-77
23. Mase K and Pentland A. Automatic Lip-Reading by Optical-Flow, Systems and Computers in Japan 199; 22(6).
24. Metaxis D and Terzopoulos D. Shape and non-rigid motion estimation through physics-based synthesis. IEEE Trans. PAMI, 1991; 22(6):67-76
25. Lorensen WE and Cline HE. Marching Cubes: A high resolution 3D surface reconstruction algorithm. Comp. Graphics 1987; 21(4):163-169
26. Kohonen T. Self-Organizing maps, Springer-Verlag, Heidelberg, 1995
27. Piegl L. The NURBS book, Springer-Verlag, Heidelberg, 1993
28. Kass M, Witkin A and Terzopoulos D. Snakes: Active contour models. Int. J. of Comp. Vision 1988; 321-331
29. Hatze H. Quantitative analysis, synthesis and optimization of human motion. Human Movement Science 3 1984; 5-25
30. Ferrigno G and Pedotti A. ELITE: A Digital Dedicated Hardware System for Movement Analysis Via Real-Time TV Signal Processing. IEEE Trans. on Biomed. Eng. 1990; 32:943-949
31. Burdea G and Coiffet P. Virtual Reality Technology, Wiley & So, 1994
32. Borghese NA, Ferrigno G and Pedotti A. 3D Movement Detection: A Hierarchical Approach. Proc. 1988 IEEE Int. Conf. Systems, Man, and Cybernetics, Vol. I, , Pergamon Press, 1988, pp 303-306
33. Ferrigno G, Borghese NA, and Pedotti A. Pattern Recognition in 3D Automatic Human Motion Analysis - ISPRS J. of Photogr. and Rem. Sens. 1990; 45:227-246
34. Borghese NA. SMART3D: Tracking of movements surveyed by a multiple set of TVcameras, Proc. X Symp. ISBS, 1992, pp108-111

35. Hu X and Ahuja N. Marching Point Features with Ordered Geometric, Rigidity and Disparity Constraints. IEEE Trans. PAMI 1994; 16(10):1041-1049

36. DeFanti TA and Sandin DJ. Final Report to the National Endowment of the Arts. Tech. Rep. Univ. Illinois, NEA R60-34-163, 1977

37. Grimes GJ. Digital Data Entry Glove Interface Device, Bell Telephone Laboratories, Murray Hill, NJ, US Patent 4,414,537, 1983

38. Kramer J and Leifer L. The Talking Glove: An Expressive and Receptive 'Verbal' Communication Aid for the Deaf-Blind, and Non-vocal, Tech. Rep., Dept. Electr. Eng., Stanford University, 1989

39. Weng J, Cohen P and Herniou M. Camera Calibration with Distortion Models and Accuracy Evaluation. IEEE Trans. PAMI 1992; 14(10):965-979

40. Wolf PR. Elements of Photogrammetry, McGraw Hill, 1983

41. Bergamasco M. The Glad-in-Art Project. Proc. VR93, Springer-Verlag, Heidelberg, 1993, pp 251-258

42. Millman PA and Colgate JE. Design of a Four Degrees of Freedom Force-Reflecting Manipulandum with a Specified Force/Torque Workspace. Proc. IEEE Robotics and Automation, vol. 2, 1991, pp 1488-1493

43. Buttolo P and Hannaford B. Pen-Based Force Display for Precision Manipulation in Virtual Environment. Proc. IEEE VRAIS95, 1995, pp 217-224

44. Hirota K, Hirose M. Surface Display: A Force Feedback System Simulating the Surface of and Object. Proc. IEEE Int. Workshop on Robot and Human Comm., 1994, pp 251-254

45. Salcudean SE and Yan J. Towards a Force-Reflecting Motion-Scaling System for Microsurgery. Proc. IEEE Robotics and Automation, vol. 3, 1994, pp 2296-2301

46. Gomi H. and Kawato M. Equilibrium-Point Control Hypothesis Examined by Measuring Arm Stiffness During Multijoint Movement. Science, 1996; 272:117-119

47. Ohshima T, Yamamoto H, and Tamura H. Gaze Directed Adaptive Rendering for Interacting with Virtual Space. Proc. IEEE VRAIS96, 1996, pp 103-110

48. Chiou GI and Hwang J. A Neural Network-Based Stochastic Active Contour Model (NNS-SNAKE) for Contour Finding of Distinct Features. IEEE Trans. on Imag. Proc. 1995; 4(10):1407-1416

49. Terzopoulos D and Waters K. Analysis and synthesis of facial image sequences using physical and anatomical models. IEEE Trans. PAMI 1993; 15(6):569-579

50. Terzopoulos D and Metaxas D. Dynamic 3D Models with Local and Global Deformations: Deformable Superquadrics. IEEE Trans. PAMI 1991; 13(7):703-714

51. Moody J and Darken CJ. Fast Learning in Networks of Locally-Tuned Processing Units. Neural Computation 1989; 1:281-294

52. Poggio T and Girosi F. Regularization Algorithms for Learning that are Equivalent to Multilayer Networks. Science 1990. 247:978-982

53. Poggio T and Girosi F. Networks for Approximation and Learning. Proc. IEEE 1990; 78(9):1481-1497

54. Marchini S and Borghese NA. Optimal local estimation of RBF parameters, Proc. ICAN94, Lawrence Erlbaum, 1994 pp 71-75

55. Watson DF. Computing the n-dimensional Delaunay tessellation with application to Voronoi polytopes. The Computer Journal 1981; 24(2):167-172

56. Fang T and Piegl L. Algorithm for Delaunay triangulation and convex hull computation using sparse matrix. Computer Aided Design 1992; 24(8):425-436

57. Cignoni P, Montani C, Perego R and Scopigno R. Parallel 3D Delaunay Triangulation. Proc. Eurographics 93, 1993, pp 129-142

58. Ruprecht D and Muller H. Image Warping with Scattered Data Interpolation. IEEE CG&A 1995; 37-43

59. Oishi Y and Sugihara K. Topology-Oriented Divide-and-Conquer Algorithm for Voronoi Diagrams. Graphical Models and Image Processing 1995; 57(4):303-314

60. Singh K, Ohya J and Parent R. Human Figure Synthesis and Animation for Virtual Space Teleconferencing. Proc. IEEE VRAIS 95, 1995, pp 118-126

61. Summerfield Q. Visual Perception of Phonetic Gestures. Modularity and the Motor Theory of Speech Perception, Mattingly G and Studdert-Kennedy M(eds), Lawrence Erlbaum Associates, Hillsdale, N.J, pp 117-137

62. Rosen JM, Laub Jr DR, Pieper SD, Mecinski AM, Soltanian H, McKenna MA, Chen D, Delp SL, Loan JP, and Basdogan C. Virtual Reality and Medicine: From Training Systems to performance Machines. Proc. IEEE VRAIS96, 1996, pp 5-13 ·

63. Platt SM and Badler NI. Animating Facial Expressions. Computer Graphics 1982; 15(3).

64. Bathe K. Finite Element Procedures in Engineering Analysis, Prentice-Hall, 1982.

65. Essa IA and Pentland AP. Coding, Analysis, Interpretation and Recognition of Facial Expressions. IEEE Trans. PAMI 1996, in press

66. Pieper S, Rosen J and Zeltzer D. Interactive graphics for plastic surgery: A task level analysis and implementation. Computer Graphics, Special Issue: ACM Siggraph, 1992; 127-134

67. Ekman P and Friesen WV. Facial Action Coding System, Consulting Psychologists Press Inc, Palo Alto, CA, 1978

68. Savarè R, Aliverti A, Ferrigno G and Pedotti A. Surface Analysis by Laser Beam Scanning and Sterephotogrammetry. Proc. 3rd European Conference on Engineering and Medicine, Firenze, 1995

69. Martinetz TM, Berkovich G and Schulten KJ. Neural-Gas Network for Vector Quantization and its Application to Times-Series Prediction. IEEE Trans. on Neural Networks 1993; 4(4):558-569

70. Fontana M, Borghese NA and Ferrari S. Image reconstruction using improved "Nerual Gas". Proc. WIRN95, 1995, pp

71. Rokita P. Generating Depth-of-Field Effects in Virtual Reality Applications. IEEE CG&A, 1995; 18-21

72. Pentland A. Computational Complexity versus simulated environments, Computer Graphics 1990; 24(2):185-192

73. Garcia-Alonso A, Serrano N, and Flaquer J. Solving the collision detection problem, Computer Graphics and Applications 1994; 14(3):36-43

74. Smith A, Kitamura Y, Takemura H and Kishino F. A Simple and Efficient Method for Accurate Collision Detection Among Deformable Polyhedral Objects in Arbitrary Motion, Proc. IEEE VRAIS95, 1995 pp 136-144

75. Gilbert H, Johnson GE and Keerth SS. A fast procedure for computing the distance between complex objects in three-dimensional space. IEEE J. of Robotics and Automation 1988; 4(2):193-203

76. Vanecek G. Back-face culling applied to collision detection of polyhedra. Tech. Rep., Purdue University, Dept. of Comp. Science, 1994

77. Kitamura Y, Smith A, Takemura H, and Kishino F. Optimization and parallelization of octree-based collision detection for real-time performance. Proc. IEICE Conference, 1994

78. Parke FI. Parametrized models for facial animation, CG&A 1982; 61-68

79. Granieri J and Badler NI. Simulating humans in VR. Virtual Reality and its Applications, Earnshaw R, Jones H, and Vince J (eds), Academic Press, London, UK, 1995

80. Foley JD. Interfaces for Advanced Computing, Scientific American 1987; 127-135

81. Oromaner J. Flat Panels Proliferate and Challenge the CRT, Electronic Products 1992; 34(8):21-24

82. Bolas M. Human Factors in the Design of as Immersive Display, IEEE CG&A 1994; 55-59

83. Sun J and Perona P. Early Computation of Shape and Reflectance in the Visual System. Natue 1996; 379:165-169

84. Gibson JJ. The Senses Considered as Perceptual Systems. Houghton-Mifflin, Boston, 1966

85. Hendrix C and Barfield W. Presence in Virtual Environment as a Function of Visual and Auditory Cues, Proc. IEEE VRAIS95, 1995, pp 74-82

86. Kolasinski EM. Simulator Sickness in Virtual Environments, Tech. Report, US Army Research Institue, Orlando FL, 1996

87. Granieri J and Crabtree N. Production and Playback of Human Figure Motion for 3D Virtual Environments, Proc. IEEE VRAIS95, 1995, pp 127-135

88. Badler N, Webber B, Becket W, Geib C, Moore M, Pelachaud C, Reich B and Stone M. Planning for animation. Interactive Computer Animation (Magnenat-Thalman N and Thalman D eds), Prentice-Hall, 1995

89. Stansfield S, Shawver D, Rogers D and Hightower R. Mission visualization for planning and training. IEEE CG&A 1995; 15(5):12-14

90. Rosen JM, Laub DR, Pieper SD, Mecinski AM, Soltanian H, McKenna MA, Chen D, Delp SL, Loan JP and Basdogan C. Virtual Reality and Medicine: From Training Systems to Performance Machines. Proc. IEEE VRAIS96, 1996, pp 5-13

91. Delp SL, Loan JP, Hoy MG, Zajac FE, Topp EL and Rosen JM. An interactive graphics-based model of the lower extremity to study orthopaedic surgical procedures. IEEE Trans. Biomed. Eng. 1990; 37(8):757

92. Delp SL, and Zajac FE. Force and Moment generating capacity of the lower-limb muscles before and after tendon lengthening. Clin. Orthopaedics and Related Res. 1992; 284:247-259

93. Barrus JW, Waters RC and Anderson DB. Locales and Beacons: Efficient and Precise Support for Large Multi-User Virtual Environments. Proc. VRAIS96, 1996, pp 204-213

Recent Results in On-line Prediction and Boosting

Nicolò Cesa-Bianchi   Sandra Panizza

DSI, Università di Milano

via Comelico 39

20135 Milano, Italy

`{cesabian,panizza}@dsi.unimi.it`

1 Introduction

The successful design of practical algorithms for solving a class of problems very often depends on the existence of a formal model where algorithmical ideas can be developed, analyzed, and compared. In the case of machine learning, a number of such formal models have been proposed in the past. Some of them have been successful in generating elegant mathematical results but, on the other hand, they have had a rather limited impact on the practical side. In this paper, we suggest that the on-line prediction model is a good source of interesting algorithmic ideas with a great potential for new applications. To this end, we will describe a simple algorithm based on "multiplicative weights," we will analyze this algorithm within the on-line prediction model, and finally we will show some of its variants and applications.

In the on-line prediction model a learner (or predictor) must predict, one by one, the elements of an unknown sequence. At the t-th round, the learner observes the t-th element y_t of the sequence and must generate a prediction $\widehat{y}_{t+1}$ for the $t + 1$st element y_{t+1}. This prediction is evaluated by means of a fixed function measuring how much $\widehat{y}_{t+1}$ differs from y_{t+1}; we call this difference the learner's loss at round t. For boolean predictions we measure the performance both of the experts and of the algorithm with the *discrete loss*, which simply counts the number of prediction mistakes. Instead, when the learner's predictions $\widehat{y}_t$ range in the continuous interval $[0, 1]$, we use the *absolute* loss $|y_t - \widehat{y}_t|$. A crucial aspect in this model is that nothing is assumed about the mechanism generating the sequence $y_1, y_2, \ldots$ of elements to predict. Of course, under this assumption we can not construct a learner that incurs a small total loss on *any* sequence. To fix this we pose a *relative* goal, that is we want to compare the loss of the learner with the loss of the best element in a given set of learners which we call *comparison class*. By best element we mean the learner in the comparison class having the smallest loss on the particular sequence that was actually observed. In this respect, our goal is to develope learning algorithms whose losses are "not much worse" than the loss of the best learner in the comparison class. To avoid confusion, we call the learners in the comparison class "experts". We use this term because our learning algorithm will use the predictions of these learners as an advice for computing his own prediction. The learner and the experts might also use some *side information* which, in our case, takes the form of a sequence $x_1, x_2 \ldots$ where each x_t is interpreted as a set of observations to which the value y_t is associated. For example, in a whether forecasting problem, x_t might be yesterday's temperature, pressure, and humidity average measurements and $y_t \in \{0, 1\}$ is 1 if and only if it rains today. If such side information is available, then, before generating their predictions for the t-th element y_t, the learner and the experts are given access to the t-th observation x_t. However, all the learners we consider in this paper do not make any direct use of side information: their predictions depend only on the experts' predictions (which, in turn, may depend on the side information). Clearly, allowing the learner to make direct use of the side information can improve the bounds presented here.

Consider an on-line prediction problem where the y_t's belong to $\{0, 1\}$ and the sequence $x_1, x_2, \ldots$ of observations is such that $x_t \in [0, 1]^n$. Possible sets of experts for this problem are:

1. The functions $e_1, \ldots, e_n$ such that $e_i(x) = 1$ if and only if $x_i \geq 1/2$. These experts compute their predictions simply by thresholding one component of the vector of observations.

2. The functions of the form $g : \{0, 1\}^k \to \{0, 1\}$. In this case we assume $\widehat{y}_{t+1} = g(y_{t-k}, \ldots, y_t)$, i.e. the prediction of the expert associated to g depends only on the last k elements observed in the sequence, whenever available. Note that these experts do not make use of the observations x_t.

3. An arbitrary set $\{f_1, \ldots, f_N\}$ of $\{0, 1\}$-valued functions defined on $[0, 1]^n$.

Different algorithms for solving the on-line expert prediction problem have been recently proposed by Littlestone and Warmuth [15], Vovk [17] and others, both for the simple setting in which all predictions and outcomes are boolean, and for the more general setting in which the advice of the experts and/or predictions of the algorithms are in the continuous interval $[0, 1]$. The main idea underlining all the proposed methods is actually quite simple. For each expert E_i and each round t, the algorithm maintains a weight, $w_{i,t}$, and these weights are used to combine the advice of the N experts on bit y_t in order to produce the algorithm's own prediction. After the correct value is revealed, the weights of those experts giving bad advice are slashed by an update factor $\beta_t \in [0, 1)$. Algorithms based on such multiplicative weighting schemes have been shown to be, in some sense, asymptotically optimal both in the boolean [3, 18] and in the continuous [2] settings.

More precisely, if in the boolean setting we denote with $M_A(y)$ and $M_{\text{BEST}}(y)$ the total discrete loss (i.e. number of mistakes) on the sequence y incurred by, respectively, algorithm A and the best expert (for that sequence) in the given comparison class, then there exists an algorithm WM using a multiplicative weigthing scheme for which we can prove bounds of the form

$$M_{\text{WM}}(y) \leq 2M_{\text{BEST}}(y) + \log N + 2\sqrt{M_{\text{BEST}}(y)\log N} \tag{1}$$

for any sequence y.

When, instead, the algorithm's predictions are allowed to range in the interval $[0,1]$, there is a variant IG of the algorithm WM whose bounds are exactly half of the right-hand-side in (1). That is, by denoting with $\overline{M}_{\text{IG}}(y)$ the total absolute loss of IG on the sequence y, we obtain

$$\overline{M}_{\text{IG}}(y) \leq M_{\text{BEST}}(y) + \frac{\log N}{2} + \sqrt{M_{\text{BEST}}(y)\log N}.$$

Thus, for algorithm IG, the additional loss term over the loss incurred by the best expert, $\overline{M}_{\text{IG}}(y) - M_{\text{BEST}}(y)$, grows at a sublinear rate as a function of $M_{\text{BEST}}(y)$.

Generalizations of the expert prediction model to different loss functions have been studied by Vovk [18], Haussler, Kivinen, and Warmuth [11], Yamanishi [19].

More recently, these results have been extended to more general comparison classes for variants of the basic multiplicative weight algorithm. In particular, Kivinen and Warmuth [14] analyzed the on-line prediction model when the comparison class consists of linear combinations (with bounded coefficients) of arbitrary experts and proved worst-case bounds for a new family of multiplicative weighting algorithms called Exponentiated Gradient (EG) algorithms. Helmbold, Kivinen, and Warmuth extended these results to handle comparison classes consisting of sigmoided linear combinations of N experts (equivalent to a neural network with one hidden layer of experts and one output neuron with logistic activation function) and proved worst-case bounds for the EG algorithm and the entropic loss [12]. In particular, they were able to prove loss bounds of the form

$$\tilde{M}_{\text{EG}}(y) \leq \frac{4}{3}\left(\tilde{M}_{\boldsymbol{u}}(y) + U^2\ln(2N)\right)$$

where $\tilde{M}_{\boldsymbol{u}}(y)$ is the total entropic loss of EG and $\tilde{M}_{\boldsymbol{u}}(y)$ is the total entropic loss of the best sigmoided neuron with weight vector $\boldsymbol{u}$ such that $\sum_{i=1}^{N}|u_i| \leq U$.

The multiplicative weighting scheme is quite general and can be applied to a wide variety of learning problems. These include gambling, repeated games and predictions of points in R^n (see [7, 1]). Perhaps the most surprising of these applications is the derivation of a new algorithm for "boosting" weak learning algorithms [7]. This important application will be analyzed in the last part of the paper.

In the next section we define the on-line prediction problem with experts and describe a basic learning algorithm for this setting. We also propose some interesting generalization of the on-line prediction problem to the case where the learner is allowed to output randomized (or continuous) predictions. We then present a general Bayesian framework for on-line boolean prediction within which previously known algorithms as well as new ones can easily be derived. Some interesting applications of the multiplicative weighting algorithms are then presented in the second part of the paper. In particular, in Section 3 we apply the algorithms developed in Section 2 to solve an adversarial version of the multi-armed bandit problem. In section 4 some closely related algorithms are applied to derive a new boosting algorithm.

2 On-line prediction with experts

We now define the on-line model of prediction with experts. We will restrict our attention to the case when the values y_ts to predict are boolean.

Let $E_1, \ldots, E_N$ be N arbitrary experts. In this model learning takes place in a sequence of rounds. On each round $t = 1, \ldots, T$

1. each expert E_i generates a prediction $\widehat{y}_{i,t} \in \{0, 1\}$ (possibly based on side information),

2. the learner observes the predictions of all the experts and makes a prediction $\widehat{y}_t \in \{0, 1\}$,

3. the learner observes the correct bit y_t.

We assume the learner's prediction $\widehat{y}_t$ at time t depends only on the observed sequence $y_1, \ldots, y_{t-1}$ and on the past experts predictions $\widehat{y}_{i,1}, \ldots, \widehat{y}_{i,t-1}$ for $i = 1, \ldots, N$.

We use $M_i(y^T) = \sum_{t=1}^{T} |y_t - \widehat{y}_{i,t}|$ to denote the number of mistakes made by expert E_i on sequence $y^T = (y_1, \ldots, y_T)$. Similarly, we write $M_A(y^T) = \sum_{t=1}^{T} |y_t - \widehat{y}_t|$ to denote the number of mistakes on sequence y^T made by learner A. The goal of the learner is to minimize the number of mistakes relative to the best expert.

2.1 Weighted Majority

A basic learning algorithm for this setting is "Weighted Majority" (WM). The idea underlying this method is very simple. For each expert E_i, WM maintains a weight $w_{i,t}$, where t denotes the round number. Initially, $w_{i,1} = 1$ for each expert E_i, $i = 1, \ldots, N$. On each round t, the learner computes the weighted sums

$$W_t^0 = \sum_{i \in \mathcal{E}_0} w_{i,t} \qquad W_t^1 = \sum_{j \in \mathcal{E}_1} w_{j,t}$$

where $\mathcal{E}_0$ is the set $\{i : 1 \leq i \leq N, \widehat{y}_{i,t} = 0\}$ of the experts predicting 0 on the t-th round and $\mathcal{E}_1$ is defined analogously. Then the learner predicts according to the largest between W_t^0 and W_t^1 (or predicts arbitrarily in case of a tie).

After receiving the actual value of y_t, the learner adjusts the weight of those experts which made a mistake according to the multiplicative rule

$$w_{i,t+1} = w_{i,t} \cdot \beta^{|y_t - \widehat{y}_{i,t}|} \tag{2}$$

where $0 \leq \beta < 1$. The essential property of this weighting scheme is that the experts making many mistakes get their weights rapidly slashed, thus reducing their influence on the majority vote.

The analysis of the Weighted Majority algorithm is also very simple. Choose $T > 0$ and fix an arbitrary sequence $y^T \in \{0,1\}^T$. Let $W_t = \sum_{i=1}^{N} w_{i,t}$ be the total weight of the experts at round t. Then $W_1 = N$, $W_{t+1} \leq W_t$ on each round t and $W_{t+1} \leq W_t \cdot (1+\beta)/2$ on those rounds t where the learner makes a mistake. Thus

$$W_{T+1} \leq \left(\frac{1+\beta}{2}\right)^m$$

holds, where $m = M_{\mathrm{WM}}(y^T)$. As

$$W_{T+1} = \sum_{i=1}^{N} w_{i,T+1} \geq \beta^{M_j(y^T)} \qquad \text{for any } 1 \leq j \leq N$$

we find that

$$M_{\mathrm{WM}}(y^T) \leq \left\lfloor \frac{\log N + M_{\mathrm{BEST}}(y^T)\log(1/\beta)}{\log\frac{2}{1+\beta}} \right\rfloor \tag{3}$$

where $M_{\mathrm{BEST}}(y^T)$ is the number of mistakes of the best expert on the same sequence y^T.

Note that (3) holds for any $T > 0$ and for any sequence $y^T \in \{0,1\}^T$. Furthermore, the number of mistakes of WM on any $y \in \{0,1\}^*$ depends linearly both on $\log N$ (where N is the number of experts) and on the number $M_{\mathrm{BEST}}(y)$ of mistakes of the best expert on y.

Knowing some information on the sequence y to predict, such as an upper bound K on $M_{\mathrm{BEST}}(y)$, allows to choose β in order to minimize the right-hand side of (3). We start by rewriting (3) as

$$\begin{aligned} M_{\mathrm{WM}}(y) \\ \leq \quad \max\{q \in \mathbf{N} : q \leq \log N + q\log(1+\beta) + M_{\mathrm{BEST}}(y)\log(1/\beta)\}. \end{aligned} \tag{4}$$

Replacing $M_{\mathrm{BEST}}(y)$ by its upper bound K and choosing $\beta = K/(q-K)$ to minimize $q\log(1+\beta) + K\log(1/\beta)$ yields

$$M_{\mathrm{WM}}(y) \leq \max\left\{q \in \mathbf{N} : q \leq \frac{\log N}{1 - H(K/q)}\right\} \tag{5}$$

where H is the binary entropy function $H(x) = -\log(x) - (1-x)\log(1-x)$. A different (and more involved) analysis of (4) yields the closed-form bound

$$M_{\mathrm{WM}}(y) \leq 2M_{\mathrm{BEST}}(y) + \log N + 2\sqrt{K \ln N}. \tag{6}$$

Notes. The Weighted Majority algorithm has been proposed by Littlestone and Warmuth [15] who proved (3) along with a number of other bounds for variants of WM. Bound (5) has been first shown by Vovk [17] (see also [3] for related results). Finally, bound (6) is a direct derivation from a bound shown in [2] for a randomized variant of WM.

2.2 Bayesian interpretations

Although Weighted Majority is applied in a setting where no assumptions are made on the sequence to be predicted, it is possible to work out a nice Bayesian interpretation for this algorithm. To get this Bayesian interpretation, we can imagine that the sequence $y^T = y_1, \ldots, y_T$, rather then being "worst-case", is generated by first selecting an expert uniformly at random and then by corrupting the expert's predictions with an i.i.d. noise process with rate $0 \leq \eta < 1/2$. In other words, each bit of the unknown sequence $y^T = y_1, \ldots, y_T$ differs from that predicted by the selected expert with a fixed probability η.

In this probabilistic setting an optimal prediction strategy is defined by the Bayes decision rule that, on each round t, simply outputs the label with the highest posterior probability (having seen y^{t-1}), i.e.

$$\hat{y}_t = \arg \max_{y \in \{0,1\}} P(y \mid y^{t-1}).$$

This rule is especially easy to compute when we express the probability $P(y \mid y^{t-1})$ as a sum over the expert set $\{E_1, \ldots, E_N\}$,

$$P(y \mid y^{t-1}) = \sum_{i=1}^{N} P_i(y \mid y^{t-1}) \cdot P(i \mid y^{t-1}).$$

The first term in each summand, $P_i(y \mid y^{t-1})$, is simply the probability, under expert E_i, that the tth bit of the sequence will be y, given that the previous $t-1$ were $y_1, \ldots, y_{t-1}$. When the sequence is generated by the i.i.d. noise process described above, this probability is easily seen to be

$$P_i(y \mid y^{t-1}) = \begin{cases} \eta & \text{if } y = E_i(y^{t-1}), \\ 1 - \eta & \text{otherwise.} \end{cases}$$

The second term, $P(i \mid y^{t-1})$, represents instead the posterior probability of expert E_i given that we have observed the prefix y^{t-1}. Taken together, the sum $\sum_{i=1}^{N} P_i(y \mid y^{t-1}) P(i \mid y^{t-1})$ represents the contributions from each expert E_i, weighted according to its posterior probability, to the confidence that the next outcome will be $y \in \{0, 1\}$. By applying Bayes rule, we can compute the posterior probability $P(i \mid y^{t-1})$ in term of the prior probability $Q(i)$ of expert E_i which, for simplicity, we assume to be uniform, and the likelihood $P_i(y^{t-1})$ of y^{t-1} under E_i, i.e.

$$P(i \mid y^{t-1}) = \frac{P_i(y^{t-1})Q(i)}{\sum_{j \in J} P_j(y^{t-1})Q(j)}. \tag{7}$$

Here $P_i(y^0)$ is defined to be 1. In our case, it is easy to see that

$$P_i(y^T) = \prod_{t=1}^{T} P_i(y_t \mid y^{t-1}) = \eta^k (1 - \eta)^{T-k},$$

where $k = M_i(y^T)$ is the number of prediction mistakes made by E_i on the sequence y^T. Since the uniform prior makes the normalization in (7) equal for all i, Bayes optimal prediction $\hat{y}_t$ can be equivalently written as

$$\hat{y}_t = \arg \max_{y \in \{0,1\}} \sum_{i=1}^{N} P_i(y \mid y^{t-1}) \cdot P_i(y^{t-1}). \tag{8}$$

According to (8) the Bayesian predictor can be thought of as having, at the beginning of each round t, a current set of distributions $\{P_i(y^{t-1})\}_{i=1}^{N}$ for the experts, each $P_i(y^{t-1})$ representing the likelihood of y^{t-1} under E_i. Then, by using these distributions, Bayes determines the conditional probabilities of each label and predicts according to the more likely one. Following each round, the distributions assigned to the experts are updated based on the experts' posterior probabilities of the bit y_t given the prefix y^{t-1}.

It is now quite intuitive how to rephrase this Bayesian predictor in term of weighted voting schemes in which the connection to the probability of the sequence can be abandoned. More precisely, when predicting the bit y_t, each expert E_i should have the weight $P_i(y^{t-1})$. Furthermore, during round t, each expert E_i should vote with a fraction $P_i(E_i(y^{t-1}) \mid y^{t-1})$ of its weight for its own prediction, $E_i(y^{t-1})$, and with the remaining fraction for the opposite prediction. After the bit y_t is revealed, each expert's weight should be updated, multiplying it by the factor $P_i(y_t \mid y^{t-1})$. It is easy to verify that this algorithm generates the same predictions and experts' weights of the WM algorithm whenever the noise rate η and WM's parameter β satisfy the relation $\eta = \beta/(1 + \beta)$.

The Bayesian framework that we have described so far is based on the simplicistic assumption that the noise process is characterized by a fixed rate $0 \leq \eta < 1/2$. As already mentioned in Section 2.1, if the algorithm knows in advance a bound K on the number of mistakes made by the best expert, then we can optimally tune the parameter β as a function of N and K, thus obtaining a better performance for the algorithm. From a Bayesian point of view, this amounts to use this additional information to create a more accurate probabilistic model of the world.

Without such strong prior knowledge, a reasonable approach is to assume a prior distribution on the noise rate η and use evidence from the sequence and the prior to dynamically estimate the (assumed fixed) noise rate η. In this case, it is natural to compute the Bayes optimal prediction (8) by using the expected value of η with respect to the posterior distribution. In particular, it is convenient to let the prior distribution on η be a Beta distribution with parameters $a, b > 0$. The corresponding density function is

$$\beta_{a,b}(\eta) = \frac{\eta^{a-1}(1 - \eta)^{b-1}}{\int_0^1 \eta^{a-1}(1 - \eta)^{b-1} d\eta},$$

and its expected value is $a/(a+b)$. Observe that by changing the setting of the parameters a and b we obtain different distributions. For instance, choices of a and b for which $a > b$ correspond to a distribution skewed to the right, and vice versa.

We can now compute the probabilities in (8) according to these new probabilistic assumptions. For the sake of simplicity, let us assume that E_i's prediction on round t is such that $E_i(y^{t-1}) \neq y$. In this case, $P_i(y \mid y^{t-1})$ is simply the posterior probability of expert E_i being incorrect on round t (given the prefix y^{t-1}). Then,

$$
\begin{aligned}
P_i(E_i(y^{t-1}) &= y \mid y^{t-1}) \\
&= \int_0^1 \hat{P}_i(E_i(y^{t-1}) = y \mid \eta, y^{t-1}) \cdot \hat{P}_i(\eta \mid y^{t-1}) d\eta \\
&= \int_0^1 \eta \cdot \hat{P}_i(\eta \mid y^{t-1}) d\eta = \frac{M_i(y^{t-1}) + a}{t - 1 + a + b}.
\end{aligned}
\tag{9}
$$

(Recall that $M_i(y^{t-1})$ is the total number of prediction mistakes made by E_i on y^{t-1}.) Equation (9) follows from the fact that when the noise process is i.i.d. the posterior distribution under the Beta prior with parameters a and b is itself a Beta distribution with parameters $a' = M_i(y^{t-1}) + a$ and $b' = t - 1 - M_i(y^{t-1}) + b$, respectively. The estimate given by (9) is usually referred to as the *mean posterior estimate*. Note, when $t = 1$, before seeing any outcomes, the estimate produced is simply $a/(a + b)$, the normalized value of the parameters a and b, which is the mean of the Beta prior. The simplicity of this method of estimation is one of the reasons Beta densities (which are a special case of the Dirichlet densities) are so attractive.

From (9) it is easy to see that the probability of y^T under expert E_i reduces to

$$
P_i(y^T) = \frac{\left[\prod_{k=0}^{M_i(y^T)-1}(k + a) \right] \left[\prod_{m=0}^{T-1-M_i(y^T)}(m + b) \right]}{\prod_{n=0}^{T-1}(n + a + b)}.
\tag{10}
$$

When the distributions in (8) are computed as described by (9) and (10), we obtain a new family of Bayesian optimal predictors. As we did before, it is now straightforward to derive from this family of Bayesian predictors a corresponding family of multiplicative weigthing algorithms, which associate a weight to each expert and predict with a weighted voting scheme of the experts' predictions. We call the algorithm derived from these Bayesian predictors Bayesian Binomial Weighting (BBW) — see Figure 1.

These algorithms, rather than using a fixed update factor for the weights as done by WM, use a variable update factor that depends on t and on the past expert's behavior, $M_i(y^t)$. The BBW algorithms are actually quite intuitive in this form: the weight of each expert E_i is split between the predictions $E_i(y^{t-1})$ and the complementary prediction. The split is based on the mean posterior estimate of the expert's probability of making an incorrect prediction on round t. Thus, on each round t, the expert's weight is more concentrated on the bit that is more likely to be observed given the past expert's behavior.

Algorithm: Bayesian Binomial Weighting
Parameters: A set $E_1, \ldots, E_N$ of experts and positive integers a and b.
Initialization: Set learner's counter m and experts' counters $k_1, \ldots, k_N$ to 0.
Repeat for $t = 1, 2, \ldots$

1. For $i = 1, \ldots, N$ let

$$w_{i,t}^{E_i(y^{t-1})} = w_{i,t} \cdot \left(1 - \frac{k_i + a}{m + a + b}\right), \quad w_{i,t}^{1 - E_i(y^{t-1})} = w_{i,t} \cdot \frac{k_i + a}{m + a + b}.$$

2. Predict with the boolean value $\widehat{y}_t$ satisfying

$$\hat{y}_t = \arg \max_{y \in \{0,1\}} \left\{ \sum_{i=1}^{N} w_{i,t}^y \right\}$$

3. Read correct bit y_t.

4. Increment learner's counter m by 1. Set the weight of each expert E_i to $w_{i,t}^{y_t}$ (i.e. to the weight with which E_i voted for y_t) and increment by 1 the counter k_i of those experts E_i that made a mistake.

Figure 1: Bayesian Binomial Weighting (BBW) algorithm.

Since we are primarily interested in a worst-case setting, we analyze the performance of the BBW algorithms without making any of the probabilistic assumptions that were used to derive them. A very general upper bound on the number of prediction mistakes made by the BBW algorithm, for integer values of the parameters a and b, is established by the following result. This result holds for a slightly modified version of the BBW algorithm that we call *conservative* BBW. The conservative BBW algorithm is just like the original algorithm, except step 4 in Figure 1 is executed only in rounds in which BBW makes a mistake.

Theorem 1 *For all positive integers a and b and for all $y^T \in \{0,1\}^T$, $T \in \mathbb{N}$, if the conservative BBW algorithm is run on y^T with parameters a and b, then*

$$M_{\mathrm{BBW}}(y^T) \ \leq \ \max \left\{ q \in \mathbb{N} : q \leq \log N + \log \binom{q + a + b - 2}{\widehat{M}_{\mathrm{BEST}}(y^T) + a - 1} \right.$$
$$\left. + \ \log \left(1 + \frac{q}{a + b - 1}\right) - \log \binom{a + b - 2}{a - 1} \right\},$$

where $\widehat{M}_{\mathrm{BEST}}(y^T) = \min_{1 \leq i \leq N} \widehat{M}_i(y^T)$ and

$$\widehat{M}_i(y^T) = \left| \{1 \leq t \leq T : E_i(y^{t-1}) \neq y_t \wedge M_{\mathrm{BBW}}(y^{t-1}) \neq y_t\} \right|$$

is the number of prediction mistakes made by E_i in the rounds in which BBW made mistakes as well.

By fixing particular values for the parameters a and b, more readable bounds can be obtained which enable us to compare the performance of some of the algorithms of the family with the performance of Weighted Majority. An interesting special case of Theorem 1 is when a and b are both chosen equal to 1. For this setting the Beta distribution becomes uniform and this corresponds to a noise rate η chosen uniformly at random in the interval $(0, 1)$. The resulting mean posterior estimate is called *Laplacian estimate*. Curiously enough, the setting $a = 1$ and $b = 2$ (prior skewed to the left) yields a bound for BBW slightly better than the one obtained using the Laplacian estimate. These two results are summarized in the corollary below.

Corollary 1 *For all $y^T \in \{0, 1\}^T$, $T \in \mathbb{N}$, if the conservative BBW algorithm is run on y^T using $a = b = 1$ (Laplacian estimate), then the number of mistakes is at most*

$$M_{\mathrm{BBW}}(y^T)$$
$$\leq \quad \max\left\{ q \in \mathbb{N} : q \leq \log N + \log\left(\frac{q}{\widehat{M}_{\mathrm{BEST}}(y^T)} \right) + \log(q+1) \right\}. \quad (11)$$

Furthermore, if BBW is run on the same sequence using $a = 1$ and $b = 2$, then the number of mistakes is at most

$$\max\left\{ q \in \mathbb{N} : q \leq \log N + \log\left(\frac{q}{\widehat{M}_{\mathrm{BEST}}(y^\ell)} \right) + \log(q+1) \right\} - 1$$

where $\widehat{M}_{\mathrm{BEST}}(y^T)$ is defined as in Theorem 1.

The above results arise two natural questions: How can one optimize the choice of a and b in the bound of Theorem 1? And how does this bound (or its special cases in Corollary 1) compare to the bound (6), which is based on the preliminary knowledge of an upper bound K on the number of mistakes made by the best expert? It turns out that bounds (6) and (11) are asymptotically equivalent. Furthermore, they are also asymptotically optimal, hence no tuning of the parameters a and b in algorithm BBW can improve, at least asymptotically, over these bounds.

Notes. Bayesian derivations for the Weighted Majority algorithm were also presented in [10]. The Bayesian Binomial Weighting algorithm has been introduced by Cesa-Bianchi *et al.* in [4], where Theorem 1, Corollary 1, and a number of other bounds for variants of BBW were proven. Definition and proofs for the asymptotical optimality of tuned WM and BBW can be found in [3] and [4].

2.3   An optimal randomized learner

An interesting generalization of the on-line boolean prediction problem with experts is the case where the learner is allowed to output randomized predictions. That is, we assume that on each round t the learner flips a coin with bias

$\widehat{p}_t \in [0, 1]$ for heads, and then assigns $\widehat{y}_t = 1$ if and only if the outcome of the coin toss is heads. Hence, the probability $\mathbf{P}\{\widehat{y}_t \neq y_t\}$ that the learner makes a mistake on round t is $|y_t - \widehat{p}_t| \in [0, 1]$. We use $\overline{M}_A(y^T) = \sum_{t=1}^{T} |y_t - \widehat{p}_t|$ to denote the expected number of mistakes (or total absolute loss) of a randomized learner A on the sequence $y^T = (y_1, \ldots, y_T)$.

In the attempt of deriving a randomized variant of the Weighted Majority algorithm described in Section 2.1, it is perhaps natural to look at the "average" $r_t = W_t^1 / W_t$ as a good choice for the bias $\widehat{p}_t$ (recall that W_t^1 is the total weight at round t of the experts predicting 1 on y_t and W_t is the total weight at round t of all experts). However, it turns out that better bounds on the total loss can be proven for learners that let $\widehat{p}_t$ be some nonlinear function of r_t. Consider the randomized learner IG that, on each round t, predicts 1 with probability

$$\widehat{p}_t = \frac{\log(1 - r_t + \beta r_t)}{\log(1 - r_t + \beta r_t) + \log(\beta(1 - r_t) + r_t)} \tag{12}$$

where $r_t = W_t^1 / W_t$. After the correct value y_t is observed, IG updates the weight of each expert E_i according to the multiplicative rule (2), i.e.

$$w_{i,t+1} = w_{i,t} \cdot \beta^{|y_t - \widehat{y}_{i,t}|} = w_{i,t} \cdot (1 - (1 - \beta)|y_t - \widehat{y}_{i,t}|)$$

where the last equation holds because $|y_t - \widehat{y}_{i,t}| \in \{0, 1\}$. We have that for any $T > 0$ and for any sequence $y \in \{0, 1\}^T$,

$$\overline{M}_{\text{IG}}(y^T) \leq \frac{\log N + M_{\text{BEST}}(y^T) \log(1/\beta)}{2 \log \frac{2}{1+\beta}}. \tag{13}$$

Some remarks are in order. First, note that the right-hand side of (13) is exactly half of the right-hand side of the corresponding bound (3) for Weighted Majority. Hence, allowing randomization improves the bound by a factor of a half.

Second, the sigmoidal function used by IG has a nice information-theoretic interpretation. According to the Bayesian framework developed in Subsection 2.2, for IG we have

$$P_i(y^{t-1}) = \frac{\beta^{M_i(y^{t-1})}}{(1+\beta)^{t-1}} \quad \text{and} \quad P(i \mid y^{t-1}) = \frac{\beta^{M_i(y^{t-1})}}{\sum_{j=1}^{N} \beta^{M_j(y^{t-1})}}$$

Thus

$$
\begin{aligned}
P(y_t = 0 \mid y^{t-1}) &= \sum_{i=1}^{N} P_i(y_t = 0 \mid y^{t-1}) P(i \mid y^{t-1}) \\
&\propto \frac{\sum_{i=1}^{N} \beta^{\widehat{y}_{i,t}} \beta^{M_i(y^{t-1})}}{\sum_{j=1}^{N} \beta^{M_j(y^{t-1})}} \\
&= \frac{\sum_{i=1}^{N} (1 - (1 - \beta)\widehat{y}_{i,t}) \beta^{M_i(y^{t-1})}}{\sum_{j=1}^{N} \beta^{M_j(y^{t-1})}} \\
&= 1 - (1 - \beta) r_{t+1}
\end{aligned}
$$

50

and (12) can be rewritten as

$$\widehat{p}_t = -\frac{\log P(y_t = 0 \mid y^{t-1})}{Z},$$

where Z is the normalization constant. The quantity $-\log P(y_t = 0 \mid y^{t-1})$ is called the *instantaneous information gain* provided by the event $y_t = 0$. Since, in this case, $\widehat{p}_t$ is the estimate of the probability that $y_t = 1$, the function (12) says that such an estimate should be proportional to the information gained by the occurrence of the complementary event $y_t = 0$. Thus, in a certain sense, each mistake is traded with a corresponding gain of information.

Third, bound (13) can be generalized to the case where the experts are randomized. That is, on round t each expert E_i computes a bias $\widehat{p}_{i,t} \in [0,1]$ corresponding to the probability of $\widehat{y}_{i,t} = 1$. The total loss of expert E_i on a sequence y^T is then computed as $\overline{M}_i(y^T) = \sum_{t=1}^T |y_t - \widehat{p}_{i,t}|$. In this case, the quantity r_t used by IG to compute the bias in round t (see (12)) becomes

$$r_t = \frac{\sum_{i=1}^N w_{i,t}\widehat{p}_{i,t}}{\sum_{j=1}^N w_{j,t}}$$

and (13) is generalized to

$$\overline{M}_{\text{IG}}(y^T) \le \frac{\log N + \overline{M}_{\text{BEST}}(y^T)\log(1/\beta)}{2\log\frac{2}{1+\beta}} \tag{14}$$

where $\overline{M}_{\text{BEST}}(y^T)$ is defined analogously to $M_{\text{BEST}}(y^T)$. Note that when experts are randomized, the on-line prediction protocol is modified so that, on each round t, the learner collects the biases $\widehat{p}_{1,t},\ldots,\widehat{p}_{N,t}$ instead of the actual experts' predictions $\widehat{y}_{1,t},\ldots,\widehat{y}_{N,t}$.

In case an upper bound K on the total loss $\overline{M}_{\text{BEST}}(y^T)$ of the best expert is available to IG at the beginning, we get the improved bounds

$$\overline{M}_{\text{IG}}(y^T) \ \le \ \max\left\{q \in \mathbb{N} : q \le \frac{\log N}{2(1 - H(K/q))}\right\} \tag{15}$$

$$\overline{M}_{\text{IG}}(y^T) \ \le \ \overline{M}_{\text{BEST}}(y^T) + \frac{\log N}{2} + \sqrt{K\ln N} \tag{16}$$

corresponding respectively to bounds (5) and (6).

Bound (16) is of special interest. By using T as a trivial upper bound on the loss of the best expert on any sequence y^T, we get

$$\frac{\overline{M}_{\text{IG}}(y^T) - \overline{M}_{\text{BEST}}(y^T)}{T} = \sqrt{\frac{\ln N}{T}} + O\left(\frac{1}{T}\right), \tag{17}$$

showing that IG's additional loss per trial over that of the best expert goes to 0 with rate $O(1/\sqrt{T})$.

The constant in front of $\sqrt{\frac{\ln N}{T}}$ can be easily improved by allowing IG to simulate an additional expert E_{N+1} with constant bias $\widehat{p}_{N+1,t} = 1/2$. With this additional expert we have $\overline{M}_{\text{BEST}}(y^T) \leq T/2$ and (17) becomes

$$\frac{\overline{M}_{\text{IG}}(y^T) - \overline{M}_{\text{BEST}}(y^T)}{T} = \sqrt{\frac{\ln(N+1)}{2T}} + O\left(\frac{1}{T}\right). \tag{18}$$

One might wonder whether (18) can be improved for all T and N. The answer is no. In fact, one can prove that for *any* randomized learner A and for any $N, T > 0$ there is a set of N experts and a sequence $y^T \in \{0,1\}^T$ such that

$$\frac{\overline{M}_A(y^T) - \overline{M}_{\text{BEST}}(y^T)}{\sqrt{(T \ln N)/2}} \geq 1 - o(1) \qquad \text{where } \lim_{N \to \infty} \lim_{T \to \infty} o(1) = 0.$$

Notes. Vovk [17] derived bounds (14) and (15) for an algorithm closely related to IG. The derivation of algorithm IG, bound (16) and the asymptotical optimality proof mentioned in the last paragraph were all presented in [2].

3 The adversarial bandit problem

In this section we show how some algorithms developed for the on-line prediction model can be successfully used to solve an adversarial version of the well studied multi-armed bandit problem, originally proposed by Robbins [16].

In this setup, a gambler must choose which of N slot machines to play. At each time step, he pulls the arm of one of the machines and receives a reward or payoff (possibly zero). The gambler's purpose is to maximize his total reward over a sequence of rounds. Since each arm is assumed to have a different distribution of rewards, the goal is to find the arm with the best expected return as early as possible, and then to keep gambling using that arm.

The problem is a classical example of the trade-off between exploration and exploitation. On the one hand, if the gambler plays exclusively on the machine that he thinks is best ("exploitation"), he may fail to discover that one of the other arms actually has a higher average return. On the other hand, if he spends too much time trying out all the machines and gathering statistics ("exploration"), he may fail to play the best arm often enough to get a high total return.

In the past, the bandit problem has almost always been studied with the aid of statistical assumptions on the process generating the rewards for each arm. In the gambling example, for instance, it might be natural to assume that the distribution of rewards for each arm is Gaussian and time-invariant. However, in general it may be difficult or impossible to determine the right statistical assumptions for a given domain, and some domains may be inherently adversarial in nature so that no such assumptions are appropriate. In this variant of the bandit problem we apply the same type of worst-case analysis as in the on-line prediction model and make *no* statistical assumptions about the generation of payoffs (except for the fact that they are chosen from a bounded range).

More formally, the adversarial bandit problem is formulated as follows. There are N arms denoted by the integers from 1 through N. On each round $t = 1, \ldots, T$

1. the player fixes a distribution P_t on $\{1, \ldots, N\}$,

2. each arm $i \in \{1, \ldots, N\}$ generates a payoff $x_{i,t} \in [0, 1]$,

3. the player observes payoff $x_{i,t}$ with probability $P_t(i)$.

We define the total expected payoff of player A on a sequence of T rounds by $\overline{G}_A = \sum_{t=1}^{T} \sum_{i=1}^{N} x_{i,t} P_t(i)$. Similarly, we define the total payoff of arm i by $G_i = \sum_{t=1} x_{i,t}$. The goal of the player is to maximize the total expected payoff relative to the total payoff $G_{\text{BEST}} = \max_{1 \leq i \leq N} G_i$ of the best arm.

To solve this problem, we first attack the following (easier) problem, where the payoffs of all arms are accessible in each round. We call this the "full information game". On each round $t = 1, \ldots, T$

1. the player fixes a distribution P_t on $\{1, \ldots, N\}$,

2. each arm $i \in \{1, \ldots, N\}$ generates a payoff $x_{i,t} \in [0, 1]$,

3. the player observes all payoffs.

As in the adversarial bandit problem, the goal of the player in the full information game is to maximize the total expected payoff relatively to the payoff of the best arm.

It is easy now to formulate this game as an on-line prediction problem. Each arm i plays the role of a randomized expert E_i whose loss at time t is $-x_{i,t} \in [-1, 0]$. The learner is constrained to predict in each round according to some expert E_{i_t} randomly chosen according to distribution P_t (i.e. the learner cannot combine predictions) thus suffering the expert's loss $-x_{i_t,t}$. The learner's goal is to minimize the total expected loss $-\sum_t \sum_i x_{i,t} P_t(i)$, which amounts to maximize the expected payoff $\sum_t \sum_i x_{i,t} P_t(i)$.

By exploiting this close relationship with the on-line prediction problem, we can solve the full information game using a variant of the Weighted Majority algorithm. Interpreting the weights generated by the multiplicative update (2) as a probability distribution, we consider the player Hedge that chooses arm i in round t with probability

$$P_t(i) = \frac{(1+\alpha)^{\sum_{s=1}^{t-1} x_{i,s}}}{\sum_{j=1}^{N} (1+\alpha)^{\sum_{s=1}^{t-1} x_{j,s}}} \tag{19}$$

where $\alpha > 0$. Note that each term $(1 + \alpha)^{x_{i,t}}$ can be written as $\beta^{-x_{i,t}}$ for $\beta = 1/(1 + \alpha) \in (0, 1]$. This is in agreement with our interpretation of payoff $x_{i,t}$ as a negative loss.

Algorithm Exp3
Parameters: Reals $\alpha > 0$ and $\gamma \in [0, 1]$.
Initialization: Initialize Hedge.
Repeat for $t = 1, 2, \ldots$ until game ends

1. Get the distribution P_t from Hedge.

2. Let the chosen arm i_t be arm j with probability
   $\widehat{P}_t(j) \doteq (1 - \gamma)P_t(j) + \gamma/K$.

3. Receive reward $x_{i_t,t} \in [0, 1]$.

4. Feed the simulated reward vector $\widehat{x}_{1,t}, \ldots, \widehat{x}_{N,t}$ back to Hedge, where

$$\widehat{x}_{j,t} = \begin{cases} \dfrac{\gamma}{N} \cdot \dfrac{x_{i_t,t}}{\widehat{P}_t(i_t)} & \text{if } j = i_t \\ 0 & \text{otherwise.} \end{cases}$$

Figure 2: Algorithm Exp3 for the adversarial bandit problem.

The performance of the player using P_t defined in (19) is

$$\sum_{t=1}^{T}\sum_{i=1}^{N} x_{i,t}P_t(i) \geq \frac{\left(\sum_{t=1}^{T} x_{j,t}\right)\ln(1 + \alpha) - \ln N}{\alpha}$$

for all arms $j = 1, \ldots, N$. By an appropriate choice of $\alpha = \alpha(T, N)$, we obtain the following

$$\overline{G}_{\text{Hedge}} = \sum_{t=1}^{T}\sum_{i=1}^{N} x_{i,t}P_t(i) \geq G_{\text{best}} - \sqrt{2T \ln N}.$$

Note that the expected total payoff per round of Hedge converges to the total payoff per round of the best arm with rate $O(1/\sqrt{T})$.

We now move to the analysis of the adversarial bandit problem. We present an algorithm Exp3 that runs the algorithm Hedge as a subroutine. The algorithm is described in Figure 2. On each round t, Exp3 receives the distribution P_t from Hedge, and selects an arm i_t according to the distribution $\widehat{P}_t$, which is a mixture of P_t and the uniform distribution. Intuitively, mixing in the uniform distribution is done in order to make sure that the algorithm tries out all N arms and gets good estimates of the rewards for each. Otherwise, the algorithm might miss a good arm because the initial payoffs it observes for this arm are low and large payoffs that occur later are not observed because the arm is not selected. After Exp3 receives the payoff $x_{i_t,t}$ associated with the chosen arm, it generates a simulated payoff vector $\widehat{x}_{1,t}, \ldots, \widehat{x}_{N,t}$ for Hedge. As Hedge requires full information, all components of this vector must be filled in, even for the arms that were not selected. For arms $j \neq i_t$ not chosen, we set $\widehat{x}_{j,t}$ to be zero. For the chosen arm i_t, we set the simulated payoff $\widehat{x}_{i_t,t}$ proportional to $x_{i_t,t}/\widehat{P}_t(i_t)$. This compensates the payoff of arms that are unlikely to be chosen and guarantees that the expected simulated payoff associated with any fixed arm j is proportional to the actual payoff of the arm, i.e., that $\mathbf{E}_{i_t}[\widehat{x}_{j,t} \mid i_1, \ldots, i_{t-1}] = \frac{\gamma}{N} x_{j,t}$ for any fixed choice of $i_i, \ldots, i_{t-1}$. The constant scaling factor γ/N is used to guarantee that the rewards fed back to Hedge are in the range $[0, 1]$ as required. From this perspective, it is necessary to mix in the uniform distribution with the distribution generated by Hedge in order to ensure that the simulated reward before scaling, $x_{i_t}/\widehat{P}_t(i_t)$, is not too large.

We now give the main theorem of this section, which bounds the regret of algorithm Exp3.

Theorem 2 *For $\alpha > 0$, $\gamma \in [0, 1]$, the expected gain of algorithm* Exp3 *is at least*

$$
\begin{aligned}
\overline{G}_{\mathrm{Exp3}} &\geq \frac{1-\gamma}{\alpha}\left(G_{\mathrm{best}}\ln(1+\alpha) - \frac{N\ln N}{\gamma}\right) \\
&\geq G_{\mathrm{best}} - \left(\gamma + \frac{\alpha}{2}\right)G_{\mathrm{best}} - \frac{N\ln N}{\alpha\gamma}.
\end{aligned}
$$

If we know an upper bound g on G_{best}, then the parameters γ and α can be chosen so that

$$
\overline{G}_{\mathrm{Exp3}} \geq G_{\mathrm{best}} - \frac{3}{\sqrt[3]{2}}g^{2/3}(N\ln N)^{1/3}. \tag{20}
$$

It can be shown that *any* player for the adversarial bandit problem has a total expected payoff that, in the worst case, is bounded from the above by

$$
G_{\mathrm{BEST}} - \Omega\left(\sqrt{G_{\mathrm{BEST}}N}\right). \tag{21}
$$

The gap between (20) and (21) has been recently tightened by showing that a slight variant of Exp3 has total expected payoff of at least $G_{\mathrm{BEST}} - \sqrt{G_{\mathrm{BEST}}N}$. Using the trivial bound $G_{\mathrm{BEST}} \leq T$ we obtain, for this variant of Exp3, that the expected payoff per round converges to the payoff per round of the best arm with rate $O\left(1/\sqrt{T}\right)$.

Notes. The material covered in this section is taken from [1].

4 Boosting

In this section we apply the multiplicative weighting technique presented in Section 2.1 to derive a new algorithm for "boosting," i.e., for converting a moderately inaccurate prediction rule into one that performs extremely well. This is, perhaps, the most surprising application of the multiplicative weighting schemes.

In the framework we adopt here, the boosting algorithm receives a set of N training examples $S = (x_1, y_1), \ldots, (x_N, y_N)$ from $X \times \{0, 1\}$, where X is a given instance space. The goal of the booster is simply to find a hyphotesis

$H : X \rightarrow \{0, 1\}$ with low empirical error relative to the training sample, that is to find a H such that $H(x_i) = y_i$ for most $(x_i, y_i) \in S$. In order to solve the problem, suppose that the booster is provided with a classifier A (e.g. a neural network) that outputs, when presented with a subsample S_t drawn according to some distribution D_t over S, a hypothesis H_t with low error $\epsilon_t = D_t(H_t(x) \neq y)$ with respect to D_t. Although the hypotheses produced by A could be very inaccurate, it is reasonable to expect them to be at least a little bit better than pure random guessing. Furthermore, the booster, by repeatedly running the classifier A on many different samples S_t, $t = 1, 2, \ldots, T$, can obtain many different hypotheses H_t. To be able to use this information to construct a hypothesis with low empirical error, the booster faces two main problems. First, how to force the classifier A to output the most useful hypotheses; second, how to combine the returned hypotheses into a single, highly accurate final hypothesis. These are in fact the main issues involved in designing a boosting algorithm.

Boosting algorithms usually work by calling the classifier A many times, each time presenting it with a different distribution D_t over S, and finally combining all of the generated hypotheses H_t into a single final hypothesis H_{FIN}. The intuitive idea underlining this technique is to alter the distribution D_t over the instances in such a way that the probability of the parts of the training set S where the previous hypotheses $H_1, \ldots, H_{t-1}$ performed worst is increased, thus forcing the classifier to generate hypotheses that are more accurate on these parts.

More formally, boosting proceeds in rounds. On round $t = 1, \ldots, T$:

- The booster defines a distribution D_t on the instances in S and passes it to the classifier A.

- The classifier A produces a hypothesis H_t with error ϵ_t with respect to D_t, where $\epsilon_t = D_t(H_t(x) \neq y_t)$.

After T rounds, the returned hypotheses $H_1, \ldots, H_T$ are combined into a final hypothesis H_{FIN}. To avoid confusion, we call the hypotheses returned by the classifier during the boosting process *weak* hypotheses and we call *final* the hypothesis returned by the boosting algorithm. We use the term "weak" to emphasize the fact that the single hypotheses H_t returned by A can be quite inaccurate (i.e. the error ϵ_t might be arbitrarily close to 1/2.)

An important practical deficiency of some boosting algorithms is that they require an a priori lower bound on the accuracy of the set of weak hypotheses produced by the classifier. Moreover, the theoretical performance of the final hypothesis depends only on the performance of the least accurate of the weak hypotheses. Clearly, not only this estimate is usually unknown in practice, but the accuracy of the hypotheses returned by A will typically vary wildly with respect to the distributions D_t.

In the rest of the section we will present a boosting algorithm that needs no prior knowledge about the accuracies of the weak hypotheses and whose performance depends on the average performance of all the weak hypotheses,

---

Algorithm: AdaBoost

Input: Training examples $(x_1, y_1), \ldots, (x_N, y_N)$, distribution D over the N examples, classifier A, integer T specifying the number of iterations.

Initialization: $w_{i,1} = D(i)$ for $i = 1, \ldots, N$.

Repeat for $t = 1, 2, \ldots, T$

1. Set $P_t(i) = w_{i,t} / \sum_{i=1}^{N} w_{i,t}$.

2. Feed the distribution P_t to A and get back a hypothesis $H_t : X \to \{0, 1\}$.

3. Compute the error of H_t: $\epsilon_t = \sum_{i=1}^{N} |H_t(x_i) - y_i| P_t(i)$ and set $\beta_t = \epsilon_t/(1 - \epsilon_t)$.

4. Set the new weight vector to $w_{i,t+1} = w_{i,t} \beta_t^{1 - |H_t(x_i) - y_i|}$.

Ouput the hypothesis

$$H_{\mathrm{FIN}} = \begin{cases} 1 & \text{if } \sum_{t=1}^{T} \left(\log \frac{1}{\beta_t} \right) H_t(x) \geq \frac{1}{2} \sum_{t=1}^{T} \log \frac{1}{\beta_t} \\ 0 & \text{otherwise.} \end{cases}$$

Figure 3: The boosting algorithm AdaBoost.

rather than just that of the weakest one. In deriving our boosting algorithm we first reduce the boosting problem to an on-line prediction problem and then we use a variant of the WM algorithm to solve the prediction problem.

We start investigating the relationship between boosting and on-line prediction. Recall that, in the on-line prediction model, we want to compute a weight vector over the expert set so that, on each round, the weight associated with an expert is increased if the expert makes few mistakes. However, in the boosting model, we want to compute a distribution over the instances x_i in S in a way that, on each round, the probability associated with an instance is increased if the instance is most likely to be misclassified by the previously generated weak hypotheses. Since we have an algorithm (WM) which computes weights over experts and we need one that computes weights over instances, the obvious solution is to reverse the roles of experts and instances and then to apply the WM algorithm to the on-line prediction problem over instances. This leads to a reduction where the experts correspond to the instances and the rounds of the on-line prediction problem are instead associated to the weak hypotheses produced by the classifier.

Using the above reduction and applying WM directly to the derived prediction problem, we obtain a boosting algorithm that maintains a set of weights $w_{1,t}, \ldots, w_{N,t}$ over the instances in the training set. On each round t, the distribution D_t presented to A is obtained by simply renormalizing these weights. Based on the distribution D_t, the classifier A then generates a weak hypothesis H_t with error ϵ_t with respect to D_t. Using the new hypothesis, the booster updates the weight vector and the process goes on. After T iterations, the booster outputs as final hypothesis H_{FIN} a weighted majority vote of the weak hypotheses $H_1, \ldots, H_T$ returned by the classifier. The resulting boosting algorithm is described in Figure 3. The algorithm becomes more intuitive when described in the following informal way.

- After each hypothesis H_t is generated, the weight $w_{i,t}$ associated with each instance x_i is decreased if H_t is correct on that instance and otherwise is increased. Thus, each distribution D_t focuses on the examples in S that are most likely to be misclassified by the previous hypotheses.

- The parameter β appearing in the weight update rule (step 4 in Figure 3) is no longer fixed ahead of time but is rather chosen as a decreasing function of the error ϵ_t of each weak hypothesis H_t. This makes intuitive sense: more accurate hypotheses cause larger changes in the generated distribution.

- Finally, the weight assigned to H_t in the final hypothesis is also a decreasing function of β_t. Thus, more accurate weak hypotheses have more influence on the outcome of the final hypothesis.

We now give the main theorem of this section, which bounds the error of the final hypothesis produced by the boosting algorithm of Figure 3 in terms of the error of the weak hypotheses returned by the classifier.

Theorem 3 *Let $\epsilon_1, \ldots, \epsilon_T$ be the errors of the weak hypotheses $H_1, \ldots, H_T$ when the classifier A is called by the boosting algorithm defined in Figure 3. Then the error $\epsilon = D(H_{\text{FIN}}(x_i) \neq y_i)$ of the final hypothesis H_{FIN} is upper bounded by*

$$\epsilon \leq 2^T \prod_{t=1}^{T} \sqrt{\epsilon_t(1 - \epsilon_t)} \tag{22}$$

When D is chosen to be the uniform distribution, i.e. $D(x_i) = 1/N$, equation (22) yields a bound on the empirical error of the final hypothesis relative to the training set S.

Some remarks are in order. First, note that the bound (22) on the error of the final hypothesis H_{FIN} depends on the error of all the weak hypotheses. It is important to point out that, for previously known boosting algorithms, the bounds on the error depended only on the maximal error of the weakest hypothesis, thus ignoring the advantage that can be gained from weak hypotheses whose error is smaller. This advantage seems to be relevant in practical applications of boosting, since there we expect the error of the classifier A to increase as the distribution D_t shift more and more away from the uniform distribution. Second, observe that Theorem 3 applies also if, for some weak hypotheses H_t, $\epsilon_t \geq 1/2$. However, if the weak hypotheses produced by the classifier consistently have error only slightly better than $1/2$, for instance $\epsilon_t \leq 1/2 - \gamma$, then the error of H_{FIN} can also be bounded by

$$\epsilon \leq \prod_{i=1}^{T} \sqrt{1 - 4\gamma^2} \leq e^{-2T\gamma^2}.$$

Thus, Theorem 3 shows that, in this case, the error of the final hypothesis H_{FIN} drops to zero exponentially fast.

To summarize, we have seen that, by formulating the boosting problem as an on-line prediction problem and by applying a variant of WM to this latter problem, we can obtain a boosting algorithm, that unlike the previous ones, does not require any prior knowledge about the accuracies of the weak hypotheses. Instead, it adapts to these accuracies and generates a weighted majority hypothesis whose error depends on the error of all the weak hypotheses. Furthermore, the algorithm is very flexible and easy to implement.

Notes. The boosting algorithm AdaBoost has been proposed by Freund and Schapire [7]. They also study some generalizations of this algorithm to the problem of learning functions whose range, rather than being binary, is an arbitrary finite set or a bounded segment of the real line. An interesting connection between game theory and boosting is analyzed in [9]. This relationship turns out to be highly relevant in the development and understanding of such learning algorithms. Some experiments on real learning problems were performed in [8]. Recently, Kearns and Mansour [13, 6] applied the boosting framework to the problem of learning decision trees. More experiments on boosting decision trees can be found in [5].

References

[1] P. Auer, N. Cesa-Bianchi, Y. Freund, and R.E. Schapire. Gambling in a rigged casino: The adversarial multi-armed bandit problem. In *Proceedings of the 36th Annual Symposium on the Foundations of Computer Science*, pages 322–331. IEEE press, 1995.

[2] N. Cesa-Bianchi, Y. Freund, D.P. Helmbold, D. Haussler, R. Schapire, and M.K. Warmuth. How to use expert advice. Technical Report UCSC-CRL-95-19, University of California at Santa Cruz, 1995. An extended abstract appeared in the Proceedings of the 25th ACM Symposium on the Theory of Computation.

[3] N. Cesa-Bianchi, Y. Freund, D.P. Helmbold, and M.K. Warmuth. On-line prediction and conversion strategies. *Machine Learning*. To appear. An extended abstract appeared in the Proceedings of the First EuroCOLT Workshop.

[4] N. Cesa-Bianchi, D.P. Helmbold, and S. Panizza. On Bayes methods for on-line boolean prediction. In *Proceedings of the 9th Annual Conference on Computational Learning Theory*, pages 314–324. ACM Press, 1996.

[5] C. Cortes and H. Drucker. Boosting decision trees. In *Advances in Neural Information Processing Systems 8*. MIT Press, 1996. To appear.

[6] T. Dieterich, M. Kearns, and Y. Mansour. Applying the weak learning framework to understand and improve C4.5. In *Proceedings of the 13th International Conference on Machine Learning*, pages 96–104. Morgan Kaufmann, 1996.

[7] Y. Freund and R. Schapire. A decision-theoretic generalization of on-line learning and an application to boosting. In *Proceedings of the 2nd Euro-COLT Workshop*, pages 23–37. Lecture Notes on Artificial Intelligence, Vol. 904, Springer-Verlag, 1995.

[8] Y. Freund and R. Schapire. Experiments with a new boosting algorithm. In *Proceedings of the 13th International Conference on Machine Learning*, pages 148–156. Morgan Kaufmann, 1996.

[9] Y. Freund and R. Schapire. Game theory, on-line prediction and boosting. In *Proceedings of the 9th Annual Conference on Computational Learning Theory*. ACM Press, 1996.

[10] D. Haussler and A. Barron. How well does the Bayes method work in on-line predictions of $\{+1, -1\}$ values? In *Proceedings of 3rd NEC Symposium*, pages 74–100. SIAM, 1993.

[11] D. Haussler, J. Kivinen, and M.K. Warmuth. Tight worst-case loss bounds for predicting with expert advice. In *Proceedings of the 2nd European Conference on Computational Learning Theory*, pages 69–83. Lecture Notes on Artificial Intelligence, Vol. 904, 1995.

[12] D.P. Helmbold, J. Kivinen, and M.K. Warmuth. Worst-case loss bounds for sigmoided neurons. In *Advances in Neural Information Processing Systems 8*. MIT Press, 1996. In press.

[13] M. Kearns and Y. Mansour. On the boosting ability of top-down decision tree learning algorithms. In *Proceedings of the 28th ACM Symposium on the Theory of Computing*, pages 459–468. ACM Press, 1996.

[14] J. Kivinen and M.K. Warmuth. Exponentiated gradient versus gradient descent for linear predictors. Technical Report UCSC-CRL-94-16, University of California at Santa Cruz, 1994. To appear in *Information and Computation*.

[15] N. Littlestone and M.K. Warmuth. The weighted majority algorithm. *Information and Computation*, 108:212–261, 1994.

[16] H. Robbins. Some aspects of the sequential design of experiments. *Bullettin of the American Mathematical Society*, 55:527–535, 1952.

[17] V.G. Vovk. Aggregating strategies. In *Proceedings of the 3rd Annual Workshop on Computational Learning Theory*, pages 372–383, 1990.

[18] V.G. Vovk. A game of prediction with expert advice. In *Proceedings of the 8th Annual Conference on Computational Learning Theory*, pages 51–60, 1995. An extended version is to appear in *Journal of Computer and Systems Sciences*.

[19] K. Yamanishi. On-line maximum likelihood prediction with respect to general loss functions. In *Proceedings of the 2nd European Conference on Computational Learning Theory*, pages 84–98. Lecture Notes on Artificial Intelligence, Vol. 904, 1995.

SECTION 3

EDUARDO R. CAIANIELLO LECTURES

Input/Output HMMs: A Recurrent Bayesian Network View

Paolo Frasconi

Dipartimento di Sistemi e Informatica. Università di Firenze

Via di Santa Marta 3 – 50139 Firenze (Italy)

paolo@mcculloch.ing.unifi.it

http://www-dsi.ing.unifi.it/~paolo

Abstract

This paper reviews Markovian models for sequence processing tasks, with particular emphasis on input/output hidden Markov models (IOHMMs) for supervised learning on temporal domains. HMMs and IOHMMs are viewed as special cases of belief networks that might be called *recurrent* Bayesian networks. This view opens the way to more general structures that could be devised for learning probabilistic relationships among sets of data streams (instead of just input and output data streams) or that might exploit multiple hidden state variables. Introducing the concept of belief network unfolding it is shown that recurrent Bayesian networks operating on discrete domains are equivalent to recurrent neural networks with higher order connections and linear units.

1   Introduction

In this paper I focus on problems involving learning temporal relationships. Examples of problems of this kind are numerous and include sequence classification (e.g., speech recognition, DNA sequences classification), sequence production (e.g., generation of actuator signals in robotics) and sequence prediction (e.g., financial time series forecasting). Learning tasks interfaced with a sequential data generation process require special architectures and algorithms. Architectures characterized by an algebraic input/output relationship (such as feedforward neural networks) are inadequate models of sequential data because the absence of adaptive memory prevents temporal context to be taken into account in a flexible way.

The neural network research community attempted quite early to overcome these limitation by allowing feedback in the connection graph, resulting in the so called recurrent neural networks (Rumelhart et al., 1986; Jordan, 1986). Thanks to cycles in their connections, recurrent networks exhibit a *dynamic* input/output behavior and can store past information for arbitrary durations. Importantly, the memory depth of recurrent networks is not fixed a priori by the architecture (as it would happen by simply inserting time delays in the connections (Waibel et al., 1989)). Instead, memory depth can be learned from data by adjusting the weights of feedback connections.

The speech recognition research community, on the other hand, had already introduced another learning device endowed with temporal memorization capabilities: The hidden Markov models (HMMs) (Levinson et al., 1983; Rabiner, 1989; Poritz, 1988). Standard HMMs assume a finite set of states with probabilistic transitions. Observations (outputs) are associated to state variables (or sometimes to state transitions). The state at time t is assumed to be a probabilistic function of the state at time $t-1$ only. This Markov probabilistic independence assumptions makes the model computationally tractable.

Standard HMMs are known to be effective for adaptive modeling of temporal contingencies and, besides speech recognition, they have been successfully applied in other domains such as molecular biology (Baldi et al., 1993). As earlier pointed out by Bridle (1989), they can be interpreted as recurrent connectionist models with linear units. A major limitation of standard HMMs is that they can only deal with one single data stream. Recently, Bengio & Frasconi (to appear) have introduced input/output HMMs (IOHMMs), a model that allows to deal with pairs of data streams, so that input sequences can be mapped to output sequences. IOHMMs are similar to HMMs, except that state transitions are not constant over time but are instead controlled by the current input; moreover, the output is a probabilistic function of both the state and the input.

In this paper, Markovian models, and in particular input/output HMMs, are reviewed from the perspective of graphical models, a.k.a. belief networks or probabilistic independence networks. (Pearl, 1988; Whittaker, 1990; Jensen et al., 1990). A belief network is a graphical representation of conditional independence assumptions among a given set of variables. Belief networks have been mainly studied in statistics and in artificial intelligence, where they provide a very interesting tool for dealing with uncertainty in expert systems (Pearl, 1988; Andersen et al., 1989), but also in statistical physics and operations research (Shachter, 1986). The machine learning community has recently got in deeper touch with graphical models, stimulated by the need of learning structure and parameters of belief networks from data (Buntine, 1996; Heckerman et al., 1995).

2 Belief Networks and Hidden Markov Model Architectures

In this section some basic notions about belief networks are reviewed and Markovian models are introduced.

2.1 Basic definitions

The following notation will be used in the following. Continuous or discrete random variables will be denoted by uppercase letters (e.g. X) and generic values for random variables will be denoted by lowercase letters (e.g., x). Subscripts will be used to identify a particular variable in a set (e.g., $X_i \in \mathcal{X}$)

and superscript will index values for discrete random variables (i.e., admissible values for discrete variable X_i will be denoted as $x_i^1, x_i^2, \ldots, x_i^{r_i}$). Multivariate variables will be denoted by boldface uppercase letters (e.g., $\boldsymbol{X}$). $P(x_i)$ will be used as a shorthand for $P(X_i = x_i)$ when the value is left unspecified and $P(x_i^k)$ will be used as a shorthand for $P(X_i = x_i^k)$.

Consider the finite universe of variables $\mathcal{V} = \{v_1, \ldots, v_n\}$ and three disjoint subsets $\mathcal{X}, \mathcal{Y}, \mathcal{Z}$. Variables in $\mathcal{X}$ are said to be *conditionally independent* of variables in $\mathcal{Y}$ *given* $\mathcal{Z}$ if $P(x|y, z) = P(x|z)$ for each $X \in \mathcal{X}, Y \in \mathcal{Y}, Z \in \mathcal{Z}$ and for each x, z such that $P(x|z) > 0$. The conditional independence relation can be expressed in a graphical form using undirected graphs or directed acyclic graphs, called *belief networks* (Pearl, 1988).

2.2 Bayesian networks

In this paper, only directed probabilistic independence graphs will be considered, which are known as Bayesian networks. The semantics of Bayesian networks is quite simple. Consider a directed acyclic graph $\mathcal{D}$ such that nodes are associated to variables in $\mathcal{V}$ by some bijection (hence, no distinction in notation will be made between nodes and variables). Then, the joint probability over $\mathcal{V}$ can be factorized as

$$P(v_1, \ldots, v_n) = \prod_{i=1}^{n} P(v_i | \mathrm{pa}(V_i)) \tag{1}$$

where $\mathrm{pa}(V_i)$ denotes the set of parents of V_i in $\mathcal{D}$. Hence, in order to fully specify a probabilistic model for which the conditional independences encoded by $\mathcal{D}$ hold, one only needs to specify the set of conditional densities $P(v_i | \mathrm{pa}(V_i))$ for each variable (node) V_i. When building Bayesian networks, parents of V_i are chosen to be all those variables which are supposed to have a direct causal impact on V_i.

Probabilistic inference in belief networks consists of updating the belief for a set *uninstantiated* variables $\mathcal{X}$, given the evidence entered in another set of *instantiated* variables $\mathcal{Y}$. The inference problem can be elegantly solved using network propagation algorithms (Pearl, 1988; Jensen et al., 1990).

2.3 Standard Hidden Markov Models

Consider a serially ordered sequence of variables $Y_1, \ldots, Y_t, \ldots$ in which t denotes a discrete time index. A standard HMM is a model of the joint distribution $P(Y_1, \ldots, Y_t, \ldots)$. As often done in nonparametric statistical models (particularly in connectionism), a complex joint density such as the above can be represented more compactly by introducing auxiliary variables which need not to be actually observed. HMMs rely on a sequence of *hidden* discrete state variables $X_1, \ldots, X_t, \ldots$. Hidden states are often justified as describing the underlying stochastic process responsible of data generation. For example, such a process corresponds to the human speech production system in applications to speech recognition (Rabiner, 1989).

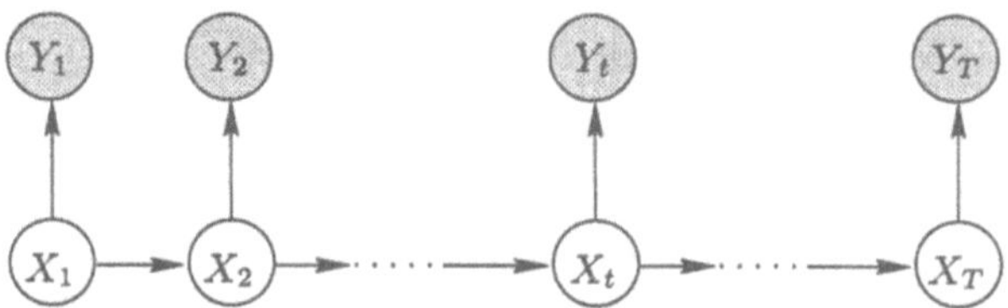

Figure 1: The time-unfolded belief network for state-emitting standard HMMs.

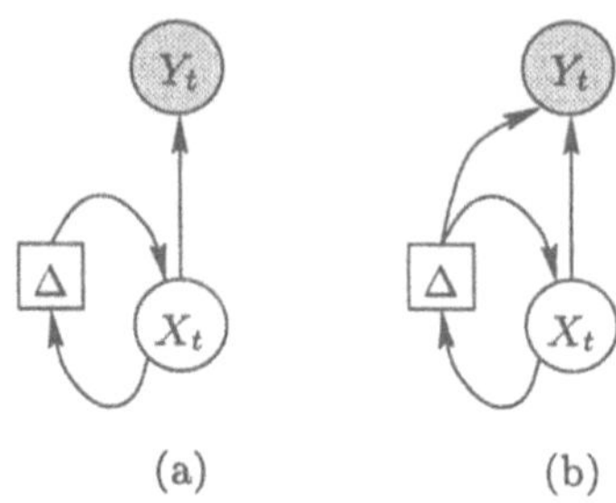

(a)   (b)

Figure 2: Recurrent belief networks for standard HMMs. a): state-emitting model; b): transition-emitting model.

States are supposed to form a Markov chain: X_t is assumed to be conditionally dependent of X_{t-1} only. Moreover, each *observation* Y_t is supposed to be conditionally dependent of X_t only. Hence, it is easy to recognize that the graph depicted in Figure 1 is a belief network for standard HMMs. Shading indicates variables observed in the data (i.e., receiving evidence during training or during recall). State nodes store the densities $P(x_t|x_{t-1})$. Assuming there are n distinguished states $x_t^1, \ldots, x_t^n$, the state transition densities are collected in $n \times n$ stochastic matrices A_t, where $a_{ij,t} = P(x_t^i|x_{t-1}^j)$. Observation nodes store the densities $P(y_t|x_t)$. In the case of discrete sequences (i.e., $y_t \in \{y_t^1, \ldots, y_t^r\}$) the observation densities are collected in $n \times r$ matrices B_t, where $b_{ik,t} = P(y_t^k|x_t^i)$.

Standard HMMs are almost always assumed to be *stationary*, i.e. both the state transition and the observation densities are assumed to be constant over time. Hence, the network of Figure 2a is a more synthetic graphical representation of HMMs. Like recurrent neural networks, Figure 2 contains a delay device Δ and nodes are supposed to be instantiated with different values at different time steps. Borrowing the terminology from recurrent neural networks (Rumelhart et al., 1986), the network shown in Figure 1 can be considered as the *time-unfolded* version of the network shown in Figure 2a.

Standard HMMs as described above are sometimes referred to as *state-emitting* models, as opposed to *transition-emitting* models in which observations are conditionally dependent of a pair of states (i.e., X_{t-1} and X_t). Transition emitting HMMs are represented in Figure 2b, in which the output density is

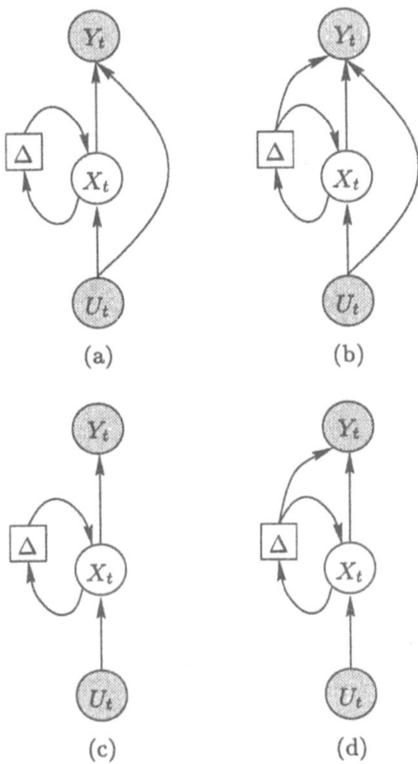

Figure 3: Recurrent belief networks for IOHMMs. a): input-dependent state-emitting model; b): input-dependent transition-emitting model; c): input-independent state-emitting model; d): input-independent transition-emitting model.

$$P(y_t | x_t, x_{t-1}).$$

2.4 Input/Output Hidden Markov Models

Consider two synchronous serially ordered sequences of variables $U_1, \ldots, U_t, \ldots$ and $Y_1, \ldots, Y_t, \ldots$ in which t denotes a discrete time index. An input/output HMM (IOHMM) is a model of the conditional distribution

$$P(Y_1, \ldots, Y_t, \ldots | U_1, \ldots, U_t, \ldots),$$

where U_t are the *input* variables and Y_t are the *output* variables. The idea for extending HMMs to IOHMMs is very simple: Just add the *exogenous* variable U_t to the network of Figure 2 so that the current state (at time t) is conditionally dependent of both the previous state (at time $t-1$) and the current input.

Four variants of IOHMMs can be defined, depending whether outputs are emitted on states or on transitions and whether the output is conditionally

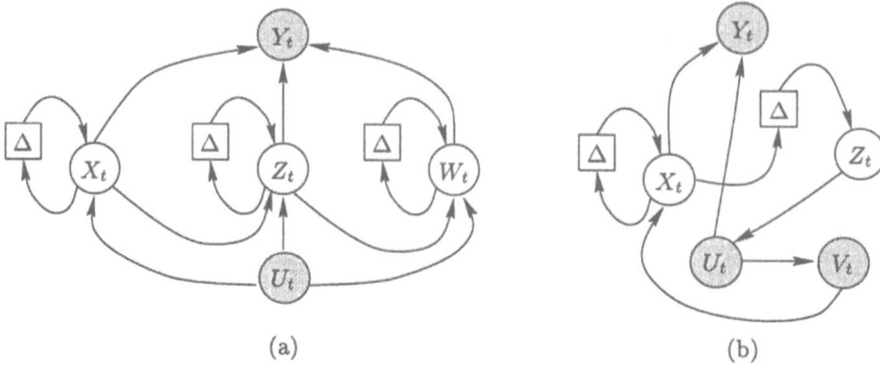

Figure 4: Some examples of recurrent Bayesian networks that generalize Markovian models. Nodes with self-loops are stationary non-homogeneous HMMs whose state transitions are controlled by their parents.

dependent of the input. All four variants are shown in Figure 3 (time-unfolded versions can be easily derived).

Note that, like HMMs, the Markov chain of IOHMMs is stationary since the conditional densities attached to each arc are constant over time. Unlike standard HMMs, however, the Markov chain of IOHMMs is not *homogeneous* since state transitions depend on time through the input variable U_t.

2.5 Beyond IOHMMs: Recurrent Bayesian Networks

One of the chief merits of graphical models is probably the synthetic expressive power they have. It is not difficult to imagine more complex structures such as those depicted in Figure 4. The semantics attached to of all these structures is related to conditional independence, as in traditional "static" Bayesian networks. However, traditional Bayesian networks are acyclic, while here cyclic directed paths are allowed provided that at least one time delay device is interposed along each cycle. Equation 1 is then generalized as follows:

$$P(v_{1,1}, \ldots, v_{n,1}, \ldots, v_{1,t}, \ldots, v_{n,t}, \ldots, v_{1,T}, \ldots, v_{n,T})$$
$$= \prod_{t=1}^{T} \prod_{i=1}^{n} P(v_{i,t} | \mathrm{pa}(V_i)_{1, t-\Delta_{i,1}}, \ldots, \mathrm{pa}(V_i)_{p_i, t-\Delta_{i,p_i}}) \tag{2}$$

where $v_{i,t}$ is the value of V_i at time t, T is the the time horizon being considered, $\mathrm{pa}(V_i)_{j,t}$ is the value of the j-th parent of V_i at time t, p_i is the number of parents of node V_i, and $\Delta_{i,j} = 1$ if there is a time delay in the arc from V_j to V_i.

A first direction for extending IOHMMs is to introduce more than just one hidden state variable. An example is shown in Figure 4a), in which the state variable is hierarchically decomposed so that multiple time scales are incorporated in the model. Using graphical models formalism, Smyth et al. (1996) have described more complex HMMs with multiple state chains. Hierarchical

HMMs with different state variables associated to different time scales have been proposed in (ElHihi & Bengio, 1996). Multiple state variables can also be justified as decomposition of the interacting factors involved in the data generation process (Ghahramani & Jordan, 1995).

A second direction for extending IOHMMs is to allow more than just two variables (i.e., input and output) to be visible (see, e.g. Figure 4b). In this way, probabilistic relationships among sets of data streams can be modeled.

A particularly important point for belief networks is the computational complexity of the inference algorithms (which also constitute the basis for the learning procedures). Complex recurrent Bayesian networks can very quickly become intractable and the complexity for a network as simple as the one depicted in Figure 4a grows exponentially with the number of hidden variables. Methods for approximating the inference algorithms, such as those proposed in (Saul & Jordan, 1996), are therefore very important.

3 Connectionist Architectures for IOHMMs

In this section some links between IOHMMs and recurrent neural networks are highlighted. In section 3.1 the case of discrete sequences is considered. Section 3.2 presents a connectionist architecture for continuous sequences.

3.1 Discrete Sequences Case: Belief Network Unfolding

In the case of discrete sequences the next/state densities $P(x_t|x_{t-1}, u_t)$ can simply be modeled as probability tables with parameters $\theta_{ijk} = P(x_t^i|x_{t-1}^j, u_t^k)$. Similarly, the output densities $P(y_t|x_t, u_t)$ can be modeled by probability tables with parameters $\vartheta_{\ell ik} = P(y_t^\ell|x_t^i, u_t^k)$. As explained in this section, parameters defined in this way can be though of as connection weight in a linear second order recurrent neural network.

In order to explain the connectionist analogy, I will introduce the concept of *state unfolding*[1] transformation of a Bayesian network, here described in the general case. Let be $\mathcal{D}$ a directed acyclic graph and assume that nodes are associated to discrete variables V_i, each taking values in $\{v_i^1, \ldots, v_i^{r_i}\}$. The unfolded network $\mathcal{D}_U$ is a higher order neural network having $\prod_{i=1}^{n} r_i$ linear nodes v_i^k, $i = 1, \ldots, n$, $k = 1, \ldots, r_i$. The activations of nodes v_i^k, $k = 1, \ldots, r_i$ will be supposed to represent the belief associated to variable V_i. For each node V_i in $\mathcal{D}$, consider the set of parents $\mathrm{pa}(V_i)$. Let $p_i = |\mathrm{pa}(V_i)|$ (the number of parents of V_i) and $m_i = \prod_{j \in \mathrm{pa}(V_i)} r_j$ (the number of configurations of the parents of V_i). Introduce m_i multipliers and connect each p_i-tuple $v_{j_1}^{k_1}, v_{j_2}^{k_2}, \ldots, v_{j_{p_i}}^{k_{p_i}}$ to one of the multipliers. Finally, connect each multiplier to each of the nodes v_i^k for $k = 1, \ldots, r_i$ with weight $p(v_i^k|\mathrm{pa}(V_i)^j)$, having denoted with $\mathrm{pa}(V_i)^j$ the j-th configuration of the parents of V_i.

---

[1] The unfolding transformation described here exploits the state dimension of variables and should not be confused with the time-unfolded representation of Markovian models described in Section 2, in which the temporal dimension is exploited.

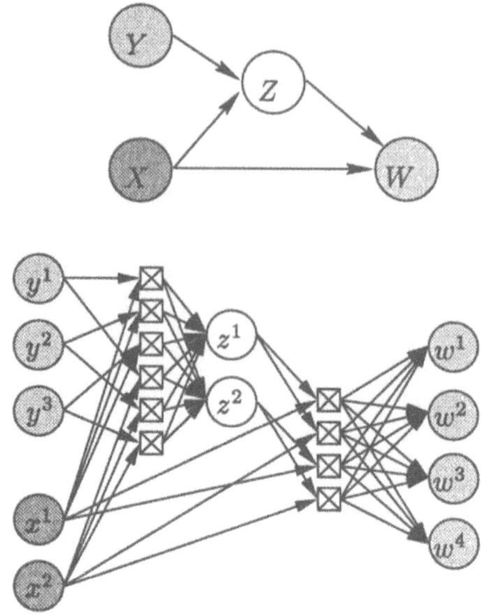

Figure 5: Example of belief network unfolding.

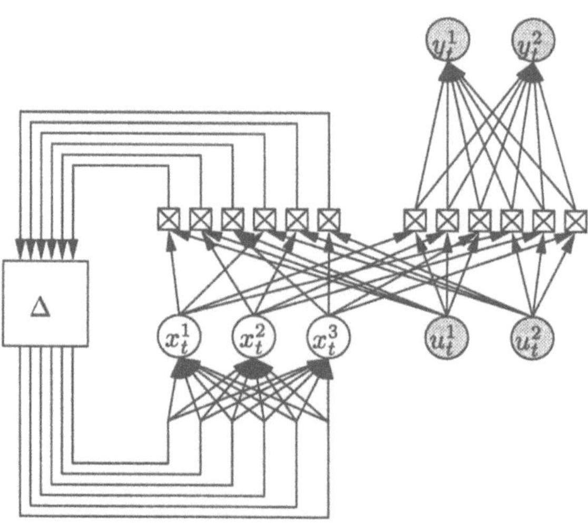

Figure 6: State-unfolding of discrete IOHMMs. The resulting graph closely resembles second order recurrent neural networks. This example refers to a model having three states and binary input and output.

An example of belief network unfolding is shown in Figure 5. When evidence is entered into each parent, the activation of each group of nodes v_j^k is either zero or one, reflecting a "one-hot" encoding of V_i's parents instantiation. It is easy to see that causal belief propagation from $\mathrm{pa}(V_i)$ to V_i (i.e., following the direction of arrows) is simply computed by forward propagation of activation signals in the unfolded network $\mathcal{D}_U$.

Unfolding the belief network of IOHMMs yields a structure very similar to second order recurrent neural networks (see Figure 6). The main difference is that units in second order neural networks (as described, e.g. in (Giles et al., 1992; Watrous & Kuhn, 1992)) are nonlinear, whereas in the present case units have a linear activation function.

It is interesting to note that the subgraph associated to state nodes in unfolded HMMs or IOHMMs corresponds to the underlying state transition graph. Such graph is complete if the Markovian model is ergodic (Rabiner, 1989). However, arcs in the state transition graph are often removed to reflect specific domain knowledge. For example, HMMs for speech recognition have transition graphs typically constrained to form a left-to-right chain (Rabiner, 1989). In the case of IOHMMs, each arc is also controlled by the current input and unfolded discrete IOHMMs are essentially stochastic finite state automata.

3.2 Continuous Sequences Case: Recurrent Mixture Model

In the case of continuous (and potentially multivariate) sequences, the next-state density $P(x_t|x_{t-1}, u_t)$ and the output density $P(y_t|x_t, u_t)$ can be modeled by feedforward neural networks. Since there are n discrete values for the hidden state variables, these models can be decomposed into $2n$ densities $P(x_t|x_{t-1}^j, u_t)$ and $P(y_t|x_t^j, u_t)$ for $j = 1, \ldots, n$. The connectionist architecture for continuous IOHMMs proposed in (Bengio & Frasconi, to appear) uses n *state subnetworks* $\mathcal{N}_1, \ldots, \mathcal{N}_j, \ldots, \mathcal{N}_n$ in charge of predicting $E[x_t|x_{t-1}^j, u_t]$ and n *output subnetworks* $\mathcal{O}_1, \ldots, \mathcal{O}_j, \ldots, \mathcal{O}_n$ in charge of predicting $E[y_t|x_t^j, u_t]$. Since the hidden states are assumed to be multinomially distributed, a softmax function (Bridle, 1989) has to be used for the last layer of each state subnetwork. The choice of the output function for the last layer of each output subnetwork, instead, should reflect the probability distribution of y_t. For example, a linear output layer should be used if outputs are supposed to be distributed according to a multivariate Gaussian density. More in general, a proper *link* function can be found for any distribution belonging to the exponential family (McCullagh & Nelder, 1989).

The resulting recurrent architecture is shown in Figure 7. It is easy to recognize that the output portion of the architecture is structured like a mixture of experts (Jacobs et al., 1991) in which each expert corresponds to a particular hidden state of the underlying Markov process. The mixing proportions are estimated by the state portion of the architecture, that therefore plays the role of a recurrent gating network. A difference with respect to a similar architecture

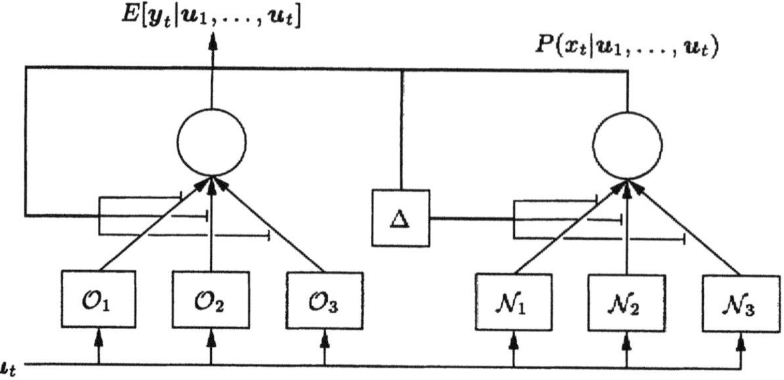

Figure 7: A connectionist architecture for continuous IOHMMs. The architecture can be interpreted as a recurrent mixture of experts.

proposed by Cacciatore & Nowlan (1994) is that the gating network itself is decomposed as a mixture model.

4 Training IOHMMs

IOHMMs can be trained by a supervised procedure based on maximum likelihood (MLE) or maximum a posteriori (MAP). The training data are supposed to be a set of input/output sequences independently sampled from the same distribution:

$$\mathcal{D} = (\mathcal{U}, \mathcal{Y}) = \{(\boldsymbol{u}_1^{(p)}, \ldots, \boldsymbol{u}_{T_p}^{(p)}, \boldsymbol{y}_1^{(p)}, \ldots, \boldsymbol{y}_{T_p}^{(p)})\}_{p=1,\ldots P}$$

where P is the total number of training examples and p is an example index. $\boldsymbol{u}_t$ and $\boldsymbol{y}_t$ can be (potentially multivariate) continuous or discrete variables. As shown above, a particular model $\mathcal{M}$ can be decomposed into $2n$ submodels for the densities $P(x_t|x_{t-1}^j, \boldsymbol{u}_t, \boldsymbol{\Theta}_{s,j})$ (with parameters $\boldsymbol{\Theta}_{s,j}$) and $P(\boldsymbol{y}_t|x_t^j, \boldsymbol{u}_t, \boldsymbol{\Theta}_{o,j})$ (with parameters $\boldsymbol{\Theta}_{o,j}$). The whole set of parameters for the model will be denoted by $\boldsymbol{\Theta}$. Parameters independence will be assumed. The training problem considered here is merely quantitative and corresponds to find the "best" parameters such that $P(\boldsymbol{\Theta}|\mathcal{D}, \mathcal{M})$ is maximum. The structural form of the model $\mathcal{M}$ (that includes the number of hidden states) is assumed to be fixed and will be omitted from notation. In the absence of priors for the parameters, the learning problem is simply solved by MLE, where the likelihood function is

$$L(\boldsymbol{\Theta}|\mathcal{D}) \propto P(\mathcal{Y}|\mathcal{U}, \boldsymbol{\Theta}) = \prod_{p=1}^{P} P(\boldsymbol{y}_1^{(p)}, \ldots, \boldsymbol{y}_{T_p}^{(p)}|\boldsymbol{u}_1^{(p)}, \ldots, \boldsymbol{u}_{T_p}^{(p)}, \boldsymbol{\Theta}).$$

The example index p will be omitted in the following to simplify notation.

The presence of the hidden state variable makes the MLE problem impossible to solve in closed form (sufficient statistics for the parameters are not observed directly from data). The problem, however, is suitable for using an expectation-maximization (EM) algorithm (Dempster et al., 1977) provided that the next-state and the output densities belong to the exponential family. The cases of discrete (symbolic) and continuous IOHMMs will be considered separately in the following. Training procedures for mixed models with continuous inputs and discrete outputs (or vice versa) can be derived straightforwardly from the methods described below.

4.1 Discrete sequences

In this case the sufficient statistics for the parameters of the next-state densities are the counts N_{ijk}, the number of times a transition from state x^j to state x^i occurs when the input symbol is u^k. Assuming a state-emitting input-dependent model[2] the sufficient statistics for the parameters of the output densities are the counts $M_{ik\ell}$, the number of times symbol y^ℓ is emitted in state x_i and the input symbol is u^k.

All these counts are not directly measurable since the state variable is hidden. In this case the EM algorithm proceeds by iterating until convergence the following two steps:

E-step compute the expected sufficient statistics

$$h_{ijk} = E[N_{ijk}|\mathcal{D}, \hat{\Theta}];$$

$$\check{g}_{ik\ell} = E[M_{ik\ell}|\mathcal{D}, \hat{\Theta}];$$

M-step update the parameters as

$$\hat{\theta}_{ijk} \leftarrow \frac{h_{ijk}}{\sum_i h_{ijk}}$$

$$\hat{\vartheta}_{ik\ell} \leftarrow \frac{g_{ik\ell}}{\sum_i g_{ik\ell}}$$

The expected sufficient statistics are easily computed by propagation of the evidence entered through the node U_t (input) and the node Y_t (supervision). Belief propagation in standard HMMs is known to be equivalent to the Baum-Welch (forward-backward) algorithm (Smyth et al., 1996). In the case of IOHMMs the forward and backward iterations have the following structure (Bengio & Frasconi, to appear):

$$\alpha_{i,t} \doteq P(y_1, \ldots, y_t, x_t^i | u_1, \ldots, u_t) = \vartheta_{i,u_t,y_t} \sum_j \theta_{ij,u_t} \alpha_{j,t-1}$$

$$\beta_{i,t} \doteq P(y_{t+1}, \ldots, y_T, x_t^i | u_t, \ldots, u_T) = \sum_j \vartheta_{j,u_t,y_{t+1}} \theta_{ji,u_{t+1}} \beta_{j,t+1} \tag{3}$$

---

[2]Modifications for transition-emitting or output-independent models are straightforward.

74

Finally:

$$
\begin{aligned}
h_{ijk} &= \sum_{p=1}^{P} \sum_{t:u_t=u^k} \gamma \vartheta_{ik,y_t} \alpha_{j,t-1} \beta_{i,t} \theta_{ijk} \\
g_{ik\ell} &= \sum_{p=1}^{P} \sum_{t:u_t=u^k,y_t=y^\ell} \gamma \alpha_{i,t} \beta_{i,t}
\end{aligned}
\tag{4}
$$

being γ normalization constants.

4.2 Continuous sequences

When using feedforward neural networks for modeling the next-state and the output densities the learning procedure is further complicated by the presence of additional hidden variables inside the networks. Training can be still solved in the expectation-maximization framework but the maximization step requires in turn an iterative approach (such as gradient descent), resulting in a global procedure with two nested loops. Assuming subnetworks without a hidden layer (i.e., generalized linear models), the iteratively reweighted least square algorithm could be used in the maximization step (Jordan & Jacobs, 1994).

Another option is to resort to a generalized EM (GEM) algorithm, in which the expected log-likelihood function Q is just *increased* instead of being maximized. GEM algorithms also converge to a local optimum of the likelihood (Dempster et al., 1977), but their convergence can be significantly slower. Assuming feedforward subnetworks with hidden layers, the GEM algorithm for IOHMMs proceeds by iterating until convergence the following two steps:

E-step compute the auxiliary function

$$
Q(\Theta, \hat{\Theta}) = E_{\mathcal{X}}[\log L(\Theta|\mathcal{D}, \mathcal{X})]|\mathcal{D}, \hat{\Theta}];
$$

M-step update the parameters following gradient ascent:

$$
\hat{\Theta} \leftarrow \hat{\Theta} + \epsilon \frac{\partial Q(\Theta, \hat{\Theta})}{\partial \Theta}
$$

The auxiliary function Q has the following cross entropy form:

$$
\begin{aligned}
Q(\Theta, \hat{\Theta}) &= \sum_{p=1}^{P} \sum_{t=1}^{T} \sum_{i=1}^{n} \hat{g}_{i,t} \log P(y_t|x_t^i, u_t, \Theta) \\
&+ \sum_{j=1}^{n} \hat{h}_{ij,t} \log P(x_t^i|x_{t-1}^j, u_t, \Theta)
\end{aligned}
\tag{5}
$$

where $\hat{g}_{i,t}$ and $\hat{h}_{ij,t}$ are computed by forward-backward recurrences similar to eqs. (3) (see (Bengio & Frasconi, to appear) for details). The gradient with respect to the connection weights of the networks is computed by simply applying backpropagation to the cost function (5).

5 IOHMMs, grammatical inference, and prior knowledge refinement

The problem of grammatical inference consists of inferring a set of rules that define a formal language from a set of labeled strings (Angluin & Smith, 1983). Grammatical inference is an interesting problem in machine learning since it can be viewed as a prototype for more complex language processing tasks (such as speech recognition and understanding). Even for the simplest class of regular grammars, for which the recognizer is a finite state automaton (FSA), the inference problem is known to be NP-complete (Angluin & Smith, 1983).

As explained in section 3.1, unfolding discrete IOHMMs along the state variable yields a stochastic finite automaton structure. In particular, if all parameters θ_{ijk} and $\vartheta_{\ell i}$ are either zero or one, an input-independent IOHMM is exactly equivalent to a Moore FSA. Frasconi & Bengio (1994) demonstrated that IOHMMs can be trained to recognize regular sets from example, with accuracy comparable to or better than recurrent second order networks. In most cases, the parameters of trained models actually converged to zero-one values. IOHMMs, however, behave like dynamical recognizers and output a continuous probability value; hence, a string will be accepted if the probability predicted by the model is greater than .5 plus some threshold. In this way, it has been experimentally observed that IOHMMs with non zero-one parameters are able to simulate machines having a number of states even superior to the number of states in the Markov chain. This behavior is theoretically confirmed by studies about the l-complexity or regular languages, which is defined as the smallest size of a linear network that accepts a language with respect to some threshold (Siegelmann & Giles, to appear).

One important issue that has been extensively investigated is the possibility of using adaptive recurrent networks to refine partial knowledge of the grammar to be inferred. Methods for mapping knowledge about finite state transitions into the connection weights have been developed for both first and second order recurrent networks (Omlin & Giles, to appear; Maclin & Shavlik, 1992; Frasconi et al., 1995). A common characteristic of all these methods is that exact FSA injection requires large connection weights (Omlin & Giles, to appear), i.e. dynamical units working in the saturated region of the sigmoidal function. Hence, the advantages of prior knowledge are somewhat tempered by the problem of vanishing gradients associated to learning long-term dependencies with dynamical attractors (Bengio et al., 1994).

Prior knowledge in the form of FSA transition rules can also be injected into discrete IOHMMs in a very natural way. The method simply exploits the standard Bayesian algorithm for quantitative training in belief networks using maximum a posteriori estimation. Suppose state-transition parameters Θ_{jk} (associated to previous state x^j and current input u^k) are Dirichlet distributed:

$$P(\theta_{jk}) = \gamma \prod_{i=1}^{n} \theta_{ijk}^{\alpha_{ijk}}$$

being γ a normalization constant and α_{ijk} the parameters of the Dirichlet distribution. After MAP learning, these parameters have the form

$$P(\theta_{jk}|\mathcal{D}) = \gamma \prod_{i=1}^{n} \theta_{ijk}^{\alpha_{ijk}^{0}+\eta_{ijk}}$$

where α_{ijk}^{0} identify the prior distribution and η_{ijk} are estimated from data after repeated iterations of the EM algorithm. The priors α_{ijk}^{0} can be obtained from known transition rules in a straightforward way. Suppose it is known that state x^i follows state x^j when the current input is u^k; then set $\alpha_{ijk}^{0} = H$ and $\alpha_{ljk}^{0} = \epsilon$ for $l \neq i$, being H a large real number and ϵ a small real real number. In this context H plays a role very similar to the hint strength described by Omlin & Giles in their algorithm for second order networks. There are, however, two interesting advantages of partially preset IOHMMs refined by MAP with respect to partially preset second order networks refined by gradient descent. First, rules introduced as Bayesian priors are naturally preserved during training, whereas unconstrained gradient descent cannot guarantee that rules are not destroyed by the training procedure[3]. Second, nearly deterministic IOHMMs (i.e. with state transition probabilities close to zero or one) do not suffer the problem of diffusion of credit during network propagation and thus are in a favorable situation for learning long-term dependencies (Bengio & Frasconi, 1995).

6 Conclusions

IOHMMs have been presented as a special case of recurrent Bayesian networks. There are, at least, three advantages in viewing IOHMMs under this perspective. First, potentially sophisticated extensions can be designed with relatively little theoretical effort. These extensions can be used to modeling relationships among a *set* of sequences using a *set* of state variables. Second, standard belief network propagation algorithms can be employed to compute predictions (during recall) and to to compute expected sufficient statistics (during training). Finally, prior knowledge can be injected using the solid theoretical background of Bayesian learning.

Acknowledgments

Input/output HMMs were developed in collaboration with Yoshua Bengio. The IOHMMs view presented in this paper, however, is under the complete responsibility of the author.

---

[3]See Frasconi et al. (1995) for a prior knowledge preservation method based on linear weight constraints in first order networks.

References

Andersen, S. K., Olesen, K. G., Jensen, F. V., & Jensen, F. (1989). Hugin—a shell for building Bayesian belief universes for expert systems. In *Int. Joint Conference of Artificial Intelligence*, Detroit. IEEE Press.

Angluin, D. & Smith, C. (1983). Inductive inference: Theory and methods. *Computing Surveys, 15*(3), 237–269.

Baldi, P., Chauvin, Y., Hunkapiller, T., & McClure, M. A. (1993). Hidden Markov Models in molecular biology: New algorithms and applications. In Hanson, S. J., Cowan, J. D., & Giles, C. L. (Eds.), *Advances in Neural Information Processing Systems*, volume 5, (pp. 747–754). Morgan Kaufmann, San Mateo, CA.

Bengio, Y. & Frasconi, P. (1995). Diffusion of context and credit information in Markovian models. *Journal of Artificial Intelligence Research, 3*, 249–270.

Bengio, Y. & Frasconi, P. (to appear). Input/output HMMs for sequence processing. *IEEE Transactions on Neural Networks*.

Bengio, Y., Simard, P., & Frasconi, P. (1994). Learning long-term dependencies with gradient descent is difficult. *IEEE Transactions on Neural Networks, 5*(2), 157–166.

Bridle, J. (1989). Probabilistic interpretation of feedforward classification network outputs, with relationships to statistical pattern recognition. In F. Fogelman-Soulie & J. Hérault (Eds.), *Neuro-computing: Algorithms, Architectures, and Applications*. New York: Springer-Verlag.

Buntine, W. (1996). A guide to the literature on learning graphical models. *IEEE Transactions of Knowledge and Data Engineering, 8*(2), 195–210.

Cacciatore, T. W. & Nowlan, S. J. (1994). Mixtures of controllers for jump linear and non-linear plants. In Cowan, J. D., Tesauro, G., & Alspector, J. (Eds.), *Advances in Neural Information Processing Systems 6*, San Mateo, CA. Morgan Kaufmann.

Dempster, A. P., Laird, N. M., & Rubin, D. B. (1977). Maximum-likelihood from incomplete data via the EM algorithm. *Journal of Royal Statistical Society B, 39*, 1–38.

ElHihi, S. & Bengio, Y. (1996). Hierarchical recurrent neural networks for long-term dependencies. In Touretzky, D. S., Mozer, M. C., & Hasselmo, M. E. (Eds.), *Advances in Neural Information Processing Systems 8*, Cambridge MA. The MIT Press.

Frasconi, P. & Bengio, Y. (1994). An EM approach to grammatical inference: Input/output hmms. In *Proc. 12th International Conference on Pattern Recognition*, Jerusalem (Israel). IEEE Press.

Frasconi, P., Gori, M., & Soda, G. (1995). Recurrent neural networks and prior knowledge for sequence processing: A constrained nondeterministic approach. *Knowledge Based Systems, 8*(6), 313–332.

Ghahramani, Z. & Jordan, M. I. (1995). Factorial hidden Markov models. Technical Report CCS 9502, Dept. of Brain & Cognitive Sciences, MIT, Cambridge, MA.

Giles, C. L., Miller, C. B., Chen, D., Chen, H. H., Sun, G. Z., & Lee, Y. C. (1992). Learning and extracting finite state automata with second-order recurrent neural networks. *Neural Computation, 4*(3), 393–405.

Heckerman, D., Geiger, D., & Chickering, D. M. (1995). Learning Bayesian networks: The combination of knowledge and statistical data. *Machine Learning, 20*, 197–243.

Jacobs, R. A., Jordan, M. I., Nowlan, S. J., & Hinton, G. E. (1991). Adaptive mixtures of local experts. *Neural Computation, 3*(1), 79–87.

Jensen, F. V., Lauritzen, S. L., & Olosen, K. G. (1990). Bayesian updating in recursive graphical models by local computations. *Computational Statistical Quarterly, 4*, 269–282.

Jordan, M. I. (1986). Attractor dynamics and parallelism in a connectionist sequential machine. In *Proceedings of the Eighth Annual Conference of the Cognitive Science Society*, (pp. 531–546)., Amherst 1986. Lawrence Erlbaum, Hillsdale.

Jordan, M. I. & Jacobs, R. A. (1994). Hierarchical mixtures of experts and the EM algorithm. *Neural Computation, 6*(2), 181–214.

Levinson, S. E., Rabiner, L. R., & Sondhi, M. M. (1983). An introduction to the application of the theory of probabilistic functions of a Markov process to automatic speech recognition. *The Bell System Technical Journal, 62*(4), 1035–1074.

Maclin, R. & Shavlik, J. (1992). Using knowledge-based neural networks to improve algorithms: Refining the Chou-Fasman algorithm for protein folding. In *Proceedings of the Tenth National Conference on Artificial Intelligence*, (pp. 165–170)., San Jose, CA.

McCullagh, P. & Nelder, J. A. (1989). *Generalized Linear Models* (2nd ed.). London: Chapman & Hall.

Omlin, C. W. & Giles, C. L. (to appear). Constructing deterministic finite-state automata in recurrent neural networks. *Journal of the ACM*.

Pearl, J. (1988). *Probabilistic Reasoning in Intelligent Systems : Networks of Plausible Inference*. Morgan Kaufmann.

Poritz, A. (1988). Hidden Markov models: a guided tour. In *Proc. Int. Conf. Acoustics, Speech, and Signal Processing*, (pp. 7–13).

Rabiner, L. R. (1989). A tutorial on hidden Markov models and selected applications in speech recognition. *Proceedings of the IEEE, 77*(2), 257–286.

Rumelhart, D. E., McClelland, J. L., & the PDP Research Group (1986). *Parallel Distributed Processing: Explorations in the Microstructure of Cognition*, volume 1. Cambridge: MIT Press.

Saul, L. K. & Jordan, M. I. (1996). Exploiting tractable substructures in intractable networks. In Touretzky, D. S., Mozer, M. C., & Hasselmo, M. E. (Eds.), *Advances in Neural Information Processing Systems 8*, Cambridge MA. MIT Press.

Shachter, R. D. (1986). Evaluating influence diagrams. *Operations Research, 34*(6), 871–882.

Siegelmann, H. T. & Giles, C. L. (to appear). The complexity of language recognition by neural networks. *Neurocomputing*.

Smyth, P., Heckerman, D., & Jordan., M. (1996). Probabilistic independence networks for hidden Markov probability models. Technical Report MSR-TR-96-03, Microsoft Research.

Waibel, A., Hanazawa, T., Hinton, G., Shikano, K., & Lang, K. (1989). Phoneme recognition using time-delay neural networks. *IEEE Transactions on Acoustics, Speech, and Signal Processing, 37*, 328–339.

Watrous, R. L. & Kuhn, G. M. (1992). Induction of finite-state languages using second-order recurrent networks. *Neural Computation, 4*(3), 406–414.

Whittaker, J. (1990). *Graphical Models in Applied Multivariate Statistics*. Chichester: Wiley.

Neural Nonlinear Controllers and Observers: Stability Results

A. Alessandri, M. Sanguineti, and R. Zoppoli
Department of Communications, Computer and System Sciences
DIST–University of Genova

T. Parisini
Department of Electrical, Electronic and Computer Engineering
DEEI–University of Trieste

Abstract

A receding-horizon (RH) optimal control scheme and a sliding–window (SW) state estimation scheme are presented and neural approximations for these two control and estimation laws are presented. Some theoretical results on the stability of the controller and the error convergence for the estimator are discussed.

1 Statement of the receding–horizon control problem

Let us consider the discrete-time dynamic system

$$\underline{x}_{t+1} = \underline{f}(\underline{x}_t, \underline{u}_t), \quad t = 0, 1, \ldots \tag{1}$$

where $\underline{x}_t \in X \subset \Re^n$ and $\underline{u}_t \in U \subset \Re^m$. The RH problem stems from the definition of the FH cost function

$$J_{FH}(\underline{x}_t, \underline{u}_{t,t+N-1}, N, a, P) = \sum_{i=t}^{t+N-1} h(\underline{x}_i, \underline{u}_i) + a\|\underline{x}_{t+N}\|_P^2, \quad t \geq 0 \tag{2}$$

where $\underline{u}_{t\tau} \triangleq \mathrm{col}(\underline{u}_t, \ldots, \underline{u}_\tau)$, $\|\underline{x}\|_P^2 \triangleq \underline{x}^T P \underline{x}$, ($a \in \Re$ is a positive scalar, $P \in \Re^{n \times n}$ is a positive–definite symmetric matrix, and N is a positive integer denoting the length of the control horizon). Let us assume that $\underline{f} \in \mathcal{C}^1 [\Re^n \times \Re^m, \Re^n]$, with $\underline{f}(\underline{0}, \underline{0}) = \underline{0}$, and that $h \in \mathcal{C}^1 [\Re^n \times \Re^m, \Re^+]$, with $h(\underline{0}, \underline{0}) = 0$. Then we can state the following

Problem 1. *At every time instant* $t \geq 0$, *find the RH optimal control law* $\underline{u}_t^{RH^\circ} = \underline{\gamma}_{RH}^\circ(\underline{x}_t) \in U$, *where* $\underline{u}_t^{RH^\circ}$ *is the first vector of the control sequence,* $\underline{u}_t^{FH^\circ}, \ldots, \underline{u}_{t+N-1}^{FH^\circ}$ *(i.e.,* $\underline{u}_t^{RH^\circ} \triangleq \underline{u}_t^{FH^\circ}$ *), that minimizes the cost (2) for the state* $\underline{x}_t \in X$.

$\square$

As will be clear in the following, the final cost $a\|\underline{x}_{t+N}\|_P^2$ plays a key role in the development of stability results concerning the RH regulator, and is essentially a penalty function related to the relaxation of the terminal state constraint[5, 4] $\underline{x}_{t+N} = \underline{0}$ (the relaxation of the terminal state has just been addressed in a continuous–time framework[6]).

The statement of Problem 1 does not impose any particular way of computing the control vector $\underline{u}_t^{RH°}$ as a function of $\underline{x}_t$. Actually, we have two possibilities:

1) *On-line computation.* Problem 1 is an open-loop optimal control problem and may be regarded as a nonlinear programming one.

2) *Off-line computation.* This approach implies that the control law $\gamma_{RH}^°(\underline{x}_t)$ has to be computed "a priori" and stored in the regulator's memory.

Clearly, the off-line computation has advantages and disadvantages that are opposite to the ones of the on-line approach. No on-line computational effort is requested from the regulator, but a very large amount of computer memory may be required to store the closed-loop control law.

2 Stabilizing properties of the optimal and the approximate RH regulator

Let us make the following assumptions:

(i) The linear system $\underline{x}_{t+1} = A\underline{x}_t + B\underline{u}_t$, obtained via the linearization of the system (1) in a neighborhood of the origin is stabilizable.

(ii) The transition cost function $h(\underline{x}, \underline{u})$ depends on both $\underline{x}$ and $\underline{u}$, and there exist two strictly increasing functions $r, s \in C[\Re^+, \Re^+]$, with $r(0) = s(0) = 0$, such that $r(\|(\underline{x}, \underline{u})\|) \leq h(\underline{x}, \underline{u}) \leq s(\|(\underline{x}, \underline{u})\|)$, $\forall \underline{x} \in X, \forall \underline{u} \in U$, where $(\underline{x}, \underline{u}) \triangleq \mathrm{col}\,(\underline{x}, \underline{u})$.

(iii) There exists a compact set $X_0 \subseteq X$, $X_0 \in \mathcal{Z}$, with the property that there exists a control horizon $M \geq 1$ such that there exists a sequence of admissible control vectors $\{\underline{u}_i \in U, i = t, \ldots, t + M - 1\}$ that yield an admissible state trajectory $\underline{x}_i \in X, i = t, t+1, \ldots, t+M$ ending in the origin of the state space (i.e., $\underline{x}_{t+M} = \underline{0}$) for any initial state $\underline{x}_t \in X_0$.

(iv) The optimal FH feedback control functions $\gamma_{FH}^°(\underline{x}_i, i), i = t, \ldots, t+N-1$, which minimize the cost (2), are continuous with respect to $\underline{x}_i$, for any $\underline{x}_t \in X$ and any finite integer $N \geq 1$.

Let us now denote by $J_{RH}^°(\underline{x}_t, N, a, P) = \sum_{i=t}^{+\infty} h(\underline{x}_i^{RH°}, \underline{u}_i^{RH°})$ the cost associated with the RH trajectory starting from $\underline{x}_t$ (i.e., $\underline{x}_t^{RH°} = \underline{x}_t$) and with the solution of the FH control problem characterized by a control horizon N and a terminal cost function $a\|\underline{x}_{t+N}\|_P^2$. Finally, let us denote by $J_{FH}^°(\underline{x}_t, N, a, P) \triangleq J_{FH}(\underline{x}_t, \underline{u}_{t,t+N-1}^°; N, a, P) = \sum_{i=t}^{t+N-1} h(\underline{x}_i^{FH°}, \underline{u}_i^{FH°}) + a\|\underline{x}_{t+N}^{FH°}\|_P^2$ the cost corresponding to the optimal N-stage trajectory starting from $\underline{x}_t$. Then the following theorem on the stability properties of the RH regulator can be proved.

Theorem 1 [8]. *If assumptions (i) to (iv) are verified, there exist a positive* scalar $\tilde{a}$ and a positive-definite symmetric matrix $P \in \Re^{n \times n}$ such that, for any $N \geq M$ and any $a \in \Re$, $a \geq \tilde{a}$, the following properties hold:

1) The RH control law stabilizes asymptotically the origin, which is an equilibrium point of the resulting closed-loop system.

2) There exists a positive scalar β such that the set $\mathcal{W}(N, a, P) \in \mathcal{Z}$, $\mathcal{W}(N, a, P) \stackrel{\triangle}{=} \{\underline{x} \in X : J^{\circ}_{FH}(\underline{x}, N, a, P) \leq \beta\}$, is an invariant subset of X_0 and a domain of attraction for the origin, i.e., for any $\underline{x}_t \in \mathcal{W}(N, a, P)$, the state trajectory generated by the RH regulator remains entirely contained in $\mathcal{W}(N, a, P)$ and converges to the origin.

$\square$

It is now worth noting that, in deriving (both on line and off line) the RH control law $\underline{u}_t^{RH^{\circ}} = \Upsilon^{\circ}_{RH}(\underline{x}_t)$, computational errors may affect the vector $\underline{u}_t^{RH^{\circ}}$ and possibly lead to a closed-loop instability of the origin. Analogously, as will be shown in the next section, errors on $\underline{u}_t^{RH^{\circ}}$ may be induced by the neural or RBF network that approximate the RH regulator. Therefore, we give following theorem, which characterizes the properties of the RH regulator when suboptimal control vectors $\hat{\underline{u}}_i^{RH} \in U$, $i \geq t$, are used in the RH control scheme, instead of the optimal ones $\underline{u}_i^{RH^{\circ}}$ solving Problem 1. Let us denote by $\hat{\underline{x}}_i^{RH}$, $i > t$, the state vectors belonging to the suboptimal RH trajectory starting from $\underline{x}_t$.

Theorem 3[9]. *If assumptions (i) to (iv) are verified, there exist a positive scalar $\tilde{a}$ and a positive-definite symmetric matrix $P \in \Re^{n \times n}$ such that, for any $N \geq M$ and for any $a \geq \tilde{a}$, the following properties hold:*

1) There exist suitable scalars $\tilde{\delta}_i \in \Re$, $\tilde{\delta}_i > 0$ such that, if $\left\| \underline{u}_i^{RH^{\circ}} - \hat{\underline{u}}_i^{RH} \right\| \leq \tilde{\delta}_i$, $i \geq t$, then

$$\hat{\underline{x}}_i^{RH} \in \mathcal{W}(N, a, P), \quad \forall i > t, \forall \underline{x}_t \in \mathcal{W}(N, a, P) \tag{3}$$

2) For any compact set $\mathcal{W}_d \subset \Re^n$, $\mathcal{W}_d \in \mathcal{Z}$, there exist a finite integer $T \geq t$ and suitable scalars $\bar{\delta}_i \in \Re$, $\bar{\delta}_i > 0$ such that, if $\left\| \underline{u}_i^{RH^{\circ}} - \hat{\underline{u}}_i^{RH} \right\| \leq \bar{\delta}_i$, $i \geq t$, then

$$\hat{\underline{x}}_i^{RH} \in \mathcal{W}_d, \quad \forall i \geq T, \forall \underline{x}_t \in \mathcal{W}(N, a, P) \tag{4}$$

$\square$

3   The neural approximation for the RH regulator

As stated before, we are mainly interested in computing the RH control law $\underline{u}_t^{RH^{\circ}} = \Upsilon^{\circ}_{RH}(\underline{x}_t)$ off line. Then, we need to derive ("a priori") an FH closed-loop optimal control law, $\underline{u}_i^{FH^{\circ}} = \Upsilon^{\circ}_{FH}(\underline{x}_i, i)$, $t \geq 0$, $i = t, \ldots, t + N - 1$, that minimizes the cost (2) for any $\underline{x}_t \in X$. Because of the time-invariances of the dynamic system (1) and of the cost function (2), we consider the control functions $\underline{u}_i^{FH^\circ} = \Upsilon_{FH}^\circ(\underline{x}_i, i-t)$, $t \geq 0$, $i = t, \ldots, t+N-1$, instead of $\underline{u}_i^{FH^\circ} = \Upsilon_{FH}^\circ(\underline{x}_i, i)$, and state the following

Problem 3. *Find the FH optimal feedback control law* $\{\underline{u}_i^{FH^\circ} = \Upsilon_{FH}^\circ(\underline{x}_i, i-t) \in U, t \geq 0, i = t, \ldots, t+N-1\}$ *that minimizes the cost (2) for any* $\underline{x}_t \in X$.

□

Then, once the solution of Problem 3 has been achieved, we consider only the first optimal control function (i.e., the one corresponding to $i = 0$) and write

$$u_t^{RH^\circ} = \Upsilon_{RH}^\circ(\underline{x}_t) \stackrel{\triangle}{=} \Upsilon_{FH}^\circ(\underline{x}_t, 0), \quad \forall \underline{x}_t \in X, t \geq 0 \tag{5}$$

As we have the possibility of computing (off line) any number of open-loop optimal control sequences $\underline{u}_t^{FH^\circ}, \ldots, \underline{u}_{t+N-1}^{FH^\circ}$ (see Problem 1) for different vectors $\underline{x}_t \in X$, we propose to approximate the function $\Upsilon_{RH}^\circ(\underline{x}_t) = \Upsilon_{FH}^\circ(\underline{x}_t, 0)$ by a function $\hat{\Upsilon}_{RH}(\underline{x}_t, \underline{w})$, to which we assign a given structure. $\underline{w}$ is a vector of parameters to be optimized. Among various possible approximating functions, multilayer feedforward neural networks and RBF networks may be convenient to be chosen. The choice is related to several aspects which depend heavily on the particular application. For instance feedforward neural networks are often convenient when the input vector dimension is large [8] whereas RBF networks are often useful in cases where input vector dimension is small but fast numeric algorithms for the training are required as the parameter vector appears linearly [2]. In the following, we denote by $\hat{\Upsilon}_{FH}^{(\nu)}(\underline{x}_t, \underline{w})$ the approximate function, where ν is the number of parameters to be tuned, and let $\hat{\Upsilon}_{RH}^{(\nu)}(\underline{x}_t, \underline{w}) \stackrel{\triangle}{=} \hat{\Upsilon}_{FH}^{(\nu)}(\underline{x}_t, \underline{w})$. Then, we can state the following

Theorem 4 [9]. *Assume that, in the solution of Problem 3, the first control function* $\Upsilon_{RH}^\circ(\underline{x}_t) = \Upsilon_{FH}^\circ(\underline{x}_t, 0)$ *(see (5)) is unique, and that it is a $C[X, \Re^m]$ function. Then, for any $\varepsilon \in \Re, \varepsilon > 0$, there exist an integer ν and a weight vector $\underline{w}$ such that*

$$\left\| \Upsilon_{RH}^\circ(\underline{x}_t) - \hat{\Upsilon}_{RH}^{(\nu)}(\underline{x}_t, \underline{w}) \right\| < \varepsilon, \quad \forall \underline{x}_t \in X \tag{6}$$

□

Theorem 4 applies to several different choices of approximating functions comprising the ones considered here[3]. From Theorem 4, we obtain

Corollary [9]. *If assumptions (i) to (iv) are verified, there exists an approximate RH regulator $\hat{\underline{u}}_t^{RH} = \hat{\Upsilon}_{RH}^{(\nu)}(\underline{x}_t, \underline{w})$, $t \geq 0$, for which the two properties in Theorem 3 hold true. The control vectors $\hat{\underline{u}}_t^{RH}$ are constrained to take their values in the admissible set $\bar{U} \stackrel{\triangle}{=} \{\underline{u} : \underline{u} + \Delta\underline{u} \in U, \|\Delta\underline{u}\| \leq \varepsilon\}$, where, for a given $\mathcal{W}_d$ and any $\underline{x}_t \in \mathcal{W}(N, a, P)$, ε is such that $\varepsilon \leq \delta_i$, $i \geq t$ (see the scalars in Theorem 3), and $\bar{U} \in \mathcal{Z}$.*

□

Now, we need to define an approximating network that differs slightly from the one considered to state Theorem 3. The new network is the parallel of m single-output neural networks of the type described previously (i.e., containing a single hidden layer and linear output activation units). Each network generates one of the m components of the control vector $\hat{\underline{u}}_t^{RH}$. We denote by

$\hat{\gamma}_{RHj}^{(\nu_j)}(\underline{x}_t, \underline{w}_j)$ the input-output mapping of the j-th of such networks, where ν_j is the number of neural units in the hidden layer and $\underline{w}_j$ is the weight vector, and we denote by $\gamma_{RHj}^{\circ}(\underline{x}_t)$ the j-th component of the vector function $\underline{\gamma}_{RH}^{\circ}$. Then, we can state the following min–max

Problem 4. *For each function* $\hat{\gamma}_{RHj}^{(\nu_j)}(\underline{x}_t, \underline{w}_j)$, *find the numbers* $\nu_1^*, \ldots, \nu_m^*$ *of neural units such that, for* $j = 1, \ldots, m$

$$\min_{\underline{w}_j} \max_{\underline{x}_t \in X} \left| \gamma_{RHj}^{\circ}(\underline{x}_t) - \hat{\gamma}_{RHj}^{(\nu_j)}(\underline{x}_t, \underline{w}_j) \right| \leq \frac{\varepsilon}{\sqrt{m}} \tag{7}$$

$\square$

The use of a min–max approach stems from the necessity of the control errors generated by $\hat{\underline{u}}_t^{RH^{\circ}} = \hat{\underline{\gamma}}_{RH}(\underline{x}_t, \underline{w}^{\circ})$ to be uniformly bounded on X by the scalar ε specified in the corollary. In the following, we shall address the possibility of computing ε. As to the numbers ν_j^*, rather a naive trial–and–error procedure for determining them is the following: for each j, increase ν_j until the term on the left–hand side of (7) is less than or equal to $\varepsilon/\sqrt{m}$. This procedure seems reasonable, as a bound (uniform on X) to $\left| \gamma_{RHj}^{\circ}(\underline{x}_t) - \hat{\gamma}_{RHj}^{(\nu_j)}(\underline{x}_t, \underline{w}_j) \right|$ can be established that is proportional to $1/\sqrt{\nu_j}$. More specifically, the following general result can be proved.

Fact. *Under suitable assumptions on the smoothness of the function* γ_{RHj}° *and denoting by* s_j *a suitable scalar measuring such a smoothness, we have that*

$$\min_{\underline{w}_j} \max_{\underline{x}_t \in X} \left| \gamma_{RHj}^{\circ}(\underline{x}_t) - \hat{\gamma}_{RHj}^{(\nu_j)}(\underline{x}_t, \underline{w}_j) \right| \propto \left(\frac{1}{\nu_j} \right)^{\frac{s_j}{n}} \tag{8}$$

$\square$

The bound (8) on the approximation error denotes the fundamental influence of the smoothnes of the function to be approximated on the rate of convergence of the approximation scheme, that is, the rate at which the approximation error decreases as the number of parameters of the approximating structure increases. Clearly, a lot of technical issues in the above fact have beed omitted here, as we want only to stress the dependence of the rate of convergence on the smoothnes index s_j.

4 Statement of the nonlinear estimation problem

Let us consider the discrete-time dynamic system (in general, nonlinear)

$$\underline{x}_{t+1} = \underline{f}(\underline{x}_t, \underline{u}_t), \quad t = 0, 1, \ldots \tag{9}$$

where $\underline{x}_t \in \Re^n$ and $\underline{u}_t \in \Re^m$ are the state and control vectors, respectively. The initial state $\underline{x}_0$ is unknown and we assume that $\underline{x}_0 \in X_0$, where X_0 is a compact subset of $\Re^n$.

The state vector is observed through the measurement channel (in general, nonlinear)

$$\underline{y}_t = \underline{h}(\underline{x}_t) + \underline{\eta}_t, \quad t = 0, 1, \ldots \tag{10}$$

where $\underline{y}_t \in \Re^p$ is the measurement vector and $\underline{\eta}_t \in H \subset \Re^r$ is a measurement random noise, where H is a given compact set. We suppose $\underline{f}$ and $\underline{h}$ to be Lipschitz functions on compact subsets of $\Re^n \times \Re^m$ and $\Re^n$, respectively.

Now, let us consider a *sliding window* observer, i.e., an observer which makes use of a limited batch of measurement and control vectors. In this respect, we introduce the following notations. At a given stage t and for a given temporal window of length N stages, we have to process the last $N + 1$ measurement vectors $\underline{y}_{t-N}, \ldots, \underline{y}_t$ and the last N control vectors, $\underline{u}_{t-N}, \ldots, \underline{u}_{t-1}$. Following a standard notation[7], let $\underline{f}^u(\underline{x}) \stackrel{\triangle}{=} \underline{f}(\underline{x}, \underline{u})$ and $\underline{F}^{u_{t-N}^i} \stackrel{\triangle}{=} \underline{f}^{u_i} \circ \cdots \circ \underline{f}^{u_{t-N}}$, $t - N \leq i \leq t$ (without loss of generality, in the following we assume $t \geq N$). We also define $\underline{y}_{t-N}^t \stackrel{\triangle}{=} \mathrm{col}\left(\underline{y}_{t-N}, \underline{y}_{t-N+1}, \ldots, \underline{y}_t\right)$ and $\underline{u}_{t-N}^{t-1} \stackrel{\triangle}{=} \mathrm{col}\left(\underline{u}_{t-N}, \underline{u}_{t-N+1}, \ldots, \underline{u}_{t-1}\right)$. Then, we can state the following

Problem 1N. *At any stage t, for any possible batch of measurements $\underline{y}_{t-N}^t$ and control vectors $\underline{u}_{t-N}^{t-1}$, find the initial state estimate $\hat{\underline{x}}_{t-N}$ such that $\hat{\underline{x}}_t = \underline{x}_t$, with $\hat{\underline{x}}_t = \underline{F}^{u_{t-N}^{t-1}}\left(\hat{\underline{x}}_{t-N}\right)$.*

$\square$

Notice that solving Problem 1N, for any $\underline{y}_{t-N}^t, \underline{u}_{t-N}^{t-1}$, means finding a function of the form $\hat{\underline{x}}_{t-N} = \underline{\gamma}_{t-N}(\underline{y}_{t-N}^t, \underline{u}_{t-N}^{t-1})$. Now, this intrinsic functional nature of Problem 1N and its relationship with some nonlinear observability concepts will be put into evidence. To this end, for the moment we assume that no noise is acting on the measurement channel, and it is useful to introduce the map $\underline{\Phi}: \Re^n \times \Re^{mN} \longrightarrow \Re^{p(N+1)}$, where

$$\underline{\Phi}\left(\underline{x}_{t-N} \mid \underline{u}_{t-N}^{t-1}\right) \stackrel{\triangle}{=} \begin{bmatrix} \underline{h}\left(\underline{x}_{t-N}\right) \\ \underline{h} \circ \underline{F}^{u_{t-N}^{t-N}}(\underline{x}_{t-N}) \\ \vdots \\ \underline{h} \circ \underline{F}^{u_{t-N}^{t-1}}(\underline{x}_{t-N}) \end{bmatrix} = \begin{bmatrix} \underline{y}_{t-N} \\ \underline{y}_{t-N+1} \\ \vdots \\ \underline{y}_t \end{bmatrix} \tag{11}$$

(the notation $\underline{\Phi}\left(\underline{x}_{t-N} \mid \underline{u}_{t-N}^{t-1}\right)$ simply means that $\underline{u}_{t-N}^{t-1}$ has to be interpreted as a vector of parameters).

The following assumption will be used in the remaining part of the paper.

Assumption 1N:[1] there exists a compact set $X \subseteq X_0$ and a compact set $U \subset \Re^m$, such that, for any $\underline{x} \in X$ and for any $\underline{u} \in U$, $\underline{f}(\underline{x}, \underline{u}) \in X$.

Later on we shall present an estimation technique to be determined off line. Once it has been stored in the observer's memory, it will enable one to estimate the state vector "instantaneously". Toward this end, we need to state

---

[1] The assumption means that X is a controlled–invariant set. In the case of systems without control, the assumption reduces simply in assuming that there exists a compact set which is left–invariant with respect to the function $\underline{f}$.

the estimation problem in a time–invariant context. Assumption 1N ensures us that, besides U, also X (the set the state vector takes its values from) is time–invariant. Moreover, Assumption 1N will play a key role in the proof of Theorem 2N on the convergence of the state estimation error.

Now, the following observability condition can be stated [7].

Definition. *The system (9) and (10) is uniformly $N + 1$–observable with respect to X and U^N if and only if there exists an integer N such that, for any $\underline{u}_{t-N}^{t-1} \in U^N$, the mapping $\underline{\Phi}\left(\underline{x}_{t-N} \,|\, \underline{u}_{t-N}^{t-1}\right) : X \longrightarrow \Re^{p(N+1)}$ is an injective immersion.*

The injectivity of the above immersion is clearly related to the global univalence of the mapping $\left(\underline{x}_{t-N}, \underline{u}_{t-N}^{t-1}\right) \longrightarrow \left(\underline{\Phi}\left(\underline{x}_{t-N} \,|\, \underline{u}_{t-N}^{t-1}\right), \underline{u}_{t-N}^{t-1}\right)$. Now, the following sufficient conditions for the univalence property to hold can be immediately obtained[1].

Theorem 1N. *Suppose that X_0 is a convex set and that, for any $\underline{u}_{t-N}^{t-1} \in U^N$, the mapping $\underline{\Phi}\left(\underline{x}_{t-N} \,|\, \underline{u}_{t-N}^{t-1}\right)$ is differentiable with a Jacobian matrix $D(\underline{x}_{t-N}, \underline{u}_{t-N}^{t-1}) \in \Re^{[p(N+1)+mN] \times n}$, $\underline{x}_{t-N} \in X$, $\underline{u}_{t-N}^{t-1} \in U^N$. Then, for any stage $t \geq N$, two different cases may be given:*

1. *If there exists an integer $N \geq n$ such that $n = p(N + 1)$, consider the smallest rectangular compact set $\bar{X}$, such that $X \subseteq \bar{X} \subseteq X_0$. Then $\underline{\Phi}$ is globally univalent on $\bar{X} \times U^N$ if $D(\underline{x}_{t-N}, \underline{u}_{t-N}^{t-1})$ is either a P–matrix or an N–matrix[2], $\forall \underline{x}_{t-N} \in \bar{X}, \underline{u}_{t-N}^{t-1} \in U^N$.*

2. *If an integer $N \geq n$ such that $n = p(N + 1)$ does not exist, consider the smallest convex compact set $\tilde{X}$, such that $X \subseteq \tilde{X} \subseteq X_0$. Then $\underline{\Phi}$ is globally univalent on $\tilde{X} \times U^N$ if there exists a matrix $A \in \Re^{n \times [p(N+1)+mN]}$ such that $C(\underline{x}_{t-N}, \underline{u}_{t-N}^{t-1}) \triangleq AD(\underline{x}_{t-N}, \underline{u}_{t-N}^{t-1})$ has the following (row) Diagonal Dominance Property: $|c_{ii}(\underline{x}_{t-N}, \underline{u}_{t-N}^{t-1})| > \sum_{j:j \neq i} |c_{ij}(\underline{x}_{t-N}, \underline{u}_{t-N}^{t-1})|, \forall \underline{x}_{t-N} \in \tilde{X}, \underline{u}_{t-N}^{t-1} \in U^N$, where c_{ij} denotes the i-th row, j-th column element of matrix C.*

$\square$

Let us now define the set $Y \triangleq \underline{h}(X)$, $Y \subset \Re^p$. In case of $N+1$–observability of the system, a dead–beat observer of order $N+1$ can be designed by solving the system of equations (11) in any point of the set $Y^{N+1} \times U^N$, i.e., by computing the function $\hat{\underline{x}}_{t-N} = \underline{\Upsilon}_{t-N}(\underline{y}_{t-N}^{t}, \underline{u}_{t-N}^{t-1})$. However, in general, $\underline{\Upsilon}_{t-N}$ cannot be derived in closed form; moreover, it could be useful to consider an asymptotic observer, in cui the current state estimate is generated by updating the estimate at the previous stage (of course, as will be clear later on, this is necessary when considering also the measurement noise). We denote by $\hat{\underline{x}}_{it}$, $i = t - N, \ldots, t$, the state estimates derived at the stage t on the basis of the measurements $\underline{y}_{t-N}^{t}$, on the control vectors $\underline{u}_{t-N}^{t-1}$, and on the "a priori prediction" $\bar{\underline{x}}_{t-N} \triangleq$

---

[2] A matrix is defined as a P–matrix (N–matrix) if all its principal minors are strictly positive (negative).

$\underline{f}(\hat{\underline{x}}_{t-N-1,t-1},\underline{u}_{t-N-1})$. Let us denote by $\mathcal{X}_p \stackrel{\triangle}{=} \underline{f}(X,U)$ the set to which all possible predictions $\bar{\underline{x}}_{t-N}$ belong to, for any $t \geq N$ (the definition of $\mathcal{X}_p$ is possible thanks to Assumption 1N). Then, let us consider the following

Problem 1N'. *At any stage t, for any possible batch of measurements $\underline{y}^t_{t-N} \in Y^{N+1}$ and control vectors $\underline{u}^{t-1}_{t-N} \in U^N$, and for any possible $\bar{\underline{x}}_{t-N} \in \mathcal{X}_p$, find the initial state estimate $\hat{\underline{x}}_{t-N,t}$ such that $\lim\limits_{t \to +\infty} \hat{\underline{x}}_{tt} = \underline{x}_t$, with $\hat{\underline{x}}_{tt} = \underline{F}^{u^{t-1}_{t-N}}\left(\hat{\underline{x}}_{t-N,t}\right)$.*

□

Clearly the asymptotic convergence to zero of the estimation error is an achievable requirement only when no noise acts on the measurement channel. On the contrary, when such a noise is present, only the boundedness of such an error could be expected.

Now, consider a possible way by which Problem 1N' can be solved, at least in principle (only requiring the boundedness of the estimation error in the noisy case). To this end, let us re-define the compact set Y as $Y \stackrel{\triangle}{=} \{\underline{y} \in \Re^p : \underline{y} = \underline{h}(\underline{x})+\underline{\eta}, \forall \underline{x} \in X, \forall \underline{\eta} \in H\}$, and introduce the following cost function, where the dependence on the previous state estimate via the prediction $\bar{\underline{x}}_{t-N}$ is explicitly taken into account.

$$J_t = \left\|\hat{\underline{x}}_{t-N,t} - \bar{\underline{x}}_{t-N}\right\|^2_P + \sum_{i=t-N}^{t} \left\|\underline{y}_i - \underline{h}\left(\hat{\underline{x}}_{it}\right)\right\|^2_Q \tag{12}$$

where P and Q are symmetric positive definite matrices. The state estimates are linked by the constraints

$$\hat{\underline{x}}_{i+1,t} = \underline{f}\left(\hat{\underline{x}}_{i,t},\underline{u}_i\right), \quad i = t-N, t-N+1, \ldots, t-1 \tag{13}$$

Then, we can state the following

Problem 2N'. *At any stage t, for any $\underline{y}^t_{t-N} \in Y^{N+1}, \underline{u}^{t-1}_{t-N} \in U^N$, and $\bar{\underline{x}}_{t-N} \in \mathcal{X}_p$, find the optimal state estimate $\hat{\underline{x}}^{\circ}_{t-N,t}$ that minimizes the cost (12) under the constraints (13).*

□

Remark. First of all, notice that, once the state estimates but $\hat{\underline{x}}_{t-N,t}$ have been eliminated by using the constraints (13), the cost function depends only on $\hat{\underline{x}}_{t-N,t}$, $\bar{\underline{x}}_{t-N}$, $\underline{u}^{t-1}_{t-N}$, and $\underline{y}^t_{t-N}$, i.e., J_t can be expressed in the form $J_t\left(\hat{\underline{x}}_{t-N,t},\bar{\underline{x}}_{t-N},\underline{u}^{t-1}_{t-N},\underline{y}^t_{t-N}\right)$ (J_t is a $\mathcal{C}^1$ function). Moreover, if $P = 0$ (in the noiseless case), it is easy to see that a solution to Problem 1N (hence to Problem 1N') can be obtained by minimizing the interpolation cost (12) under the constraints (13) and a dead-beat observer is obtained. On the other hand, if $P \neq 0$ (i.e., we look for a recursive state observer), in order the solution of Problem 2N' could lead to a solution to Problem 1N', the problem of the convergence of the estimates to the true state needs to be addressed explicitly, and from intuition it is easy to realize that matrices P and Q may have a significant influence on the convergence of the estimation error. In the following theorem such an influence will be put into evidence.

Before the statement of the theorem, some notations and useful quantities should be introduced. First, we introduce the set $\mathcal{X}^{\circ}$ to which the optimal solutions $\hat{\underline{x}}^{\circ}_{t-N,t}$ belong to. Moreover, given a generic symmetric positive definite matrix A, $\lambda_{\min}(A)$ and $\lambda_{\max}(A)$ denote the minimum and maximum eigenvalues of A, respectively; given a generic matrix B, $\|B\| \triangleq \|B\|_{\max} = \sqrt{\lambda_{\max}(B^T B)}$ and $\|B\|_{\min} = \sqrt{\lambda_{\min}(B^T B)}$. Furtherly, consider the smallest open convex set $\mathcal{X}$ such that $\mathcal{X} \supseteq X \bigcup \mathcal{X}_p \bigcup \mathcal{X}^{\circ}$ and denote by $\bar{\mathcal{X}}$ the closure of $\mathcal{X}$ (i.e., $\bar{\mathcal{X}}$ is the smallest compact convex set containing $X \bigcup \mathcal{X}_p \bigcup \mathcal{X}^{\circ}$). k_h and k_f are global Lipschitz constants of the functions $\underline{h}$ and $\underline{f}$, restricted on $\bar{\mathcal{X}}$. Finally, the following further assumption is needed:

Assumption 2N: The system $\underline{x}_{t+1} = \underline{f}(\underline{x}_t, \underline{u}_t)$, $\underline{y}_t = \underline{h}(\underline{x}_t)$ is uniformly $N+1$–observable.

Then, we can state the following results.

Theorem 2N. *Suppose that Assumptions 1N and 2N are verified. Moreover, denote by $\hat{\underline{x}}^{\circ}_{t-N,t}$ the solution of Problem 2N' at stage $t \geq N$ and denote by $\underline{\xi}_{t-N} \triangleq \underline{x}_{t-N} - \hat{\underline{x}}^{\circ}_{t-N,t}$ the estimation error at stage $t-N$. Finally, consider the largest closed ball $N(r_\xi)$ with radius r_ξ such that $\underline{\xi}_0 \in N(r_\xi)$, and define the scalar $r_\eta \triangleq \max\limits_{\underline{\eta}_{t-N} \times \cdots \times \underline{\eta}_t \in H^{N+1}} \|\mathrm{col}(\underline{\eta}_{t-N}, \ldots, \underline{\eta}_t)\|$. Then:*

1. *Under suitable conditions specified in Part 2 below, there exist symmetric positive definite matrices P and Q such that the sequence of the error vectors $\{\underline{\xi}_{t-N}, t \geq N\}$ remains bounded for $t \to +\infty$. Moreover, its decaying rate is governed by the law*

$$\|\underline{\xi}_{t-N}\| \leq c^{t-N} \|\underline{\xi}_0\| + r_\eta \, \tilde{k} \, \frac{\lambda_{\max}(Q)}{\lambda_{\max}(P)} \sum_{k=1}^{t-N} c^k ,$$

$$\forall t > N , \, \forall \underline{\xi}_0 \in N(r_\xi) \qquad (14)$$

where $c < 1$ and $\tilde{k} > 0$ are suitable scalars specified in Part 2 below, too.

2. *The matrices P and Q, and the scalars c and $\tilde{k}$ can be characterized as follows. The scalar $\tilde{k}$ is given by $\tilde{k} = \Delta / k_f$, where $\Delta \triangleq \arg \max\limits_{\underline{x}_{t-N} \in \bar{\mathcal{X}}, \underline{u}_{t-N}^{t-1} \in U} \|D(\underline{x}_{t-N}, \underline{u}_{t-N}^{t-1})\|$. The scalar c can be written as*

$$c \triangleq 2 k_f \frac{\lambda_{\max}(P)}{\lambda_{\min}(P) + \delta^2 \, \lambda_{\min}(Q)} \qquad (15)$$

where $\delta \overset{\triangle}{=} \arg \min_{\underline{x}_{t-N} \in \bar{\mathcal{X}}, \underline{u}_{t-N}^{t-1} \in U} \|D(\underline{x}_{t-N}, \underline{u}_{t-N}^{t-1})\|_{\min}$. Moreover, consider the scalars $\bar{k}_i > 0$ such that

$$\left\| \frac{\partial \underline{h}}{\partial \underline{x}_{t-N}}(\underline{x}'_{t-N}) - \frac{\partial \underline{h}}{\partial \underline{x}_{t-N}}(\underline{x}''_{t-N}) \right\| \leq \bar{k}_1 \left\| \underline{x}'_{t-N} - \underline{x}''_{t-N} \right\|$$

$$\left\| \frac{\partial \underline{h} \circ \underline{F}^{u_{t-N}^{t-N}}}{\partial \underline{x}_{t-N}}(\underline{x}'_{t-N}) - \frac{\partial \underline{h} \circ \underline{F}^{u_{t-N}^{t-N}}}{\partial \underline{x}_{t-N}}(\underline{x}''_{t-N}) \right\| \leq \bar{k}_2 \left\| \underline{x}'_{t-N} - \underline{x}''_{t-N} \right\|$$

$$\vdots$$

$$\left\| \frac{\partial \underline{h} \circ \underline{F}^{u_{t-N}^{t-1}}}{\partial \underline{x}_{t-N}}(\underline{x}'_{t-N}) - \frac{\partial \underline{h} \circ \underline{F}^{u_{t-N}^{t-1}}}{\partial \underline{x}_{t-N}}(\underline{x}''_{t-N}) \right\| \leq \bar{k}_{N+1} \left\| \underline{x}'_{t-N} - \underline{x}''_{t-N} \right\|$$

$\forall \, \underline{x}'_{t-N}, \underline{x}''_{t-N} \in \bar{\mathcal{X}}$; $\forall \underline{u}_{t-N}^{t-1} \in U^N$ (the above scalars $\bar{k}_i$ do exist as the function D is globally Lipschitz on compact sets). Moreover, let us define the scalar $\bar{k} \overset{\triangle}{=} \sqrt{\bar{k}_1^2 + \ldots + \bar{k}_{N+1}^2}$ Then, we have that (14) holds with $c < 1$ if it is possible to choose some symmetric positive definite matrices P and Q such that

$$\lambda_{\max}(P) \geq \lambda_{\max}(Q) \frac{r_\eta}{r_\xi} \frac{2\tilde{k}}{\sqrt{1 + \frac{8r_\eta}{\bar{k}r_\xi^2}} - 1} \tag{16}$$

and

$$\lambda_{\max}(P) < \frac{1}{2k_f} \left[\lambda_{\min}(P) + \delta^2 \lambda_{\min}(Q) \right] \tag{17}$$

$\square$

Let us now address the case of approximate solution of Problem 2N coming from the neural approximation, that is let us consider the vectors $\hat{\underline{x}}_{t-N}$, $t \geq N$ approximately solving Problem 2N. Let us denote by $\underline{\varepsilon}_{t-N} \overset{\triangle}{=} \hat{\underline{x}}^\circ_{t-N} - \hat{\underline{x}}_{t-N}$ the above approximation error, by $\varepsilon \overset{\triangle}{=} \max(\|\underline{\varepsilon}_{t-N}\|)$ the largest possible magnitude of such an error and by $\hat{\underline{\xi}}_{t-N} \overset{\triangle}{=} \underline{x}_{t-N} - \hat{\underline{x}}_{t-N}$ the state estimation error when approximate solutions of Problem 2N are considered. Then, we can prove the following

Theorem 3N. *Suppose that Assumptions 1N and 2N are verified and use the same notations as for Theorem 2N. Then:*

1. *Under suitable conditions specified in Part 2 below, there exist symmetric positive definite matrices P and Q such that the sequence of the error*

vectors $\{\hat{\underline{\xi}}_{t-N}, t \geq N\}$ remains bounded for $t \to +\infty$. Moreover, its decaying rate is governed by the law

$$\|\hat{\underline{\xi}}_{t-N}\| \leq c^{t-N} \varepsilon + c^{t-N} r_\xi + \left(r_\eta \, \tilde{k} \, c \, \frac{\lambda_{\max}(Q)}{\lambda_{\max}(P)} + 3\varepsilon \right) \sum_{k=0}^{t-N-1} c^k ,$$

$$\forall t > N \tag{18}$$

where $c < 1$ and $\bar{k}, \tilde{k} > 0$ are the same scalars defined in Theorem 2N. Then, we have that (18) holds with $c < 1$ if it is possible to choose some symmetric positive definite matrices P and Q such that

$$\lambda_{\max}(P) \geq \lambda_{\max}(Q) \, \frac{2\tilde{k}\, r_\eta}{8\varepsilon + 2r_\xi} \, \frac{1}{\sqrt{1 + \dfrac{32 r_\eta}{\bar{k}(8\varepsilon + 2r_\xi)^2}} - 1} \tag{19}$$

and

$$\lambda_{\max}(P) < \frac{1}{2k_f} \left[\lambda_{\min}(P) + \delta^2 \lambda_{\min}(Q) \right] \tag{20}$$

$\square$

References

[1] J. M. Fitts. On the observability of nonlinear systems with applications to non-linear regression analysis. *Information Sciences*, 4:129–156, 1972.

[2] F. Girosi. Regularization theory, radial basis functions and networks. In V. Cherkassky, J.H. Friedman, and H. Wechsler, editors, *From Statistics to Neural Networks. Theory and Pattern Recognition Applications.* Springer-Verlag, Subseries F, Computer and Systems Sciences, 1994.

[3] K. Hornik, M. Stinchombe, and H. White. Multilayer feedforward networks are universal approximators. *Neural Networks*, 2:359–366, 1989.

[4] S. S. Keerthi and E. G. Gilbert. Optimal infinite-horizon feedback laws for a general class of constrained discrete-time systems: stability and moving-horizon approximations. *Journal of Optimization Theory and Applications*, 57:265–293, 1988.

[5] D. Q. Mayne and H. Michalska. Receding horizon control of nonlinear systems. *IEEE Trans. on Automatic Control*, 35:814–824, 1990.

[6] H. Michalska and D. Q. Mayne. Robust receding horizon control of constrained nonlinear systems. *IEEE Trans. on Automatic Control*, 38:1623–1633, 1993.

[7] P. E. Moraal and J. W. Grizzle. Observer design for nonlinear systems with discrete–time measurements. *IEEE Trans. on Automatic Control*, 3:395–404, 1995.

[8] T. Parisini, M. Sanguineti, and R. Zoppoli. Nonlinear stabilization by receding-horizon neural regulators. In *Proc. 34th IEEE Conf. on Decision and Control (also to appear in Int. J. of Control)*, pages 2433–2441, New Orleans, LA, 1995.

[9] T. Parisini and R. Zoppoli. A receding-horizon regulator for nonlinear systems and a neural approximation. *Automatica*, 31:1443–1451, 1995.

SECTION 4

MATHEMATICAL MODELS

Neural Network Fuzzification: a Critical Review of the Fuzzy Learning Vector Quantization Model

Andrea Baraldi and Flavio Parmiggiani

Mailing address: IMGA-CNR, Via Emilia Est 770, 41100 Modena (Italy)
Ph.: +39-59-362388; Fax: +39-59-374506; e-mail: andrea@katabatic.bo.cnr.it

Abstract

In this work, the definition of fuzzy clustering mechanism, adopted in the field of neural networks, is related to the definition of the soft-max adaptation rule developed in the field of data compression. In order to discuss the functional meaning of a fuzzy clustering mechanism, one neural network model, the Fuzzy Learning Vector Quantization, FLVQ, supposedly featuring fuzzy properties, is reviewed on a conceptual basis. This analysis shows that the FLVQ design is affected by some inconsistencies which make this model unable to develop a robust soft-max adaptation strategy.

1 Introduction

A clustering Neural Network (NN) model performs an unsupervised detection of statistical regularities in a random sequence of input patterns. Among the fuzzy clustering NN models which are found in the literature, the Fuzzy Learning Vector Quantization, FLVQ [1] (which was first called Fuzzy Kohonen Clustering Network [2]), is well known. In this paper, some general definitions of interest about the topic of fuzzy clustering are discussed. Then, FLVQ is reviewed on a theoretical basis. We will conclude that FLVQ does not develop an effective fuzzy clustering mechanism.

2 Definitions

In the definition presented in [3], [4], it is stated that an artificial NN model performs *fuzzy clustering* when it allows a pattern to belong to multiple categories to different degrees depending on the neurons' ability to recognize the input pattern. Since a pattern *belongs* to multiple categories to different degrees, all these categories must be able to simultaneously *acquire* (learn) the same input pattern to different degrees. This approach is well known in the traditional field of coding techniques for data compression. The c-means algorithms feature a cost function (discretization error) characterized by many local minima. To avoid confinement of reference vectors to one of these many local minima, data compression techniques modify c-means algorithms by replacing the Winner-Takes-All (WTA) strategy with a *soft-max adaptation rule*. This rule is defined as a learning strategy that not only adjusts the winning cluster but also affects all cluster centers depending on their proximity to the input pattern [5].

To summarize, we believe that the fuzzy clustering definition provided above is equivalent to the definition of the soft-max adaptation rule traditionally employed in the field of data compression. This equivalence leads to several consequences, which also affect the terminology employed in this paper:

1. *The necessary and sufficent condition to be satisfied by a clustering NN in order to be termed fuzzy is that it adopt a soft-max (non-crisp) adaptation rule.*

2. Whenever a NN model employs a WTA (crisp) adaptation strategy, it cannot be termed fuzzy even when it employs any mathematical tool derived from fuzzy set theory (e.g., membership function, fuzzy set operations, etc.).

3.A fuzzy clustering NN, i.e., a NN which develops a soft-max adaptation rule (see point 1. above), does not necessarily exploit mathematical tools provided by fuzzy set theory.

3 Critical Analysis of FLVQ

Let us analyze, from a theoretical perspective, the degree of fuzziness of the FLVQ model [1], [2]. FLVQ tries to combine the on-line Kohonen weight adaptation rule with the clustering mechanisms proposed by the batch (non-sequential) Fuzzy c-Means (FCM) algorithm [6].

3.1 Review of FCM

The objective function of the FCM algorithm to be minimized is:

$$J_m(U,V,X) = \sum_{i=1}^{c} \sum_{k=1}^{n} (u_{i,k})^m \|x_k - V_i\|^2, \tag{1}$$

where $X = [x_k]$ is the input data set made of input patterns x_k, $U = [u_{i,k}]$ is a fuzzy c-partition of X (whose definition is provided below), $V = (V_1, V_2, ..., V_c)$ is a vector of unknown cluster centers, c is the number of categories, n is the number of input patterns, and $m > 1$, equal to the "amount of fuzziness" [2], is a weighting exponent on each fuzzy membership $u_{i,k}$. In FCM, (1) is differentiated with respect to $u_{i,k}$ (for fixed V_i) and to V_i (for fixed $u_{i,k}$), and, by applying the condition $(\sum_{i=1}^{c} u_{i,k} = 1)$, we obtain

$$u_{i,k} = \frac{\left(\frac{1}{\|x_k - V_i\|^2}\right)^{\frac{1}{(m-1)}}}{\sum_{j=1}^{c} \left(\frac{1}{\|x_k - V_j\|^2}\right)^{\frac{1}{(m-1)}}}, \qquad i = 1, 2, ..., c; \ k = 1, 2, ..., n, \tag{2}$$

$$V_i = \frac{1}{\sum_{k=1}^{n} \alpha_{i,k}} \cdot \sum_{k=1}^{n} \alpha_{i,k} \cdot x_k = \sum_{k=1}^{n} \beta_{i,k} \cdot x_k, \qquad i = 1, 2, ..., c, \tag{3}$$

where

$$\alpha_{i,k} = (u_{i,k})^m, \tag{4}$$

$$\beta_{i,k} = \frac{\alpha_{i,k}}{\sum_{k=1}^{n} \alpha_{i,k}}. \tag{5}$$

Equation (2) is a membership function because it satisfies the three conditions required to state that c fuzzy subsets are a fuzzy c-partition of X. These conditions are [3]: i) $u_{i,k}$ belongs to [0,1], $\forall i$, $\forall k$; ii) $\sum_{i=1}^{c} u_{i,k} = 1$, $\forall k$; and iii) $0 < \sum_{k=1}^{n} u_{i,k} < n$, $\forall i$. Equation (2) provides the degree of compatibility of x_k with cluster center V_i. As was the case with $u_{i,k}$, $\alpha_{i,k}$ and $\beta_{i,k}$ also belong to [0,1], $\forall i$, $\forall k$. Equation (3) is such that, at the end of the current iteration of the batch FCM algorithm (i.e., after all input patterns have been processed), a cluster center, whose index is i, is updated by means of a weighted sum of the input patterns. The weight of a generic pattern k related to cluster i is identified by $\beta_{i,k}$, which consists of the coefficient $\alpha_{i,k}$ normalized with respect to the sum of the $\alpha_{i,k}$ coefficients collected by neuron i during the processing of the n input patterns.

It can be demonstrated (e.g., see [2], p. 758) that

$$\lim_{m \to \infty} \{u_{i,k}\} = 1/c, \quad \forall i, \forall k, \tag{6}$$

$$\lim_{m \to 1} \{u_{i,k}\} = \begin{cases} 1, & \text{if neuron } i \text{ is the winning neuron, } \forall k; \\ 0, & \text{otherwise;} \end{cases} \tag{7}$$

Equation (6) shows that if $m \to \infty$, every neuron features the same memberhip value whatever input k and category i may be. Equation (7) shows that if $m \to 1$, equation (2) describes a crisp label vector situation where only one neuron out of c instances features a membership value > 0.

3.2 Review of FLVQ

FLVQ employs $m = m_t$, which is substituted in (2) to compute $u_{i,k,t}$. The m_t expression is:

$$m_t = m_0 - (t \cdot \Delta m) \qquad \text{where} \qquad \Delta m = (m_0 - m_f)/t_{max} \tag{8}$$

where t_{max} is the number of maximum epochs, and $m_0 > m_f > 1$ are two user-defined parameters. Thus: i) m_t belongs to $[m_f, m_0] \subseteq [1^+, m_0]$ (therefore, m_t satisfies the constraint required by (1)); and ii) m_t is a monotone decreasing function of time (when $t = 0$, then $m_t = m_0 > m_f$, and when $t = t_{max}$, then $m_t = m_f > 1$).

FLVQ combines (3), which is the non-sequential adaptation rule of the cluster center developed in FCM, with the sequential Kohonen weight adaptation rule:

$$\begin{aligned} V_{i,t} &= V_{i,t-1} + \frac{1}{\sum_{k=1}^{n} \alpha_{i,k,t}} \cdot \sum_{k=1}^{n} \alpha_{i,k,t}(x_k - V_{i,t-1}) \\ &= V_{i,t-1} + \sum_{k=1}^{n} \beta_{i,k,t}(x_k - V_{i,t-1}), \quad i = 1, 2, ..., c, \end{aligned} \tag{9}$$

where, analogously to (4) and (5),

$$\alpha_{i,k,t} = (u_{i,k,t})^{m_t}, \tag{10}$$

$$\beta_{i,k,t} = \frac{\alpha_{i,k,t}}{\sum_{k=1}^{n} \alpha_{i,k,t}}. \tag{11}$$

Equation (9) shows that $\beta_{i,k,t}$, rather than $\alpha_{i,k,t}$, as erroneously suggested in [2], p. 760, is the learning rate coefficient employed by FLVQ at time t allowing category i to acquire input pattern k. For $u_{i,k,t}$, $\alpha_{i,k,t}$ and $\beta_{i,k,t}$ the same properties discussed in Subsection 3.1 hold true.

3.3 Discussion of FLVQ

We believe that several drawbacks affected the development of FLVQ.

1. FLVQ does not minimize any known objective function. In [2], p. 757, it was claimed that FLVQ usually terminates in such a way that (1) is approximately minimized. In [1], p. 738, contrary to what is stated in [2], it is asserted that FLVQ does not optimize a fixed objective function. We accept this second statement on the basis of the following consideration: even though FLVQ exploits (2) as does FCM, the hypothesis employed by FCM to extract (2) from (1) does not hold true for FLVQ. In particular, equation (2) is extracted while assuming that m is kept constant, but this parameter changes with t in FLVQ where $m = m_t$. Therefore, the exploitation of (2) in FLVQ does not guarantee the minimization of (1). To some degree, this may explain why in many FLVQ experiments (e.g., see [2]) the measured J_m increases with t after reaching a minimum.

2. In [1] and [2], FLVQ is compared with VQ and the Self-Organizing Map (SOM), respectively. However, these comparisons are not consistent because neither VQ nor SOM are implemented correctly. For example, in [1], given α_t as the learning rate coefficient computed at time t, $V_{j,t}$ as the connection weights of category j at time t, t_{max} as the termination time and n as the number of input patterns, the suggested VQ algorithm is:

Initialize $t = 0$.
Initialize $V_{j,0}, j = 1, 2, ..., c$.
for $(t = 1, 2, ..., t_{max})$
    Compute the learning rate coefficient α_t.
    for $(k = 1, 2, ..., n)$
        Detect the winning neuron $V_{i,t-1}$ for x_k.
        Update the winner according to Kohonen's weight adaptation rule:

$$V_{i,t} = V_{i,t-1} + \alpha_t \cdot (x_k - V_{i,t-1}). \tag{12}$$

    endfor
endfor

where α_t is defined as a monotone decreasing function of time. In this algorithm, since the k-cycle iterates within the t-cycle, the different $V_{i,t}$ values that may be computed by (12) for the same category i are sequentially overwritten! This problem is related to the fact that the Kohonen algorithms proposed in [1] and [2] do not account for the following feature: while the time variable in

FLVQ is a global counter, representing the number of epochs for every neuron in the net, both VQ and SOM, performing on-line learning, must be provided with the following two data structures: i) a neuron-based (local) time counter, describing the number of assignments per neuron, which affects the neuron's plasticity (learning rate); and ii) the epoch counter, which is a global attribute of the whole net and which affects the network's termination. Defining T as the number of epochs, T_{max} as the maximum number of epochs to reach termination, t_i as the number of assignments per neuron i and $\alpha_i(t_i)$ as the learning rate coefficient of neuron i at time t_i, the on-line Kohonen VQ algorithm should be expressed as [7]:

Initialize $t_j = 1, j = 1, 2, ..., c.$
Initialize $V_j(1), j = 1, 2, ..., c.$
for $(T = 1, 2, ..., T_{max})$
    for $(k = 1, 2, ..., n)$
        Detect the winning neuron $V_i(t_i)$ for x_k.
        Compute the learning rate coefficient $\alpha_i(t_i)$.
        Update the winner according to Kohonen's weight adaptation rule:

$$V_i(t_i + 1) = V_i(t_i) + \alpha_i(t_i) \cdot (x_k - V_i(t_i)). \tag{13}$$

        Update $t_i = t_i + 1.$
    endfor
endfor

In this procedure, $\alpha_i(t_i)$ is a monotone decreasing function of time (e.g., $\alpha_i(t_i)$ can be defined as $exp(-(t_i)/\tau)$, where τ is a user-defined parameter). We wish to stress that the behavior of (13) is quite different from that of (12). In particular, (13) is such that: i) it does not overwrite any adjusted value of a winning prototype; and ii) it allows different prototypes to feature different values of time and, as a consequence, of their learning rate.

3. FLVQ exploits a fuzzy set membership function, but inhibits fuzzy clustering mechanisms. This aspect requires careful discussion. The following equation, found in [2], p. 760:

$$\lim_{m_t \to \infty} \{\alpha_{i,k,t}\} = 1/c. \tag{14}$$

is false. Due to (4) and (6), (14) is substituted by

$$\lim_{m_t \to \infty} \{\alpha_{i,k,t}\} = \lim_{m_t \to \infty} \{(u_{i,k,t})^{m_t}\} = \lim_{m_t \to \infty} \{(1/c)^{m_t}\} = 0. \tag{15}$$

Equation (15) is implicitly suggested in [1], p. 737, where it is asserted that if m_t is large, then $(u_{i,k,t})^{m_t} \approx (1/c)^{m_t}$ tends to be very small. Applying (15) to (11), one also obtains:

$$\lim_{m_t \to \infty} \{\beta_{i,k,t}\} = \lim_{m_t \to \infty} \frac{(1/c)^{m_t}}{n \cdot (1/c)^{m_t}} = 1/n, \qquad \forall i, \forall k. \tag{16}$$

Substituting (16) in (9) we obtain

$$\lim_{m_t \to \infty} V_{i,t} = V_{i,t-1} + \frac{1}{n} \sum_{k=1}^{n} (x_k - V_{i,t-1}), \qquad i = 1, 2, ..., c. \tag{17}$$

Vice versa, when $m_t \to 1$, by applying (7) in (10) and (11) we obtain

$$\lim_{m_t \to 1} \{\alpha_{i,k,t}\} = \begin{cases} 1, & \text{if neuron } i \text{ is the winning neuron, } \forall k; \\ 0, & \text{otherwise;} \end{cases} \tag{18}$$

$$\lim_{m_t \to 1} \{\beta_{i,k,t}\} = \begin{cases}]0,1], & \text{if neuron } i \text{ is the winning neuron, } \forall k; \\ 0, & \text{otherwise;} \end{cases} \tag{19}$$

In our opinion, in [1] and [2] the inconsistency of (14) is combined with the wrong perception of $\alpha_{i,k,t}$ as the learning rate coefficient (in place of $\beta_{i,k,t}$). It is known that the choice of m_0 is very important for FLVQ [1]. The problem is that whatever value of m_0 is chosen by the user, FLVQ does not fully exploit fuzzy learning, i.e, FLVQ does not allow several neurons to simultaneously acquire the same input pattern. In detail, let us consider the following two situations.

- The user sets m_0 close to 1, i.e., $m_0 = (1 + \epsilon)$ where $0 < \epsilon \ll 1$. Then, $1 < m_f \le m_t \le (1 + \epsilon)$ is such that FLVQ provides crisp label vectors because of (7), (18) and (19), i.e., simultaneous adjustment of several cluster centers is not allowed for any t.

- The user sets $m_0 \gg 1$. In this case, m_t values are $\gg 1$ during the first iterations of the FLVQ procedure, just when neuron plasticity should be maximized to allow coarse adjustments of connection weights. The large values of m_t are such that: i) according to (6) and (15), for any given k, the $\alpha_{i,k,t}$ terms are the same for every neuron i with $i = 1, 2, ..., c$, i.e., these coefficients are uniformly distributed across the c nodes (in agreement with [1], p. 739); however, all these terms are approximately null; and ii) according to (17), every cluster center, which is attracted by every input pattern, is not affected by any contextual behavior of its neighboring neurons, i.e., no fuzzy (contextual) learning mechanism is employed .

To summarize, when the m_0 user-defined parameter is chosen as either $m_0 \gg 1$ or $m_0 \approx 1^+$, FLVQ does not exploit any contextual effect in clustering input patterns either when $t \to t_{max}$ (then $m_t \approx m_f \approx 1^+$) or when $t \to 0$ (then $m_t \approx m_0$). This conclusion is in line with what was suggested in [1], p. 738, where the heuristic choice $1.1 < m_f < m_0 < 7$ is recommended. This conclusion also means that, whatever m_0 is, when t is low (i.e., during the initial iterations of FLVQ), neuron fuzzy clustering mechanisms are actually inhibited because either (16) or (19) tends to be true. For these reasons we believe that FLVQ does not pursue a successful combination of the FCM model with Kohonen's constraints. Due to some drawbacks in its design, FLVQ does not develop an effective soft-max adaptation rule, i.e., this model does not satisfy the fuzzy clustering definition provided in Section 2, although it does compute a fuzzy set membership function. Indeed, as observed in [1], learning rate behavior in

FLVQ is roughly opposite to the usual behavior imposed on them by other NN models, like SOM [7] and Neural Gas [5]. In these models, all c learning rates decrease towards 0 as t increases, but, in FLVQ, (16) and (19) show that the (only one) winner learning rate increases to 1 (starting at $t = 0$ from a value which may be close to zero) when t increases, while the other $(c-1)$ rates tend towards, or remain close to, zero.

4 Conclusions

FLVQ fails to develop an effective soft-max adaptation rule although it computes a fuzzy set membership function. On the contrary, Kohonen's SOM exploits a soft-max version of the WTA policy employed by a c-means clustering algorithm [5], [7]. When the processing elements of SOM lose contextual sensitivity (i.e., the radius of the resonance domain decreases to zero), then SOM becomes equivalent to a hard on-line c-means algorithm [5]. Our paradoxical conclusion is that SOM can be termed fuzzy according to the definition provided in Section 2, although SOM has never claimed to feature fuzzy properties; on the other hand, FLVQ, originally develop to improve SOM with fuzzy logic mechanisms, fails in developing an effective soft-max (fuzzy) adaptation rule. The fuzziness of SOM stresses the fact that a network of processing elements can verify the definition of fuzzy clustering when it exploits cooperative/competitive local rules identified in advance by neurophysiological studies. In the science of artificial complex systems [8], where neurophysiological evidences are employed in a multi-disciplinary framework, mathematical tools derived from fuzzy set theory should be exploited to improve contextual sensitivity of artificial cooperative/competitive local mechanisms.

References

1. Bezdek J. C, Pal N. R. Two soft relatives of learning vector quantization. Neural Networks 1995; 8(5):729-743.
2. Tsao E. C, Bezdek J. C, Pal N. R. Fuzzy Kohonen clustering network. Pattern Recognition 1994; 27(5):757-764.
3. Pao Y. Adaptive pattern recognition and neural networks. Addison-Wesley, Reading, 1989.
4. Simpson P. K. Fuzzy min-max neural networks - Part 2: clustering. IEEE Trans. Fuzzy Systems 1993; 1(1):32-45.
5. Martinetz T, Berkovich G, Schulten K. Neural-Gas network for quantization and its application to time-series predictions. IEEE Trans. Neural Networks 1993; 4(4):558-569.
6. Bezdek J. C. Pattern recognition with fuzzy objective function algorithms. Plenum, New York, 1981.
7. Kohonen T. The self-organizing map. Proceedings of the IEEE 1990; 78(9):1464-1480.
8. Serra R, Zanarini G. Complex systems and cognitive processes. Springer-Verlag, Berlin, 1990.

Cultural Evolution in a Population of Neural Networks

Daniele Denaro
Department of Computer Science University of Ancona
Ancona Italy

Domenico Parisi
Institute of Psychology National Research Council
Rome Italy

Abstract

Genetic algorithms are computational models of the evolution of good solutions to problems based on the selective reproduction of the best variants and the constant addition of random variability to the population of variants. In biological evolution variants are inherited genotypes that are transmitted from parents to offspring. In cultural evolution behavioral variants are trasmitted from one individual to another because one individual (the learner) imitates another individual (the teacher). If the two individuals belong to successive generations, reproduction of teachers is selective, and random noise in added to the cultural transmission of behaviors from teachers to learners, good solutions to problems can evolve by cultural rather than biological evolution.

We describe a model of imitation learning (inspired by Hutchins and Hazelhurst, 1995) according to which the learner learns via backpropagation using the output of the teacher in response to some shared input as its teaching input. Cultural transmission of behaviors from one generation to the next via imitation learning leads to the progressive deterioration of performance across generations. However, if only the best individuals of each generation function as teachers and, furthermore, the teaching input provided by the teacher is modified by noise before it is used by the learner, not only culturally transmitted behaviors do not deteriorate but initially nonexistent behavioral capacities can emerge evolutionarily via pure cultural transmission as they can be shown to emerge via genetic transmission.

1. Introduction

Genetic algorithms are computational models of biological evolution (Holland, 1975; Goldberg, 1989; Mitchell and Forest, 1994). They may be applied to populations of neural networks that control the behavior of artificial organisms living in some particular environment to evolve genetically transmitted behaviors well adapted to the environment. An initial population of "genotypes" is created randomly. Each genotype maps into the phenotypical neural network that controls the behavior of one particular individual. The reproductive chances of each individual in the population are dependent on the individual's behavior in the environment. Each individual will tend to behave differently from all other individuals because of the random generation of the initial genotypes. Although the individuals of the initial population will behave rather inefficiently, some individuals will behave more efficiently than others and these individuals will be more likely to reproduce. Reproduction can be agamic or sexual. Agamic reproduction implies that a single individual generates one or more "offspring", i.e., copies of its genotype. Sexual reproduction requires that two different individuals mate and generate a new combination of parts of the genotype of the "father" and parts of the genotype of the "mother". In both cases random mutations are added to the offspring genotype. The differential reproduction of individuals based on the quality of their behavior and the constant addition of variability to the genetic pool via sexual recombination and genetic mutations cause an evolutionary increase in the average quality of the behavior exhibited by successive generations of organisms. In other words, a genetic algorithm can be used to create some particular behavior in a population of organisms that initially lacks the behavior (Figure 1). (For a more detailed description of the organisms and of the task, see the next Section.)

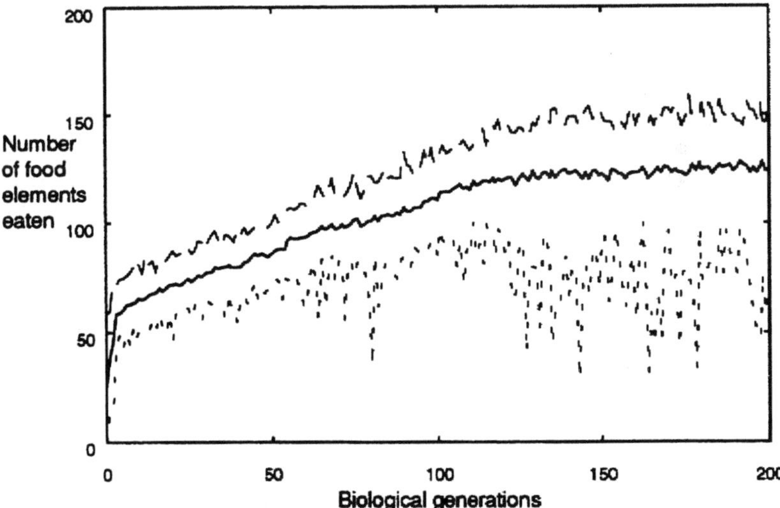

Figure 1. Evolutionary increase in the ability to find (and eat) food elements in the environment across 200 generations in a population that reproduces selectively with genetic mutations. Reproduction is agamic. The three curves represent the behavior of the best individual (top curve), average individual (middle curve), and worst individual (bottom curve) of each generation.

In the organisms we have just described the behavior which is exhibited is completely determined by the inherited genotype. In other simulations, the inherited genotype has phenotypical plasticity, that is, the phenotypical neural network that is constructed under the control of the genetic instructions contained in the genotype is sensitive to the particular environment in which the individual organism happens to live and, therefore, the same inherited genotype can result in different phenotypical networks according to the particular environment in which the genotype maps into the phenotypical network. In this case the behavior is partly genetically transmitted and partly learned in response to the particular environment (cf. Nolfi, Miglino, & Parisi, 1994). In still other simulations the phenotypical network is constructed under complete control of the genotype but then it is changed on the basis of the "experience" of the individual organism in the environment using some learning algorithm such as backpropagation (Cf. Belew, McInerney, & Schraudolph, 1992; Nolfi, Elman, & Parisi, 1994.)

In some organisms, however, a second system of transmission of behaviors is observed besides the genetic transmission system from parents to offspring. In organisms that live socially the other individuals are part of the environment for any particular individual. An individual, therefore, can learn some behavior by interacting with the other individuals. Assuming that the other individuals already possess some particular behavior a newborn individual can learn the behavior from them. This type of transmission of behaviors from one individual to another which is based not on genetic inheritance but on learning from others is called cultural transmission. As we will see, using the backpropagation procedure to model learning from others we can study the cultural transmission of behaviors in a population of networks as we can study genetic transmission.

However, biological evolution is not just genetic transmission. As we have observed, the selective reproduction of the best individuals and the addition of variability through sexual recombination and genetic mutations cause the gradual (or, in some cases, sudden) emergence of some behavioral capacity that was initially absent. Can cultural transmission also result in cultural evolution? Can the transmission of behaviors from one generation to the next via learning from others cause the emergence of some behavioral capacity that is initially absent? In this paper we present a model of cultural evolution that provides an affirmative answer to this question.

2. A model of learning from others

Following Hutchins and Hazelhurst (1995), we adopt the following model of learning from others (imitation learning). Imagine two neural networks with the same network architecture. One network is the "teacher". The teacher possesses a set of connection weights that allow it to execute efficiently some task, i.e., it knows how to map a set of inputs into the appropriate outputs. (We will ask how the teacher has come to possess these weights later on.) The second network is the "learner". The learner has random connection weights and, therefore, it is initially unable to execute the task efficiently. The learner learns the task by imitating the teacher. In any given learning trial both the teacher and the learner are exposed to the same input. Both respond to this input with some output. At the beginning of learning the two outputs will tend to be rather different. The learner learns by backpropagation. It uses the teacher's output as its teaching input. In each trial the learner modifies its connection weights in such a way that the discrepancy (error) between the teacher's output and its own output in response to the same input tends to decrease. After a certain number of trials the learner will respond to some input in the same way as the teacher. (Actually, not exactly in the same way because in backpropagation learning the error rarely or never goes to zero and, in real life, no one can imitate perfectly his or her model. This is an important point to which we will return in a moment.) The learner will have learned from the teacher. The teacher's behavior has been transmitted to the learner (Figure 2).

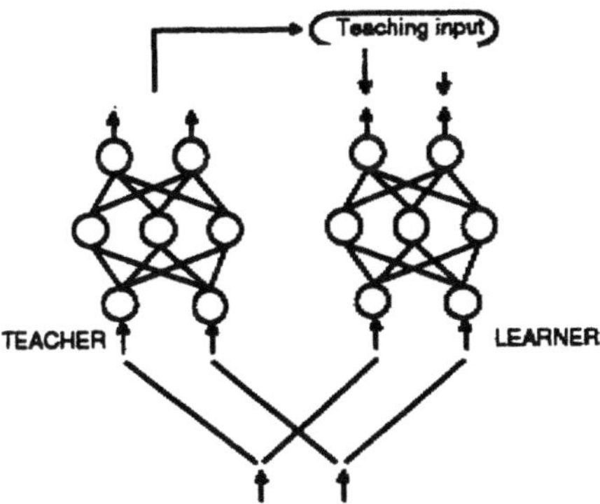

Figure 2. The network on the left is the teacher. The network on the right is the learner. Both teacher and learner are exposed to the same input. The learner learns by using the teacher's output as its teaching input in backproprgation learning.

Imagine a succession of networks with each individual network acting as the teacher for the network that follows it in the sequence. The first network in the sequence learns the Exclusive OR (XOR) task as usual, i.e., by being exposed to the correct teaching input. When the network has learned the task (say, after 10,000 epochs, when the error has reached a very low stable value) its learning is stopped and the network acts as the teacher for the second network. The second network learns the XOR task not by being exposed to the correct teaching input but by using the first network's output as its teaching input. The second network is stopped when its error with respect to the teacher also has reached a stable state. Then the second network acts as the teacher for the third network, and so on.

What is the error of each network at the end of learning if the error is calculated not with respect to the network's teacher but with respect to the correct teaching input? The results are shown in Figure 3.

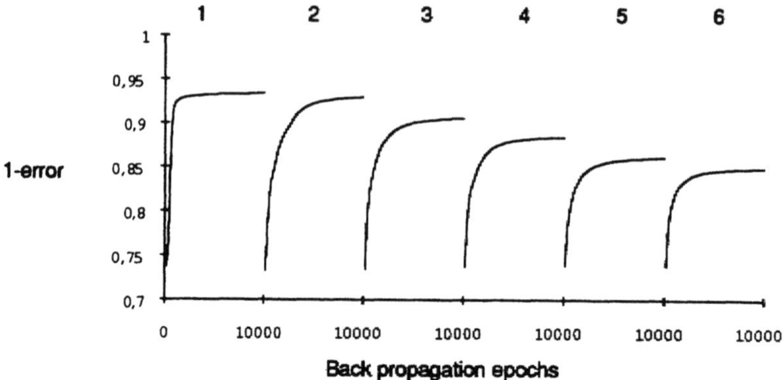

Figure 3. Imitation chain of 6 successive networks. The first network learns the XOR task from the correct teaching input. Each successive network learns the task from the preceding network (teacher) by using the teacher's output as its teaching input. The six curves show the error of each successive network with respect to the correct teaching input. (For presentation purposes learning performance is shown as 1 minus error.)

From figure 3 it is clear that culturally transmitted knowledge dissipates gradually. Since the learning error never goes to zero in backpropagation learning (and presumably in any kind of learning) the behavior of the learner at the end of learning won't be identical to the behavior of the teacher but it will be slightly worse. Therefore, the behavior which is offered by each successive teacher as a model for its "student" to imitate will be slightly worse than the behavior of its teacher and the deterioration of the behavior will accumulate across the successive generations of teachers.

Similar results are obtained if we use not an abstract task such as the XOR task but an ecological task such as finding food in an environment (Parisi, Cecconi, & Nolfi, 1990). The neural network controls the behavior of an organism in an environment that contains food elements. The network's input units encode the location of the nearest food element and the output units encode the movements of the organism. More precisely, the input units encode direction and distance of the currently nearest food element, both mapped in the interval between 0 and 1 and the output units encode angle of turning of the organism and speed of movement forward after turning, also mapped in the interval between 0 and 1. When an organism happens to step on a food element, the food element disappears (it is eaten). An individual's performance is measured as number of food elements found (eaten) during a fixed lifetime. The population is composed by a succession of generations each with a fixed number of 100 individuals. We start from a generation of individuals that are able to eat about 130 food elements during their life on the average. The next generation is composed by 100 new networks with randomly assigned connection weights each of which learns from a different individual of the preceding generation. (Since the individuals of each generation do not inherit anything genetically from the individuals of the preceding generation, these are purely cultural generations.) During learning both teacher and learner are exposed to the same input (position of nearest food) and they respond by some movement. (Since in ecological networks the input of any given trial is partly determined by the network's motor output in the preceding trial, we imagine that the learner is carried on the teacher's shoulders during the entire period of learning.) The learner learns by backpropagation using the movement encoded in the teacher's output units as its teaching input. At the end of learning the learners steps down from the teacher's shoulders and it is allowed to live an entire life all alone in its own environment.

Figure 4 shows the performance of 6 successive cultural generations of organisms with each generation learning how to find food elements from the individuals of the preceding generation. The average eating ability decreases from about 130 food elements eaten by the first generation to about 10 food elements eaten by the individuals 5 generations later.

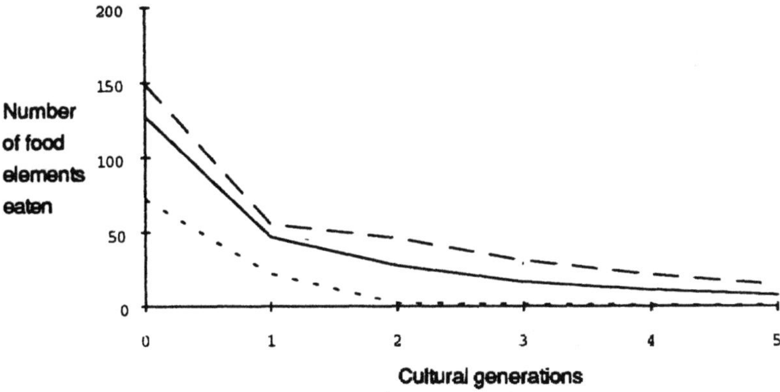

Figure 4. Eating performance of 6 successive cultural generations of individuals learning to eat by imitating the individuals of the preceding generation. The performances of the average, best, and worst individual in each generation are shown.

3. Adding selective reproduction and random variability

In the preceding Section we have seen that cultural trasmission by imitating others results in the gradual loss of culturally inherited information. Each individual that learns by imitating its teacher won't learn perfectly and, as a consequence, performance deterioration will accumulate generation after generation until any culturally transmitted ability virtually disappears. How can we explain that that is not what appears to be the case in real organisms that have cultural transmission of behaviors? Another limitation of the results reported in the preceding Section is that we start with initial teachers that already possess some behavioral capacity, either the XOR or the eating capacity. Where this behavioral capacity comes from? In real organisms cultural evolution appears to be able not only to maintain some

106

already existing capacity but also to create some capacity that is initially absent. Can our model capture this creative aspect of cultural evolution?

The model of learning by imitation described in the preceding Section refers only to the mechanism of cultural transmission of behaviors from one individual to another. If we want to study cultural evolution, and not just cultural transmission, we cannot ignore that biological evolution is not just a mechanism for the genetic transmission of behaviors and other phenotypical traits. What we should ask is the following: What is the equivalent in cultural evolution of selective reproduction and the constant addition of variability through sexual recombination and genetic mutations in biological evolution?

The results reported in Figure 4 refer to a succession of cultural generations in which each individual in one generation functions as teacher for one individual of the next generation. In other words, all the individuals that constitute one generation function as teachers for the individuals of the next generation. But let us assume that there is selective reproduction of teachers, i.e., selective cultural reproduction. Only the best individuals in each generation, i.e.,the individuals that eat the largest number of food elements, act as teachers for the individuals of the next generation. The culturally nonreproducing individuals have no "students". What would happen in these circumstances? The results for our population of food eaters are shown in Figure 5.

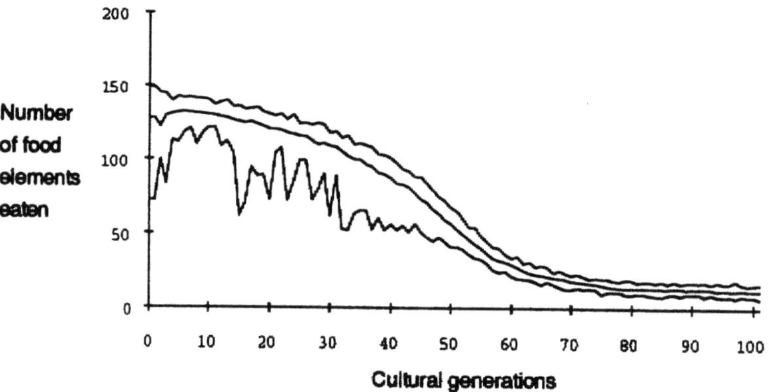

Figure 5. Eating performance of a succession of 100 cultural generations with selective reproduction of teachers. The 20 best individuals in each generation act as teachers for the 100 individuals of the next generation. Each of the 20 teachers is assigned 5 learners, i.e., 5 networks with initial random weights. Average, best, and worst performance in each generation are shown.

It is clear that selective cultural reproduction is able for a while to contrast the cumulative degradation of eating ability in successive generations. Without selective reproduction of teachers eating performance is near zero after 2 or 3 generations (cf. Figure 3). With selective reproduction performance deteriorates much more slowly but after 60 or 70 generations it again approaches zero.

Therefore, selective cultural reproduction may have some role in maintaining culturally trasmitted behaviors, but by itself is unable to prevent cultural deterioration.

Let us turn to the second mechanism that appears to be critical in biological evolution: the constant addition of variability to the genetic pool. What could be the equivalent of this mechanism in cultural evolution? We hypothesize that the equivalent of genetic mutations in cultural transmission might be the addition of random noise to the teaching input provided by the teacher to the learner before the teaching input is actually used by the learner to modify its connection weights.

Let us apply this idea to the cultural transmission of eating ability starting from a generation of teachers that already possess the ability to eat. There is selective reproduction of teachers but, in addition, in each learning trial some random noise is added to the teaching input provided by the teacher. More specifically, the activation value of each of the teacher's output units is changed by adding a quantity randomly selected from a normal distribution with a mean of 0 and a range of -0,5/+0,5. (The activation value of the teacher's output units can vary from 0 to 1.) The learner uses this modified output as its teaching input. Hence, what the learner is trying to approximate (to imitate) is not its teacher but some random deformation of the teacher. The results are shown in Figure 6.

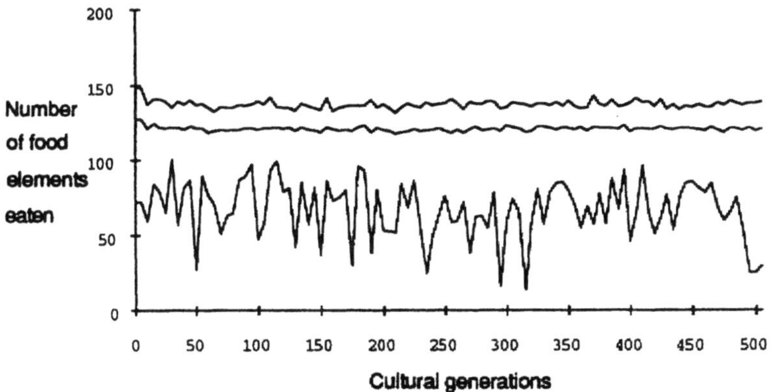

Figure 6. Eating performance of 500 cultural generations starting from a generation of individuals that already know how to eat. Noise is added to the teaching input provided by teachers to learners. The three curves represent the performance of the best, average, and worst individual in each generation.

It seems then that if we adopt selective cultural reproduction and, furthermore, we add noise to the teaching input in imitation learning, it is possible to completely eliminate the progressive deterioration of performance observed in all cases sofar in cultural transmission. Without selective reproduction of teachers and without noise in cultural transmission, eating performance quickly deteriorates after a few generations (cf. Figure 4). With selective reproduction of teachers but no noise performance deterioration is much slower but inevitable (cf. Figure 5). With both selective reproduction and noise added to cultural transmission after 500 generations the average eating ability has the same level as in the first generation (cf. Figure 6).

While selective cultural reproduction and noise added to cultural transmission apparently are sufficient to contrast the cumulative deterioration of performance due to the fact that learners do not perfectly imitate their teachers, we still have to solve the problem of moving from cultural transmission to cultural evolution. Selective reproduction together with noise can maintain some already existing ability but can cultural transmission create a nonexisting ability?

Let us start from an initial population of neural networks with randomly assigned connection weights. The average eating performance of these individuals obviously is extremely limited. At the end of life the 20 individuals with the best eating performance are selected as teachers for the next cultural generation. To each teacher are assigned 5 learners, i.e., 5 networks with initial random weights. Each learner learns by imitating its teacher but some noise is added to the teacher's output before it is used by the learner as its teaching input. At the end of learning (i.e., when the error in the backpropagation procedure reaches a stable low value) the 100 learners become the second generation and their eating performance is measured by allowing them to live in the environment. (Notice that these networks have two "performances": one is learning to respond to the input like the teacher, and the error in the backpropagation procedure is a measure of the goodness of this performance; the other one is finding food in the environment, and the number of food elements eaten is a measure of this second performance.) The process of cultural evolution is continued for 200 generations. The results are shown in Fig. 7.

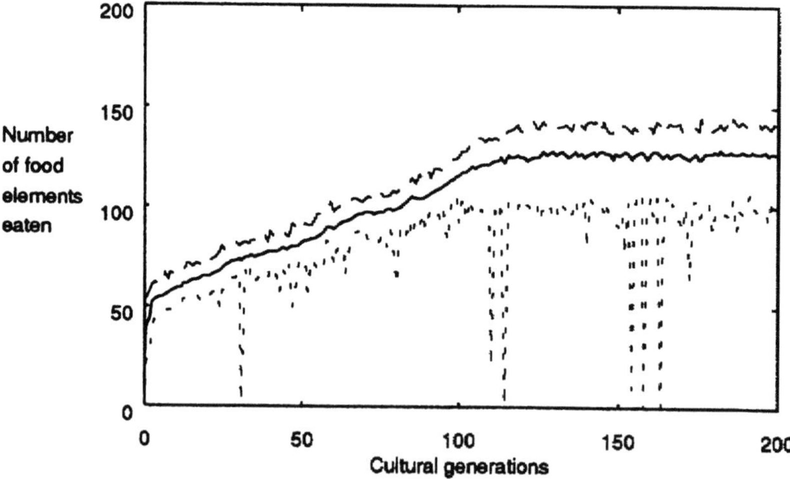

Figure 7. Evolutionary increase in the ability to find food elements in the environment across 200 generations in a population that reproduces culturally by selecting as teachers for the next generation the best individuals in each

generation and by adding random noise to cultural transmission. The three curves refer to the performance of the best, average, and worst individual in each generation.

The similarity of Figure 7 with Figure 1 is striking. Our cultural algorithm yields the same results of the genetic algorithm. Cultural transmission of behaviors through learning from teachers replaces the inheritance of genetic information from parents to offspring. Selection of teachers from among the best individuals in each generation replaces selective genetic reproduction. Random noise added to the teaching input provided by teachers replaces genetic mutations. In both cases evolution causes the emergence of a new behavioral ability that was initially absent.

4. Conclusion

Selective cultural reproduction and noise added to cultural transmission are able to maintain an already existing ability by contrasting its progressive deterioration due to imperfect transmission from teachers to learners and they are also able to create an initially nonexisting ability. These two mechanisms of cultural evolution are parallel to the two mechanisms of selective genetic reproduction and genetic mutations in biological evolution. (Sexual recombination can have a cultural analogue in a learner network learning from the output of various teacher networks.) In both cases evolution is based on randomly generating variants of already good solutions and selecting for reproduction the (rare) variants that represent improvements on an already good solution while discarding the (frequent) variants that are less good than the original solution. (Interesting mathematical models of cultural transmission and evolution can be found in Cavalli-Sforza & Feldman, 1981, and Boyd & Richerson, 1985).

It is interesting that evolution will succeed even if the improved variants are very rare. Genetic mutations change randomly some of the inherited genetic information and in rare cases these random changes result in a phenotype that performs better than its parent. However, mutations are one-shot events and it is understandable that in some rare cases they may have lucky phenotypical consequences. On the other hand, the addition of random noise to the teaching input in our model of cultural transmission occurs in all learning trials and one may wonder why the successive random changes in the teaching input do not just neutralize each other without ever resulting in cultural offspring that are better or worse than their teachers.

Figure 8 shows that this is not the case. The figure contains 3 learning curves on the left and 3 curves on the right. The 3 curves on the left refer to a set of 100 networks with initial random weights that learn the XOR task based on the correct teaching input. The 3 curves show the learning performance of the average, best , and worst individual in the set, respectively. The 3 curves on the right refer to another set of 100 individuals that learn using the output of the first set of individuals at the end of learning as their teaching input. Noise is added to this teaching input in each learning trial.

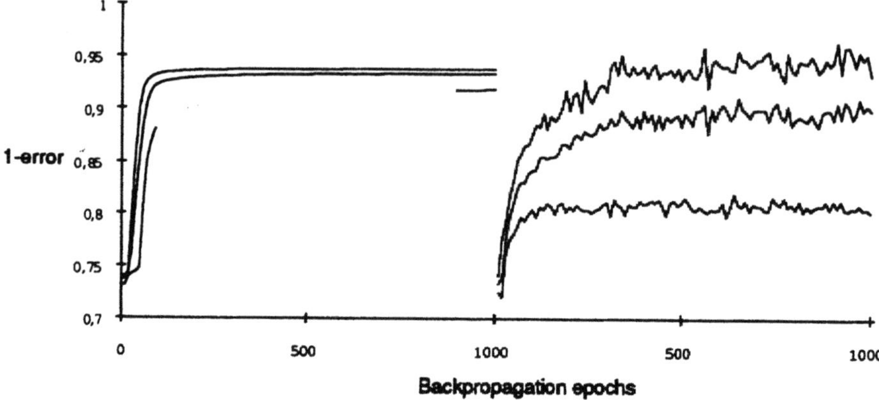

Figure 8. Learning curves for the XOR task for 100 networks learning on the basis of externally provided correct teaching input (left) and for other 100 networks that learn using the output of the first set of networks at the end of learning as their teaching input (right). Noise is added to the teaching input in this second case. The curves show the performance of the average, best, and worst individual in each set.

It is clear that adding noise to the teaching input in cultural transmission increases the variability of the observed performances. Cultural transmission without noise results in a slow and progressive deterioration of performance (cf. Figure 3). Cultural transmssion with noise may produce individuals that have a much lower performance level than the worst individual in the preceding generation but at the same time it can produce some rare individuals that have a slightly higher performance level than the best individual of the preceding generation. This appears to be sufficient to trigger cultural evolution. The selective cultural reproduction of the best individuals of each cultural generation will insure that the rare and probably slight improvements produced by noise will be retained in the population of cultural variants while the frequently produced deteriorated variants will be discarded. This process becomes cumulative and it not only is able to neutralize any global tendency to deterioration but can also result in the evolutionary cultural emergence of initially nonexistent capacities.

References

1. Belew, R.K., McInerney, J., & Schraudolph, N.N. Evolving networks: using the genetic algorithm with connectionist learning. In C.G. Langton et al. (eds.) Artificial Life II. Reading, Mass., Addison-Wesley 1992.

2. Boyd, R. & Richerson, P.J. Culture and the Evolutionary Process. Chicago, Chicago Press, 1985.

3. Cavalli-Sforza, L.L. & Feldman, M.W. Cultural Transmission and Evolution. A Quantitative Approach. Princeton, Princeton university Press, 1981.

4. Goldberg, D.E. Genetic Algorithms in Search, Optimization, and Machine Learning. Reading, Mass., Addison-Wesley, 1989.

5. Holland, J.H. Adaptation in Natural and Artificial Systems. Ann Arbor, University of Michigan Press, 1975 (reprinted by MIT Press, Cambridge, Mass., 1992).

6. Hutchins, E. & Hazelhurst, B. How to invent a lexicon: the development of shared symbols in interaction. In N. Gilbert & R. Conte (eds.) Artificial Societies: The Computer Simulation of social life. London, UCL Press, 1995.

7. Mitchell, M. & Forest, S. Genetic Algorithms and Artificial Life. Artificial Life. 1994, 1, 267-290.

8. Nolfi, S., Elman, J.L., & Parisi, D. Learning and evolution in neural networks. Adaptive Behavior, 1994, 3, 5-28.

9. Nolfi, S., Miglino, O. & Parisi, D. Phenotypic plasticity in evolving neural networks. In D.P. Gaussier & J.-D. Nicoud (eds.) From Perception to Action. Los Alamitos, CA, IEEE Computer Society Press, 1994.

10. Parisi, D., Cecconi, F., & Nolfi, S. Econets: networks that learn in an environment. Network, 1990, 1, 149-168.

A New Incremental Learning Technique

Nick Dunkin, John Shawe-Taylor
Department of Computer Science
Royal Holloway College
University of London
U.K.
{nickd, john}@dcs.rhbnc.ac.uk

Pascal Koiran
LIP
ENS-Lyon
46 Allée d'Italie
Lyon
France
koiran@lip.ens-lyon.fr

Abstract

We present a new type of constructive algorithm for incremental learning. The algorithm overcomes many of the problems associated with standard back propagation such as speed and optimum network size. We investigate the ability of the network to learn and test the resulting generalisation of the network.

keywords : *Incremental learning, neural networks, back propagation*

1   Introduction

The back propagation method of training neural networks, although popular, has many disadvantages. The most significant of these is the time that the back propagation algorithm takes to train a network; it is often necessary to run the algorithm for many hundreds of epochs before a minimum is reached.

The other major disadvantage of the back propagation algorithm has to do with the size of the network. The back propagation algorithm slows down considerably as the size of the network increases so an over estimation of the initial network size can lead to drastic increases in learning time.

There are also the problems of *over* and *under* fitting. If a network does not have enough connections then the network will not be able to learn the desired function, if however the network possesses too many connections then the architecture may appear to learn the desired function but the resulting generalisation properties of the neural network will be poor. Such a network has *over-fitted* the examples and is acting more like a look-up table.

In response to these learning problems researchers have begun to investigate constructive network architectures [2, 3, 5, 1]. Constructive neural network algorithms build a network as part of the learning process, the aim being to speed up the learning process and let the size of the network be determined from the data. Unfortunately, many constructive algorithms have a tendency to add too many nodes and end up over-fitting the training data.

Such is the case with the popular cascade correlation architecture which adds nodes incrementally as new singular layers. Once a unit has been trained its weight connections are fixed, so if this unit has learned a *feature* that turns out to be incorrect through later training it must be compensated for by later units. This can lead to over-fitting and poor generalisation properties.

In this paper we describe a new incremental learning algorithm that constructs a single sigmoidal *hidden* layer during learning in such a way that *over-fitting* is avoided, resulting in good generalisation properties and short learning times.

2 A New Incremental Algorithm

The new incremental algorithm uses two principles which combine to produce quick training times and good generalisation properties.

Firstly, at each stage only one node in the network is being trained and hence this can be done very efficiently. Secondly, the algorithm bounds the contribution of each added node which limits the expressibility of the network and thus gives desirable generalisation properties.

The initial network contains one unit in the hidden layer with input size dictated by the training data. The internal units have a sigmoidal activation function with the output node producing a simple linear function of its inputs. Every node in the internal layer has a threshold value associated with it.

Training begins with the single internal node and progresses until the reduction of the total squared error of the network is less than some constant β (usually 0.0001) over ω iterations, where ω is a constant (usually 500).

When the network has reached this *boredom threshold*, training stops and the connection weight values of the network are frozen. At this time a new unit is added to the internal layer of the network. Before training can recommence it is necessary to factor down the contribution of the previously trained network. All the previously trained connections in the output layer are factored by $(1-\bar{\alpha})$ where $\bar{\alpha}$ has a value of $1/n$. The contribution of the new node in the network is multiplied by a value of $\bar{\alpha}$. At any time during training only one sigmoidal node is being adapted. In the next section we show how, at each stage of the incremental process, the function learned by the single sigmoid node will work towards a general solution.

2.1 Theoretical Result

The following theorem was first presented by Koiran [4] and shows that an algorithm exists for training a network with n hidden nodes incrementally, that is in terms of the functionality, as good as training a fixed network with n hidden nodes. The incremental training will be of course, much easier

Theorem 1.0. *Let G be a subset of a Hilbert space H, with $||g|| \leq B$ for each $g \in G$.*
Let $m = \inf_{F \in co(G)} ||F - f||$, where $f \in H$ and $co(G)$ is the convex hull of G.
Then for every $c > \sqrt{B^2 + m^2}$,

$$||f_n - f|| \leq m + \frac{c}{\sqrt{n}},$$

where $f_n, n = 1, 2, \ldots$ is any sequence chosen to satisfy

$$||f_1 - f||^2 \leq \inf_{g \in G} ||g - f||^2 + \epsilon_1,$$

and, for $n \geq 2$,

$$||f_n - f||^2 \leq \inf_{g \in G} ||\alpha_n f_{n-1} + \bar{\alpha}_n g - f||^2 + \epsilon_n,$$

where $\alpha_n = (n-1)/n$, $\bar{\alpha}_n = 1 - \alpha_n = 1/n$, and $\epsilon_n \leq (c^2 - B^2 - m^2)/n^2$.

Theorem 1.0 shows that a solution to some problem can be derived incrementally. Let us take G as the set of possible functions realisable by a single sigmoidal unit, whose output is multiplied by a constant value, at most B,

$$G = \{A\sigma(<w, x>)|A < B\}$$

where $\sigma(x)$ is the sigmoid function and $<w, x>$ denotes the inner products of vectors w and x.

The function f_n is the convex combination of n functions from G; the state of the frozen network with n internal nodes,

$$f_n = (g + (n-1)f_{n-1})/n,$$

where g is some function from G. This means that at each addition stage of the algorithm the training of the new node is equivalent to finding g_0 given f_{n-1} such that,

$$||\alpha f_{n-1} + \bar{\alpha} g_0 - f||^2 < \inf_{g \in G} ||\alpha f_{n-1} + \bar{\alpha} g - f||^2 + \epsilon_n,$$

where $\alpha + \bar{\alpha} = 1$, $\bar{\alpha} = 1/n$. f is the target and $\epsilon_n = 1/n^2$

The function g_0 is determined by the weight vector w and output multiplier a. These are the values we must compute in order to find g_0. The weight vector is found by performing back propagation on the one node network. The output multiplier is found once this weight vector has been determined.

Training continues, using standard back propagation, until the desired error level is achieved. Back propagation is often disregarded due to its slow training times but our algorithm is only ever training one internal node at a time so back propagation is more than satisfactory as a training algorithm.

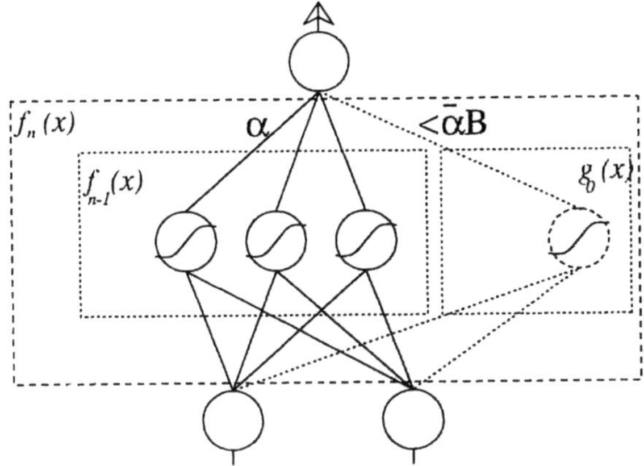

Figure 1: The network $f_{n-1}(x)$ with the addition of a new node ($g_0(x)$) giving the new network $f_n(x)$

3 Testing and Evaluation

In this section we describe the testing of the new algorithm and compare it with another popular incremental algorithm; Fahlman's cascade correlation [2].

3.1 Function Approximation

The algorithm was tested on function approximation problems, with both noisy and clean data.

3.2 Function One

The first function was a simple three dimensional function of the form,

$$f(x, y, z) = (x * \cos(y - z)) + \frac{x^2}{2} * \sin(2(y - z))$$

The complete training set consisted of 125 sets of example presentations. This training set was divided into 10 separate training sets which contained 20, 30, 40, 50, 60, 70, 80, 90, 100 and 125 training examples. These training sets were used to train ten networks. Each network converged to an acceptably low error. Each of the ten networks were then presented with a test set that contained 25 previously unseen examples. The results of these tests are presented in the next section.

3.3   Function Two

The second function to be approximated was a two dimensional function,

$$f(x, y) = (exp(cos(4 * (x + y))))/2$$

Eleven test sets were generated at random, from a normal distribution between 0 and 1 on the x and the y axis. The number of examples in each test set were 10, 20, 30, 40,50, 60, 70, 80, 90, 100 and 125. Eleven networks were trained with the test sets and then each network was tested on a test set of 25 previously unseen examples drawn at random from the a normal distribution in the same way as before.

This second experiment was then repeated with the addition of random noise on the input data. The randomly introduced noise was in the range $(-0.05 - +0.05)$.

3.4   Comparison with Cascade Correlation

One of the most popular constructive algorithms is the Cascade Correlation architecture described by Fahlman and Labiere. The function approximation problems were tested on a cascade correlation network and the results are presented in the next section.

It can be seen from the graph that the new incremental algorithm generalises well on the unseen test set. The plot flattens out to a constant low error of approximately 0.04 TSE(Total Squared Error). The plot for the cascade correlation architecture shows erratic behaviour, sometimes performing well on the unseen examples and other times performing badly. The conjecture is that although the cascade correlation architecture learns the training sets within an allowable error, the resulting generalisation properties of the trained network are unsatisfactory.

Cascade correlation performs badly on the training set with additional noise, displaying poor error rates on the test set and non uniform behaviour as the number of examples in the original training set increases. There is also an indication that over-fitting occurs in noisy data. As the number of noisy examples increases towards a 125 the performance of cascade correlation deteriorates, indicating that over-fitting may be occurring.

4   Results of Experimental Work

4.1   Comparative Results

For each experiment a graph was plotted to show how cascade correlation and the new incremental algorithm performed on the previously unseen test set with an increasing number of examples in the training set.

The performance of cascade correlation in the first experiment is particularly poor with out new algorithm producing a much more uniform behaviour pattern.

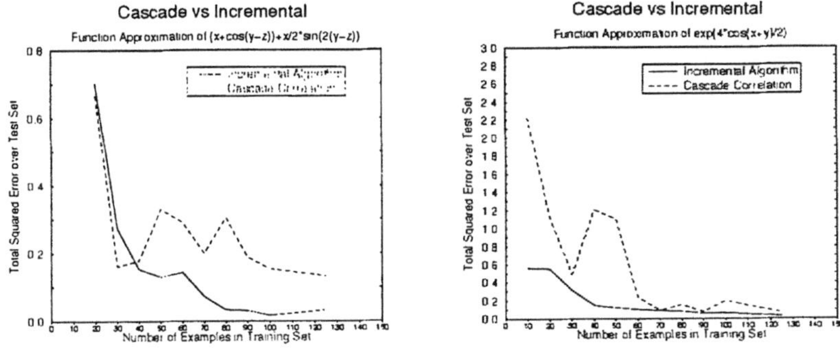

Figure 2: Results for Function One and Function Two

In the second experiment, on the two dimensional data, cascade correlation begins to perform more satisfactorily, however there are some problems apparent at the lower end of the training sets.

The training set with additional noise provides cascade correlation with some problems. There is indication of over-fitting at the far right of the graph and throughout the test sets the performance on the unseen data is unpredictable. The trace produced by the new algorithm is very similar to the trace produced for the clean data, indicating that generalisation properties of this new algorithm are good and that the addition of noise has not resulted in over-fitting of the training data.

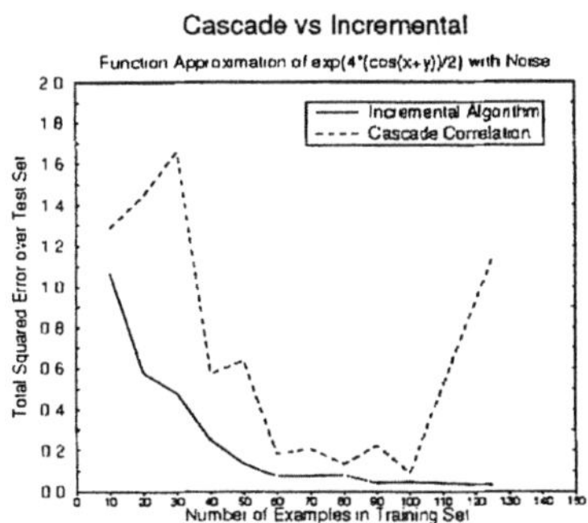

Figure 3: Function Two - with additional noise

118

4.2　The Approximated Manifolds

It is interesting to look at the manifold described by the network generated in
the second function experiment. The following approximation, from the incremental algorithm, was with 60 training points and the addition of six internal
nodes.

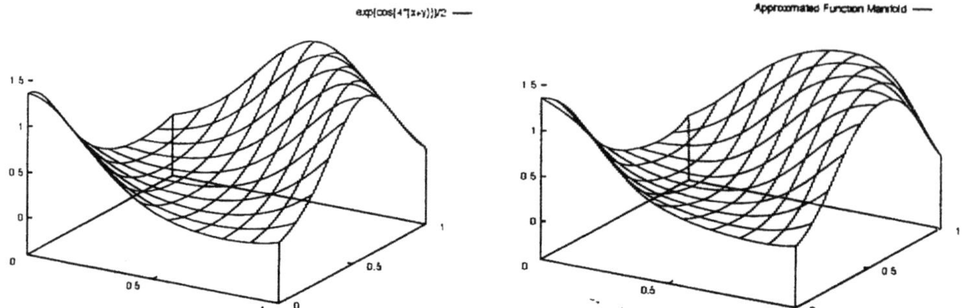

Figure 4: Target and approximated functions

5　Discussion and Further Work

In this paper we have shown, both theoretically and experimentally that this
new algorithm performs well and results in a network whose generalisation
properties are comparable with that of a neural network with a fixed *a priori*
size.

Further comparisons need to be done, first with cascade correlation and
then with other incremental algorithms.

References

[1] D.DeMers & G.Cottrel (1992) Non-Linear Dimensionality Reduction, University of California 1992

[2] Scott Fahlman & Christian Lebiere (1991). The Cascade-Correlation
Learning Architecture, CMU-CS-90-100, Technical Report

[3] Christian Jutten & Rachida Chentouf(1995), A New Scheme for Incremental Learning, Neural Processing Letters

[4] Pascal Koiran (1994) Efficient Learning of Continuous Neural Networks,
COLT 94 Techniques in Neural Learning, NC-TR-95-036, Technical Report

[5] Tin-Yau Kwok & Dit Yan Yeung(1995), Constructive Feedforward Neural
Networks for Regression Problems : A Survey, HKUST-CS95-43, Technical
Report

Solving algebraic and geometrical problems using neural networks

Mario Ferraro[1] and Terry Caelli[2]

[1]Dipartimento di Fisica Sperimentale, Universita' di Torino,
via Giuria 1, 10125 Torino, Italy,
tel. 39-11-6707376, fax 39-11-6691104, e-mail:ferraro@ph.unito.it
[2]Department of Computer Science, Curtin University of Technology,
Perth, Western Australia. 6019

Keywords: Neural computation, model-based neural networks

1 Introduction

Neural networks have been used over many years to help solve a large range of problems in pattern recognition, control and general signal processing. In this sense "neural computing" has generally been restricted to these types of domains. In contrast, here we investigate how neural networks can be adapted to actually solve numerical problems in algebra and geometry by modelling connections and transducer functions.

In classical numerical analysis solutions to problems are either obtained via approximation theory, linear programming or search methods, the former being analytic, the latter being based upon enumeration techniques which satisfy specific cost functions [2]. Our proposed approach combines both aspects of numerical computation but also restricts itself to perceptron-like computational units, in the spirit of "neural computing". However, of central importance to this work is the notion that parameterising the connections, weights, etc., constrains the associated search problem and so results in solutions which are guaranteed to satisfy the problem constraints. In this way, we note some duality between data and model parameters and also note how many problems can be formulated in terms of either estimating parameters to fit data or estimating data to fit parameters.

To this stage we (Caelli, Squire Wild, [1]: Squire and Caelli, [3]) have shown how by creating specific types of connections and parameterisations of the weighting functions we can obtain invariant pattern recognition with much fewer variables than traditional Neural Networks (NNs). In this paper we show how general this idea is by investigating how quite different mathematical computations can be solved within standard perceptron-like neural architectures.

---

[1] This project was funded by a grant from the Australian Research Grants Committee. Editorial correspondence should be addressed to Mario Ferraro.

2 Preliminary computations

There are three main aspects involved in neural computing: representation, direct computation and estimation. The first two aspects two aspects determine the ways the net is modelled where, in particular, different representations entail different net architectures. For instance, consider two vectors $\mathbf{x}$, $\mathbf{y}$, and consider the problem of designing a neural net with non-linearities, denoted by $\mathcal{SP}$, that computes the scalar product $\mathbf{x} \cdot \mathbf{y}$. The components of one vector, say $\mathbf{x}$, can be considered as input to the net, and the components of $\mathbf{y}$ can be considered as weights by setting $w_{ij} = y_j \delta_{ij}$. Then $\mathcal{SP}$ is a simple one-layered net , whose output is $O = \sum_i x_i y_i$. An appropriate representation of inputs and weights can simplify the computation considerably. However, it must be noted that such a simplification may not be possible for all problems.

A network similar to $\mathcal{SP}$, which will be denoted by $\mathcal{DT}$ can be used to compute the determinant of a 2×2 matrix. Let $[a_{ij}]$ a 2×2 matrix, and let $\mathbf{x}$, $\mathbf{y}$, be vectors defined by setting $x_1 = a_{11}$, $x_2 = -a_{12}$, $y_1 = a_{22}$, $y_2 = a_{21}$. To compute the determinant of $[a_{ij}]$ it is then enough to apply $\mathcal{SP}$ to these vectors. It is not difficult to design a net computing the determinant of a $n \times n$ matrix by using $\mathcal{DT}$ as basic subnet elements for the normal "Laplace development" (cofactors, minors, [4]). Finally, the trace of a matrix can be calculated in a trivial way, and, for future reference, let $\mathcal{T}$ be the net performing this calculation.

So far we have considered computations that can be carried out without training the net. However, in most problems some type of procedure is needed which forces the weights to assume values that guarantee the solution.

Consider the problem of solving a set of linear homogeneous equations, $A\mathbf{x} = 0$, where $A = [a_{ij}]$ is a square matrix, using a neural net that will be called $\mathcal{H}$. Assume as input to the net the elements of the coefficients matrix and define the weights between a_{ij} and a vector of hidden units h_i to be $w_{ij} = \delta_{ij} x_j$. Let the transduction function be $h_i(z) = z^2$, and suppose that all hidden units have unitary weight connections to the output unit O. Then the output of the net is

$$O = \sum_i \left[\sum_j a_{ij} x_j \right]^2 .$$

Set $O = 0$, then the solution can be found minimising O with some search technique, including backpropagation.

It must be observed that in many cases networks for solving complex problem can be formed by considering a set of simple nets $\mathcal{N}_i$ performing computations in parallel, and such a net will be denoted by $\bigcup_i \mathcal{N}_i$ or a concatenation of simple nets, such that the output of $\mathcal{N}_i$ is the input to $\mathcal{N}_{i+1}$; the concatenation will be denoted by $\mathcal{N}_i \to \mathcal{N}_{i+1}$. We will consider, in the following sections, two less obvious applications of this approach.

3 Polynomials: Representation and operations

Let $\mathbf{a}$, with components $a_i, i = 1, ..n$, be a set of inputs units, and let the weights be given by the rule $w_i = x^i$, where x is any real number. Next suppose that the input unit a_i is connected only with the weight w_i, and the output is given simply by $P = \sum a_i w_i$. In other words a polynomial $\mathcal{P}(x, n)$ can be represented by a simple net in the sense that its values are computed by changing the values of the weight x; that is, here we consider the input as coefficients and the weights as power functions of the data. More generally, this type of net can provide an engine to compute the approximation of any differentiable function. This treatment can be extended to polynomials of two variables, $\mathcal{P}(x, y, n)$, by considering an input matrix $[a_{ij}]$ and weights w_{ij} given by $w_{ij} = x^i y^j$, where x, y are real numbers. The output P is, in this case,

$$P = \sum_i \sum_j a_{ij} w_{ij},$$

Such neural nets which represent polynomials will be denoted by $\mathcal{P}$.

Different operations can be performed on $\mathcal{P}(x, y; n)$, for instance, partial derivatives of a polynomial with respect to x and y. The partial derivative of $\mathcal{P}(x, y; n)$, with respect to x can be computed by a net $\mathcal{DP}$, whose architecture is the same as $\mathcal{P}$, and the weights are $w_{ij} = ix^{i-1}y^j$. Thus we can compute, in parallel, the values of the the polynomial and its derivatives.

Finding the roots of a polynomial is also of interest . Consider a polynomial $\mathcal{P}(x, n)$ and assume that the coefficients a_i are such that there exist n roots of the polynomial, that is, n solutions to the equation $\mathcal{P}(x, n) = 0$, where these roots need not be distinct. Note that, obviously, $\mathcal{P}(x, n)$ and $\mid \mathcal{P}(x, n) \mid$ have the same roots.

The roots of $\mathcal{P}(x, n)$ can be found using a network, denoted in the following by $\mathcal{RP}$, composed of n subnets, which compute serially the roots of $\mathcal{P}(x, n)$. Computation by the k-th subnet is constrained by the results obtained by the subnets $1, 2, ...k - 1$ and, in turn, it constrains the calculations of subnets $k + 1, ...n$. A sketch for just three subnets is shown in Figure 1.

Every subnet has the same set of input units, $a_i^{(k)}$, where k denotes the subnet at which such inputs are applied, and $a_i^{(k)} = a_i^{(m)} = a_i$ for all k, m (again, each subnet receives the same external input). The input units of every subnet are connected to units $b^{(k)}$, with initial weights $w_j^{(k)} = x^j$, so that $w_j^{(k)} = w_j^{(m)}$ for all k, m. For the first subnet $b^{(1)} = P^{(1)}$, where $P^{(1)}$ is the output of subnet 1. The input units $a_1^{(k)}$ have connections to $n - k + 1$ set of units $\{g_i^{(k+1)}\}, ... \{g_i^{(n)}\}$, where, in each set, i ranges form 1 to $k - 1$. The connecting weights are $\pi_i^k = -x$, that is, they are the same for every subnet and every unit $g_i^{(k)}$. Furthermore, every $a_1^{(k)}$ is connected to the set $\{g_i^{(k)}\}$, with weights $\rho_i^{(k)} = x$. Each set $\{g_i^{(k)}\}$ is connected to an unit $h^{(k)}$, with weights equal to -1, $h^{(k)}$ also receives an input from $b^{(k)}$, with unitary connection. Finally $h^{(k)}$ is

122

connected with a unit weight to the output $P^{(k)}$ of the k-th unit. Let $I_g^{(k)}$, $I_b^{(k)}$, $I_h^{(k)}$, be the input of units of $\{g_i^{(k)}\}$, $\{b^{(k)}\}$, $\{h^{(k)}\}$, respectively; the outputs are

$$g_i^{(k)}(I_g^{(k)}) = \lg | I_g^{(k)}/a_1 |, \qquad b^{(k)}(I_b^{(k)}) = \lg(| I_b^{(k)} |), \qquad h^{(k)}(I_h^{(k)}) = \exp(I_h^{(k)}).$$

To illustrate how the net works suppose $n = 3$. The process starts by setting $P^{(1)} = 0$ and then a search procedure (for example, backpropagation) is applied until weight values $w_j^{(1)} = \xi_1^j$ are found which satisfy the condition $P^{(1)} = 0$, then ξ_1 is a root of the polynomial. Note that now $\pi_i^{(1)} = -\xi_1$, and all other weights $\pi_i^{(k)}$, $k > 1$, have not changed. Then $g_1^{(2)}$ receives from $a_1^{(1)}$ the input $-a_1\xi_1$ and from $a_1^{(2)}$ the input xa_1, so that the total input is $I_g^{(2)} = (x - \xi_1)a_1$, and the output is $g_1^{(2)}\left(I_g^{(2)}\right) = \lg | x - \xi_1 |$. The input to $b^{(2)}$ is $I_b^{(2)} = \sum a_i x^i$. Then the input received by $h^{(2)}$ is $I_h^{(2)} = \lg | b^{(2)} | - \lg | x - \xi_1 |$. Finally the output $P^{(2)}$ is given by

$$P^{(2)} = h(I_h^{(2)}) = \exp(I_h^{(2)}) = \frac{| b^{(2)} |}{| x - \xi_1 |}.$$

Note that $b^{(2)}$ has the same value as $P^{(1)}$ had before it was set to zero, to compute the first root; then the process taking place after the first root has been found factors out the term $(x - \xi_1)$ from the polynomial representation $P^{(1)}$, so that

$$P^{(2)} = \frac{| P^{(1)} |}{| (x - \xi_1) |}.$$

Note that $P^{(k)}$ and $| P^{(k)} |$ are zero for the same values of $w_j^{(k)}$. After the first root has been factored out a new one can be found, setting $P^{(2)} = 0$ and using the same search techniques as above. Let ξ_2 be this root. The process is repeated for the third subnet, leading to

$$P^{(3)} = \frac{| P^{(1)} |}{| x - \xi_1 || x - \xi_2 |},$$

as is illustrated in Figure 1.

4 Minkowski and Riemann metrics.

In this section we shall show how neural networks can be modelled to compute the distance between two points, using a Minkowski metric and also calculate the Riemann tensor, that is the coefficients of the first fundamental form, of a two-dimensional surface S.

Consider the problem of computing a distance between two points $\mathbf{x} = (x_1, x_2)$, $\mathbf{y} = (y_1, y_2)$. In this case the net has four input units $(x_1, x_2, x_3 =$

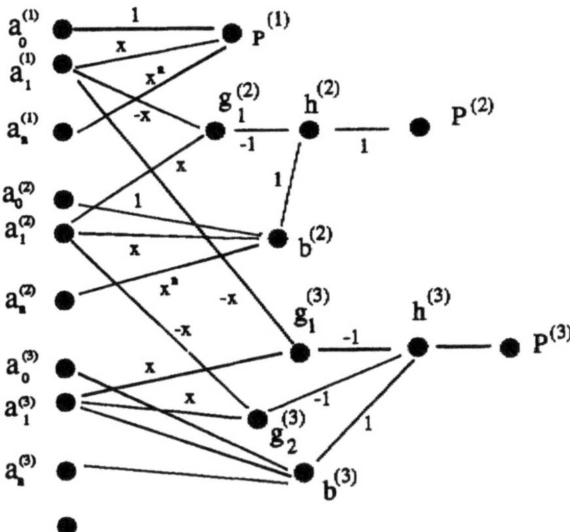

Fig. 1. Shows the neural network structure for computing the roots of a polynomial, $\mathcal{P}(x,n)$. Here, a_i refers to the input and coefficients of the polynomial; $1, x, .. x^2$ refer to the initial weights (and data values) of the net which are updated to attain the roots. The functions b, g, h refer to transduction functions which are required for the roots. P is the output which is set to zero during computation (see text for details).

$y_1, x_4 = y_2$), two hidden units $h_i, i = 1, 2$ and an output unit O. Input-output relations are given by

$$h_i = \phi(\Sigma_j w_{ij} x_j),$$
$$O = \psi(\Sigma_j k_i h_i),$$

where w_{ij}, k_i are weights and ϕ, ψ transduction functions.

A distance can be now computed, in any Minkowski metric l^p, simply by setting $w_{11} = w_{22} = 1$, $w_{13} = w_{24} = -1$, $w_{ij} = 0$, for all other cases and $k_i = 1$; the transduction functions are $\phi(z) = z^p$ and $\psi(z) = z^{1/p}$.

The results established in the previous sections also provide the basis for the computation, through a net $\mathcal{G}$, of the metric tensor g_{ij} for any surface. Let S be a surface. A parametric representation of S is a map from an open and connected parameter plane (u, v) to $\mathbf{R}^3$,

$$\mathbf{r} : (u, v) \to \{x(u, v), y(u, v), z(u, v)\},$$

and $r(u, v) = (x(u, v), y(u, v), z(u, v)) \in S$.

Each component of the vector function $\mathbf{r}$ can be approximated by a polynomial $\mathcal{P}(u, v; n)$, then the tangent vectors to S,

$$\mathbf{r}_1 = (\partial x/\partial u, \partial y/\partial u, \partial z \partial u), \quad \mathbf{r}_2 = (\partial x/\partial v, \partial y/\partial v, \partial z \partial v),$$

can now be computed by using, for each component x, y, z, a net $\mathcal{DP}_i, i = 1, 2, 3$. Let $\bigcup_i \mathcal{DP}_i$ denote the set of of these nets that can be assumed to perform parallel computations. The outputs of $\bigcup_i \mathcal{DP}_i$ are now the inputs of a set of networks $\bigcup_{ij} \mathcal{SP}_{ij}$ which calculate, in parallel, the scalar products

$$g_{ij} = \mathbf{r}_i \cdot \mathbf{r}_j.$$

Thus the net $\mathcal{G}$ is given by the concatenation of nets $\bigcup_i \mathcal{DP}_i$ with $\bigcup_i \mathcal{SP}_i j$,

$$\mathcal{G} = \bigcup_i \mathcal{DP}_i \to \bigcup_{ij} \mathcal{SP}_{ij}.$$

5 Conclusions

In this paper we have discussed and illustrated the idea of using perceptron-like neural network architectures to solve a variety of computational problems which often occur in numerical computing. We have shown that by constraining, selecting and modelling the connections in different ways we can control the types of information flow and solution obtainable with neural network architectures. Further to this, we, hopefully, have raised issues about more general application of neural computing, which, in turn, possibly poses new approaches to old problems, as, for instance, we see in the example of roots of polynomials.

References

1. Caelli TM, Squire DM, Wild TPJ. Model-Based Neural Network,Neural Networks 1993 6:613-625.
2. W.H. Press, B.P. Flannery, S. A. Teukolsky, W.T. Vetterling. Numerical Recipes in C. Cambridge University Press, Cambridge, 1988.
3. Squire DM, Caelli TM. Shift, rotation and scale invariant signatures for two-dimensional contours, in a neural network architecture. Annals of Mathematics and Artificial Intelligence 1995 (in press).
4. Korn GA, Korn TM. Mathematical Handbook for Scientists and Engineers. McGraw-Hill, New York, 1968.

This article was processed using the LaTeX macro package with LLNCS style

A Bayesian Framework for Associative Memories

Edwin R. Hancock [1] and Marcello Pelillo [2]

[1] *Department of Computer Science, University of York*
YO1 5DD York, England

[2] *Dipartimento di Matematica Applicata e Informatica, Università di Venezia*
Via Torino 155, 30173 Venezia Mestre, Italy

Abstract

We develop a Bayesian model of associative memory and demonstrate how some well-known associative memory networks can be regarded as particular instances of ours, thereby providing an empirical understanding of their behavior. Specifically, these are the exponential correlation associative memory (ECAM) of Chiueh and Goodman, and the Hopfield model. A connection with the Hamming net and the Aleksander Boolean model is also established. The framework for our study is a novel relaxation method which involves direct probabilistic modelling of the pattern corruption mechanism. The parameter of this model is the memoryless probability of error on nodes of the network. This bit-error probability is not only important for the interpretation of the ECAM and the Hopfield models, but allows also us to understand some more general properties of Bayesian pattern reconstruction by relaxation.

1 Introduction

Since Hopfield's influential work [1], there has been an explosion of interest in the study of neural network models of associative memory. These are parallel computational networks in which idealised versions of corrupted patterns are recalled by locally updating individual nodes in the network to optimise a global cost function. The original model proposed by Hopfield, however, was soon recognised to have severe limitations associated with its restricted storage capacity, and this has then motivated a number of investigators to develop alternative memory models in an attempt to overcome this fundamental drawback. One of the most successful such attempts has recently been reported by Chiueh and Goodman [2] who developed an exponential correlation associative memory (ECAM) model which not only turns out to have large storage capacity, but can also be easily implemented in VLSI hardware.

Viewed as systems that achieve global configurational optimisation via local iterative computations, these associative memories have many features in common with relaxation algorithms widely employed in the field of computer vision and pattern recognition [3, 4]. Our aim in this paper is to demonstrate not only that there is a strong conceptual similarity between the two approaches, but that the ECAM and the Hopfield models can be regarded as particular instances of discrete relaxation [4], providing them with a direct Bayesian interpretation. The analysis provides an empirical understanding of the dynamical properties of these models.

Basic to the methodology described in this paper is the direct modelling of the pattern corruption mechanism which is objective yet simple [4]. We assume that each node in the network is subject to memoryless corruption, and adopt as our criterion for global labelling the joint probability of the binary node configuration on the network. This turns out to be a compound exponential function of the Hamming distances to the idealised stored patterns, the parameter of these exponentials being the memoryless probability of bit-errors acting on the binary patterns. Node updating takes place so as to perform gradient ascent on the global configurational probability. The ECAM, too, implements an exponential update rule, and we will show how our relaxation scheme is identical to the ECAM model when the base of its exponentials is chosen as a certain function of the bit-error probability. This corresponds also to the best possible choice in terms of storage capacity. When the bit-error probability approaches $\frac{1}{2}$, our relaxation model becomes equivalent to the Hopfield update rule. Conversely, when the bit-error probability becomes asymptotically small, the relaxation approach implements a type of look-up operation, selecting the global pattern of minimum Hamming distance. This is exactly the operation performed by Lippmann's Hamming network [5], and is similar in function to the Boolean network proposed by Aleksander [6] and extensively analysed by Wong and Sherrington [7].

2 Configurational probability and relaxation

We are interested in networks that can be represented in terms of configurations of binary variables. These networks are composed of N computational nodes, and at any particular stage of updating each node will be in one of the binary states denoted by $\Omega \equiv \{-1, 1\}$. The particular realisation of the labelling of the node indexed j is denoted by s_j. With this notation, the global state of a network is represented by the configuration of binary values $S = \{s_j \in \Omega, \forall j = 1 \ldots N\}$.

Suppose that we have access to a set of training patterns. Typically, these would be configurations of binary labels which we want to recover from an initial inconsistent state of the network. Assume that there are Z such global patterns denoted by $\Lambda^\mu = \{\lambda_j^\mu \in \Omega, \forall j = 1 \ldots N\}$. According to this notation, μ is the pattern index and λ_j^μ is the binary value assigned to the site indexed j by the μth training pattern. The basic information relevant to the "consistency" of the configuration S is conveyed by the set of Hamming distances to the prototype patterns Λ^μ, $\mu = 1 \ldots Z$. With the binary node variables defined in the previous section, the Hamming distance to the prototype pattern indexed μ can be written as

$$H_\mu = \frac{1}{2}\left(N - \sum_{i=1}^{N} s_i \lambda_i^\mu\right) . \tag{1}$$

The configurational relaxation procedure is based on maximising the joint probability of the binary label configuration, i.e., $P(S)$. It is therefore necessary to find a way of enumerating $P(S)$ when the label configuration is highly inconsistent, i.e., when there are no prototype patterns for which the Hamming distance is zero. The approach is to adopt a Bayesian viewpoint in which it is assumed that only prototype patterns are legal and have uniform non-zero *a priori* probabilities of occurrence $P(\Lambda^\mu)$. Other configurations do not occur *a priori* but are the corrupted realisations of the prototype patterns. This idea is realised by applying the axiomatic property of joint probability to expand $P(S)$ over the space of consistent configurations

$$P(S) = \sum_{\mu=1}^{Z} P(S|\Lambda^\mu) P(\Lambda^\mu) . \tag{2}$$

To develop this idea into a useful update rule requires a model of the bit corruption process, that is of the conditional probabilities of the inconsistent configurations given the Z possible prototypes $P(S|\Lambda^\mu)$. We adopt a very simple viewpoint: bit errors are assumed to be memoryless and to occur with uniform probability $p \leq \frac{1}{2}$.

The first consequence of the assumed absence of memory is that the errors are independent. As a result we can factorise the conditional probabilities over the individual nodes in the network, i.e.,

$$P(S|\Lambda^\mu) = \prod_{i=1}^{N} P(s_i|\lambda_i^\mu) . \tag{3}$$

Our next step is to propose a model for the bit corruption mechanism at each node in the network. Again, taking recourse to the memoryless assumption, the probability of error on individual nodes is independent of the class of label. This leads us to the following assignment of probability

$$P(s_i|\lambda_i^\mu) = \begin{cases} 1 - p & \text{if } s_i = \lambda_i^\mu \\ p & \text{otherwise} \end{cases} \tag{4}$$

The model components given in (3) and (4) naturally lead to the following expression for $P(S)$ in terms of the set of Hamming distances to the prototype patterns

$$P(S) = \frac{\beta}{Z} \sum_{\mu=1}^{Z} \alpha^{H_\mu} \tag{5}$$

where $\alpha = \frac{p}{1-p}$ and $\beta = (1-p)^N$. We can re-write the above expression for the configurational probability in terms of the more familiar exponential function

$$P(S) = \frac{\beta}{Z} \sum_{\mu=1}^{Z} \exp\left(H_\mu \ln \frac{1}{\alpha}\right) . \tag{6}$$

Our aim in making node updates is to locate an optimum value of the configurational probability $P(S)$. The optimisation procedure is based on a conventional gradient ascent approach. When rather than being binary values the node variables are real-valued quantities in the range $[-1, 1]$, the rule for updating s_j is

$$\frac{\partial P(S)}{\partial s_j} = \frac{\beta}{Z} \ln \alpha \sum_{\mu=1}^{Z} \alpha^{H_\mu} \frac{\partial H_\mu}{\partial s_j} \tag{7}$$

where, from (1), we have $\frac{\partial H_\mu}{\partial s_j} = -\frac{1}{2}\lambda_j^\mu$. In the case of binary values, gradient ascent yields

$$s_j = \operatorname{sgn}\left\{ \sum_{\mu=1}^{Z} \lambda_j^\mu \alpha^{H_\mu} \right\} \tag{8}$$

which is the update rule of our Bayesian binary relaxation network.

3   Bayesian interpretation of the ECAM model

The ECAM is an instance of a more general associative memory model which Chiueh and Goodman [2] called the recursive correlation associative memory (RCAM). The dynamical behaviour of an RCAM is governed by the following updating rule

$$s_j = \operatorname{sgn}\left\{ \sum_{\mu=1}^{Z} \lambda_j^\mu \, f\left(\sum_{i=1}^{N} s_i \lambda_i^\mu \right) \right\} \tag{9}$$

for all $j = 1\ldots N$, where $f(\cdot)$ is an appropriate weighting function. In [2] it was shown that the dynamical system described by equation (9) is asymptotically stable both in the asynchronous and synchronous update modes, provided that the weighting function f is continuous and monotone nondecreasing over the interval $[-N, N]$. This was proved by showing that the system has an associated Liapunov (or "energy") function that governs its dynamical behavior. In addition, Chiueh and Goodman [2] showed how some known associative memory models can be viewed as a special case of the RCAM by just choosing a suitable weighting function. For example, if f is chosen to be the identity function, i.e.,

$$f(x) = x \tag{10}$$

then the model becomes essentially identical to the original Hopfield memory [1], with the weights initialised by the Hebb rule. Of particular interest is the case when the weighting function has an exponential form, i.e.,

$$f(x) = a^x \tag{11}$$

where a is a predetermined constant greater than unity. This class of memories constitutes the ECAM original model. This was shown to have large storage capacity and turns out to be especially suited for VLSI implementation [2].

Now, observe that from (1) our updating rule (8) is equivalent to

$$s_j = \text{sgn} \left\{ \sum_{\mu=1}^{Z} \lambda_j^\mu \alpha^{-\frac{1}{2} \sum_{i=1}^{N} s_i \lambda_i^\mu} \right\} \tag{12}$$

which is identical to the ECAM update rule defined by (9) and (11) when the base of exponentiation a is taken to be $\alpha^{-1/2}$ or, in other words, when

$$a = \left(\frac{1-p}{p} \right)^{\frac{1}{2}} . \tag{13}$$

Note that under the assumption that the error probability p is less than $\frac{1}{2}$, the base constant a is greater than unity as required. This provides, therefore, a direct Bayesian interpretation for the ECAM model. When the base of the exponentials a is chosen as in formula (13), the ECAM can be viewed as a method for maximising the configurational probability $P(S)$, thereby acting as a type of Bayesian pattern reconstruction process.

It may be of some interest to discuss how our Bayesian framework relates to the storage capacity result derived by Chiueh and Goodman in [2]. Our analysis here will be rather informal; a more rigorous discussion concerning the storage capacity of the ECAM can be found in [9]. Consider an input pattern S that is k bits away from the nearest memory pattern from which S is assumed to be derived through a process of memoryless bit corruption. Let $k = \kappa N$, where κ is the fraction of incorrect bits in the pattern S, and define κ' as $\kappa + \frac{1}{N}$. In [2], it was shown that when $\kappa' \geq (1 + a^2)^{-1}$ the asymptotic storage capacity of the ECAM is the largest possible for an associative memory, as determined in [8]. Observe that, under our memoryless corruption assumption, the bit-error probability p represents the expected fraction of incorrect bits in any input pattern, and is therefore intimately related to the parameter κ defined above. Specifically, κ is the actual realisation of a random variable whose expected value is p. Now, from (13), we obtain $p = (1 + a^2)^{-1}$ and thus $p' \equiv p + \frac{1}{N} > (1 + a^2)^{-1}$ for all N, which means that Chiueh and Goodman's condition for meeting the largest possible capacity is expected to hold, at least in a statistical sense. We can therefore conclude that our Bayesian framework naturally leads to a choice of the exponentiation base a which is (on the average) the best possible: it allows us to achieve the ultimate upper bound for the asymptotic storage capacity of an associative memory.

4 Bayesian interpretation of the Hopfield network

We are now interested in comparing our update rule with its counterpart for the Hopfield network model [1]. This network consists of an arrangement of nodes in which the inter-nodal connections are characterised by continuous-valued weights which store the training patterns. If w_{ij} is the interconnection weight between the nodes indexed i and j, then the network optimises the criterion $C(S) = \sum_{i,j} w_{ij} s_i s_j$, provided that the weights are symmetric, i.e., $w_{ij} = w_{ji}$. Training is frequently performed using the following Hebbian learning rule

$$w_{ij} = \frac{1}{N} \sum_{\mu=1}^{Z} \lambda_i^\mu \lambda_j^\mu .$$

(14)

By contrast with our exponential probability criterion, the Hopfield network effectively minimises a quadratic function of the Hamming distances. Operating in pattern recall mode, the network performs gradient descent on $C(S)$ with stepsize proportional to H_μ. From (1), the rule for updating the discrete node variables can be written as

$$s_j = \text{sgn} \left\{ \sum_{\mu=1}^{Z} \lambda_j^\mu \left(1 - \frac{2H_\mu}{N} \right) \right\} .$$

(15)

This clearly differs from the update rule for configurational relaxation in that it is linear in Hamming distance rather than exponential.

Under conditions in which α is close to unity, the exponentials appearing in (8) can be approximated in a linear way using the Taylor expansion:

$$s_j = \text{sgn} \left\{ \sum_{\mu=1}^{Z} \lambda_j^\mu \left[1 - (1 - \alpha) H_\mu + \dots \right] \right\} .$$

(16)

This is identical to the Hopfield dynamical rule when $\alpha = 1 - \frac{2}{N}$. The Hopfield approach can therefore be regarded as a valid approximation to configurational relaxation when $p \simeq \frac{1}{2}$. Note that in the limit when $p = \frac{1}{2}$, the update rule becomes $s_j = \text{sgn} \left\{ \sum_{\mu=1}^{Z} \lambda_j^\mu \right\}$. This means that the update decision is completely devoid of the contextual information conveyed by the global pattern configuration; it means that each node is assigned its most frequent value from the set of training patterns. In other words, the Hopfield memory only draws extremely weakly upon the available contextual information.

It is worth considering the limit of the relaxation approach when the bit-error probability p is close to zero. In this case the update rule simply selects the bit corresponding to the pattern which is closest in Hamming distance to the configuration of the network, i.e.,

$$s_j = \lambda_j^\mu, \quad \text{such that} \quad H_\mu = \min_{\nu=1}^{Z} H_\nu .$$

(17)

This is exactly the operation performed by the Hamming net described by Lippmann [5]. Moreover, this update rule is close in spirit to the operations performed by the Boolean network proposed by Aleksander [6], which have been thoroughly analysed by Wong and Sherrington [7]. In this type of network the binary-node values are used to construct an address to a memory location which contains an update value. This type of system could clearly handle the minimum Hamming distance criterion described above.

5 Conclusions

We have demonstrated how Chiueh and Goodman's ECAM model can be entirely derived within a Bayesian framework, and can therefore be interpreted as performing a type of Bayesian pattern reconstruction process. The basis for this work has been a novel relaxation scheme developed in the context of pattern recognition and computer vision, which involves direct probabilistic modelling of the pattern corruption process. We have additionally shown that the Hopfield memory can be given a Bayesian interpretation, too. Viewed from the standpoint of a pattern-space model, this interpretation corresponds to the case when the probability of individual bit errors on the network is approximately $\frac{1}{2}$. In terms of pattern recovery this is almost the worst possible case; it corresponds to complete overlap between the Hamming distance distribution for corrupt and competing stored patterns. In the limit of small bit-error probability our relaxation scheme becomes equivalent to Lippmann's Hamming network and the Boolean network proposed by Aleksander. A thorough analysis of the dynamical behavior of our relaxation model can be found in [9].

References

[1] J. J. Hopfield, "Neural networks and physical systems with emergent collective computational abilities," *Proc. Natl. Acad. Sci. USA*, vol. 79, pp. 2554–2558, 1982.

[2] T. D. Chiueh and R. M. Goodman, "Recurrent correlation associative memories," *IEEE Trans. Neural Networks*, vol. 2, pp. 275–284, 1991.

[3] S. Geman and D. Geman, "Stochastic relaxation, Gibbs distributions and the Bayesian restoration of images," *IEEE Trans. Pattern Anal. Machine Intell.*, vol. 6, pp. 721–741, 1984.

[4] E. R. Hancock and J. Kittler, "Discrete relaxation," *Pattern Recognition*, vol. 23, pp. 711–733, 1990.

[5] R. P. Lippmann, "An introduction to computing with neural nets," *IEEE ASSP Magazine*, pp. 4–22, April 1987.

[6] I. Aleksander, "The logic of connectionist systems," in I. Aleksander, Ed., *Neural Computing Architectures: The Design of Brain-Like Machines*. London, UK: North Oxford Academic Publishers, 1989, pp. 133–155.

[7] K. Y. M. Wong and D. Sherrington, "Theory of associative memory in randomly connected Boolean neural networks," *J. Phys. A*, vol. 22, pp. 2233–2264, 1989.

[8] P. A. Chou, "The capacity of the Kanerva associative memory is exponential," in D. Z. Anderson, Ed., *Neural Information Processing Systems*. New York, NY: American Institute of Physics, 1988, pp. 184–191.

[9] E. R. Hancock and M. Pelillo, "An analysis of the exponential correlation associative memory," *Proc. 13th Int. Conf. Pattern Recognition*, Vienna, Austria, 1996.

Proof of the Universal Approximation of a Set of Fuzzy Functions

Maria Marinaro
INFM - Department of Theoretical Physics - University of Salerno
Via S. Allende - 84081 Baronissi (SA) - Italy
IIASS - Via Pellegrino 19 - Vietri sul Mare (SA) - Italy
fax : + 39 89 761189

Francesco Masulli
INFM - Department of Physics - University of Genoa
Via Dodecaneso 33 - 16146 Genova - Italy
email : masulli@ge.infm.it

Domenico Oricchio
INFM - Department of Physics - University of Genoa
Via Dodecaneso 33 - 16146 Genoa - Italy
IIASS - Via Pellegrino 19 - Vietri sul Mare (SA) - Italy
email : iiass@home.mycrosys.it

Introduction.

In the last few years neuro-fuzzy systems have been investigated by many researchers [2,3].

The possibility of combine linguistics and numerical knowledges make those models very attractive.

Morever the universal function approximation property (i.e. the possibility of approximate any given real continuous function on a compact set) that holds for some fuzzy logic system [4,7,1] permits to use those models in many applications, including control [2,12], time series prediction [1], Bayes classification [6].

In this paper the Universal Approximation Theorem is proven for the fuzzy system, already presented in [5], that could be also obtained as a Sugeno Fuzzy Model [10,9].

Universal Approximation Theorem.

Let Z be the set of fuzzy functions obtained by the "Center Of Area" (COA) method and the membership function of the following form:

$$v^i(\vec{x}) = \frac{e^{-\alpha_i^2 - \beta \sum_{l \leq m} \gamma_{lm}^i (x_l - x_l^i)(x_m - x_m^i)}}{\sum_{j=1}^N e^{-\alpha_j^2 - \beta \sum_{l \leq m} \gamma_{lm}^j (x_l - x_l^j)(x_m - x_m^j)}} \qquad (1)$$

γ_{lm}^i are positive definite.

We want to prove that the uniform closure of the set Z consists of all continuous functions defined in a closed domain U (**Universal Approximation Theorem**).

We can use the Stone-Weierstrass theorem [8] to prove the Universal Approximation Theorem. To this end we must show:

1. Z is an algebra.
2. Z separates points on U.
3. Z vanishes at no point of U.

We will show that Z satisfies those three properties.

Property 1.1

Let be $f_1, f_2 \in Z$

$$f_1(\vec{x}) = \frac{\sum_{j_1=1}^N f_{j_1} e^{-\alpha_{j_1}^2 - \beta \sum_{l \leq m} \gamma_{lm}^{j_1}(x_l - x_l^{j_1})(x_m - x_m^{j_1})}}{\sum_{j_1=1}^N e^{-\alpha_{j_1}^2 - \beta \sum_{l \leq m} \gamma_{lm}^{j_1}(x_l - x_l^{j_1})(x_m - x_m^{j_1})}} \qquad (2)$$

$$f_2(\vec{x}) = \frac{\sum_{j_2=1}^N f_{j_2} e^{-\alpha_{j_2}^2 - \beta \sum_{l \leq m} \gamma_{lm}^{j_2}(x_l - x_l^{j_2})(x_m - x_m^{j_2})}}{\sum_{j_2=1}^N e^{-\alpha_{j_2}^2 - \beta \sum_{l \leq m} \gamma_{lm}^{j_2}(x_l - x_l^{j_2})(x_m - x_m^{j_2})}} \qquad (3)$$

then:

$$f_1(\vec{x}) + f_2(\vec{x}) = \frac{\sum_{j_1, j_2=1}^N \left(f_{j_1} + f_{j_2} \right) e^{-\alpha_{j_{12}}^2 - \beta \sum_{l \leq m} \gamma_{lm}^{j_{12}}(x_l - x_l^{j_{12}})(x_m - x_m^{j_{12}})}}{\sum_{j_1, j_2=1}^N e^{-\alpha_{j_{12}}^2 - \beta \sum_{l \leq m} \gamma_{lm}^{j_{12}}(x_l - x_l^{j_{12}})(x_m - x_m^{j_{12}})}} \qquad (4)$$

Eq. (4) follows from the fact that the sum of two quadratic forms defined positive

134

is a quadratic form defined positive. Then:

$$\left(\vec{x}-\vec{x}'\right)^{T}\underline{\underline{\gamma'}}\left(\vec{x}-\vec{x}'\right)+\left(\vec{x}-\vec{x}'\right)\underline{\underline{\gamma^{j}}}\left(\vec{x}-\vec{x}'\right)=\left(\vec{x}-\vec{x}^{ij}\right)\underline{\underline{\gamma^{ij}}}\left(\vec{x}-\vec{x}^{ij}\right)+\left(\beta^{ij}\right)^{2} \tag{5}$$

Therefore $f_{1}, f_{2} \in Z$.

Property 1.2

Likewise, by considering the product $f_{1} \, and \, f_{2}$, one obtains:

$$f_{1}(\vec{x}) f_{2}(\vec{x}) = \frac{\sum_{j_{1}, j_{2}=1}^{N}\left(f_{j_{1}} f_{j_{2}}\right) e^{-\alpha_{j_{12}}^{2}-\beta \sum_{l \leq m} \gamma_{lm}^{'12}\left(x_{l}-x_{l}^{'12}\right)\left(x_{m}-x_{m}^{'12}\right)}}{\sum_{j_{1}, j_{2}=1}^{N} e^{-\alpha_{j_{12}}^{2}-\beta \sum_{l \leq m} \gamma_{lm}^{'12}\left(x_{l}-x_{l}^{'12}\right)\left(x_{m}-x_{m}^{'12}\right)}} \tag{6}$$

that is in the same form of Eq. (4), and therefore $f_{1} f_{2} \in Z$.

Property 1.3

At last, let consider the product of f by a scalar

$$c f(\vec{x}) = \frac{\sum_{j=1}^{N}\left(c f_{j}\right) e^{-\alpha_{j}-\beta \sum_{l \leq m} \gamma_{lm}^{'}\left(x_{l}-x_{l}^{'}\right)\left(x_{m}-x_{m}^{'}\right)}}{\sum_{j=1}^{N} e^{-\alpha_{j}-\beta \sum_{l \leq m} \gamma_{lm}^{'}\left(x_{l}-x_{l}^{'}\right)\left(x_{m}-x_{m}^{'}\right)}} \tag{7}$$

then $c f \in Z$.

From the properties 1.1, 1.2 and 1.3 follows that Z is an algebra.

Property 2

We choose:

$$v(\vec{x}) = \frac{e^{-\beta \sum_{l \leq m} \gamma_{lm}^{'}\left(x_{l}-x_{l}^{'}\right)\left(x_{m}-x_{m}^{'}\right)}}{\sum_{j=1}^{N} e^{-\beta \sum_{l \leq m} \gamma_{lm}^{'}\left(x_{l}-x_{l}^{'}\right)\left(x_{m}-x_{m}^{'}\right)}} \tag{8}$$

and we consider the function:

$$f(\vec{x}) = \frac{f_1 e^{-\beta \sum_{l \leq m} \gamma^1_{lm}(x_l - x'_l)(x_m - x'_m)} + f_2 e^{-\beta \sum_{l \leq m} \gamma^2_{lm}(x_l - x^2_l)(x_m - x^2_m)}}{e^{-\beta \sum_{l \leq m} \gamma^1_{lm}(x_l - x'_l)(x_m - x'_m)} + e^{-\beta \sum_{l \leq m} \gamma^2_{lm}(x_l - x^2_l)(x_m - x^2_m)}} \qquad (9)$$

$if\ \vec{x}_1 \neq \vec{x}_2\ and\ \gamma^1_{lm} = \gamma^2_{lm} = \gamma_{lm}\ then:$

$$f\left(\vec{x}^1\right) = \frac{f_1 + f_2 e^{-\beta \sum_{l \leq m} \gamma_{lm}(x'_l - x^2_l)(x'_m - x^2_m)}}{1 + e^{-\beta \sum_{l \leq m} \gamma_{lm}(x'_l - x^2_l)(x'_m - x^2_m)}} \qquad (10)$$

$$f\left(\vec{x}^2\right) = \frac{f_1 e^{-\beta \sum_{l \leq m} \gamma_{lm}(x^2_l - x'_l)(x^2_m - x'_m)} + f_2}{e^{-\beta \sum_{l \leq m} \gamma_{lm}(x^2_l - x'_l)(x^2_m - x'_m)} + 1} \qquad (11)$$

by defining $\theta = 1\Big/\left(1 + e^{-\beta \sum_{l \leq m} \gamma_{lm}(x^2_l - x'_l)(x^2_m - x'_m)}\right)$, one obtains:

$$f\left(\vec{x}^1\right) = \theta f_1 + (1-\theta) f_2 \qquad (12)$$

$$f\left(\vec{x}^2\right) = (1-\theta) f_1 + \theta f_2 \qquad (13)$$

$if\ \vec{x}_1 \neq \vec{x}_2\ then\ e^{-\beta \sum_{l \leq m} \gamma^1_{lm}(x^2_l - x'_l)(x^2_m - x'_m)} \neq 0$, and $\theta \neq 1 - \theta$. Therefore, if we assume, from Eq. (12) and (13) it follows that $f\left(\vec{x}^1\right) = 1 - \theta \neq f\left(\vec{x}^2\right) = \theta$.

Thus Z separates points of U.

Property 3

The function $f \in Z$:

$$\forall \vec{x} \in U\ f(\vec{x}) = \frac{\sum_{j=1}^N f_j e^{-\alpha^2_j - \beta \sum_{l \leq m} \gamma^j_{lm}(x_l - x'_l)(x_m - x'_m)}}{\sum_{j=1}^N e^{-\alpha^2_j - \beta \sum_{l \leq m} \gamma^j_{lm}(x_l - x'_l)(x_m - x'_m)}}\ and\ f_j \geq 0\ \forall j \qquad (14)$$

vanishes at no point of U.

References

1. F. Casalino, F. Masulli, R. Caviglia, and L. Papa. Atmospheric pressure wave forecasting through fuzzy systems. In *Proceeding of WIRN '95 -VII Italian Workshop on Neural Nets*, Vietri (Italy), 1995. World Scientific. (in press).
2. J. S. R. Jang and C. T. Sun. Neuro-fuzzy modeling and control. *Proceedings of the IEEE*, 83:378-406, 1995
3. H. M. Kim and J. M. Mendel. Fuzzy basis functions: Comparisons with other basis functions. *IEEE Trans. on Fuzzy Systems*, 3:158-168, 1995
4. B. Kosko. Fuzzy systems as a universal approximators. *IEEE Transactions on Computers*,43:139-33, 1994
5. M. Marinaro and D. Oricchio. Function approximation by a neuro-fuzzy method. In A. Bonarini, D. Mancini, F. Masulli and A. Petrosino, editors, *New Trends in Fuzzy Logic*, Singapore, 1996. World Scientific. (in press)
6. F. Masulli and F. Casalino. A neuro-fuzzy system for Bayesian classification. In *Proceedings of EUFIT '95 Third European Congress on Intelligent Techniques and Soft Computing*, pages 1425-1429, Aachen (Germany), 1995. Verlag und Druck Mainz
7. J. M. Mendel. Fuzzy logic systems for engineering: A tutorial. *Proceedings of the IEEE*, 83:345-377, 1995
8. W. Rudin. *Principles of Matematical Analysis*. McGraw-Hill, Inc, New York, 1960.
9. M. Sugeno and G. T. Kang. Structure identification of fuzzy model. *Fuzzy Sets and Systems*, 28:15-33, 1988
10. T. Takagi and M. Sugeno. Fuzzy identification of systems and its applications to modeling and control. *IEEE Trans. on Systems, Man, and Cybernetics*, 15:116-132, 1985
11. L. Wang and J. M. Mendel. Fuzzy basis functions, universal approximation, and orthogonal least-squares learning. *IEEE Trans. on Neural Networks*, 5:807-14, 1992
12. L. Wang and J. M. Mendel. Generating fuzzy rules by learning from examples. *IEEE Trans. On Systems, Man, and Cybernetics*, 22:1414-1427, 1992
13. Li-Xin Wang. *Adaptive fuzzy systems and control: Design and stability analysis*. Prentice Hall.

The Stone-Weierstrass Theorem [8]

Let Z be a set of real continuous functions on a compact set U. If:

1. Z is an algebra, that is, the set Z is closed under addition, multiplication and scalar multiplication;
2. Z separates points on U, that is, for every $\vec{x}, \vec{y} \in U$, there exist $f \in Z$ such that $f(\vec{x}) \neq f(\vec{y})$;
3. Z vanishes at no point of U, that is, for each $\vec{x} \in U$ there exist $f \in Z$ such that $f(\vec{x}) \neq 0$;

Then the uniform closure of Z consists of all real function on U.

The Computational Neural Map and Its Capacity

Francesco Palmieri

Dipartimento di Ingegneria Elettronica

Università degli Studi di Napoli "Federico II"

Napoli, Italy

Davide Mattera

Dipartimento di Ingegneria Elettronica

Università degli Studi di Napoli "Federico II"

Napoli, Italy

Abstract

We see a neural network as a device that transforms its sources (they can be anywhere in the net) into the new representation given by the collection of the node activations (feature map). The issue of the network capacity and generalization is addressed in reference to the capability of a single neuron (linear or affine with a sigmoidal function) to tap into such a representation. The binary capacity is generalized to the continuous-value case also in comparison to the generalization abilities of the system.

1   Introduction

The concept of a "feature map," proposed in various forms in the literature [3], and in fact embedded in most networks used in the applications, corresponds to the basic idea of seeing the network node activations as providing an "unfolded" representation of the input space (embedding space). The natural question on which criteria should be used to build such a representation can only be answered in reference to how the map is going to be used. Typically, the representation can be tapped into for selecting features for pattern recognition, or more generally, for multi-dimensional approximation. If we assume that the tapping device is linear (or affine), except for the sigmoidal function that can easily taken into account, we are again using the linear perceptron as the basic computation unit. A classical multi-layer perceptron, for example, can be seen as a device that prepares a feature map for the perceptrons of its last layer. Note that a framework that breaks down a system into two parts is also common to most transform methods, where combinations of the transform values are used for analysis or mapping. This paradigm has also been referred to as "support vector machine" [5]. In this context, considering a generic network topology, we can think of each neuron as a linear tap from a feature map constituted by all the node activations that feed it. Simultaneously each neuron provides a component to the feature map of the other neurons. In this scenario is clear that the key role is played by: (a) the criterion on which the

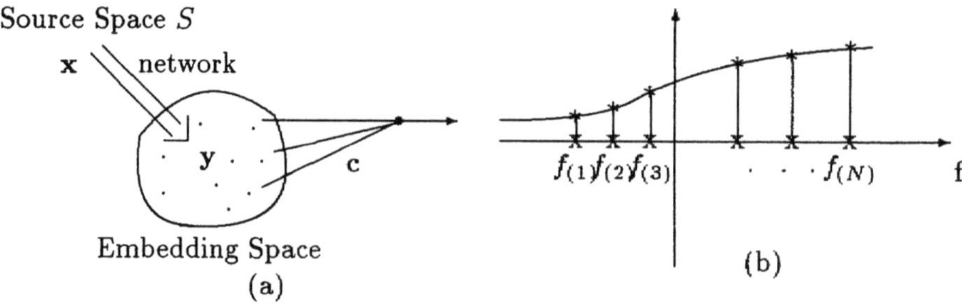

Figure 1: (a) The computational map; (b) N data points ordered on a real line.

feature map is built; (b) the capacity of a single neuron (perceptron). We have addressed the first issue with an "information-preserving" criterion, i.e. the transformation provided by the network, guarantees the best one-to-one correspondence on each set of activations with the set of current source values. This criterion [4] will not be discussed here, but we will concentrate on the question of capacity and generalization of the single perceptron as embedded in the network architecture. Let us refer to the schematic diagram of Figure 1(a). The sources $\mathbf{x} \in S \subseteq R^d$ are mapped by the network into a new space (*embedding space* or *computational map*) $\mathbf{y} \in Y \subseteq R^n$ made up of all the node activations. The nature of the mapping from $\mathbf{x}$ to $\mathbf{y}$ provided by the network is in general nonlinear, while the mechanism to access it is to be linear, or more precisely is that of a single perceptron. In other words, the computational map has to prepare the node activations to form a "desired output" through a simple linear (or affine) transformation. The set Y of node activations could be provided by the most diverse connectivity patterns and algorithms.

The binary capacity, which is based on a dichotomy counting argument, is generalized in this paper to an "ordering capacity" obtained by counting the possible rankings induced by a function. In fact, except for a one-dimensional monotonic function, we maintain that the essential capacity of a multidimensional function is related to its capability of rearranging the input points on the real line. Cover [2], after his famous paper on dichotomies for a binary perceptron [1], derived also a formula for computing the number of orderings of a linear map. We build on his findings to give the reader a comprehensive picture of the issue and propose an extension of the analysis of the generalization abilities of the system for the continuous-value case.

2 The Binary Capacity of the Map

The binary capacity of the map has to be related to its capability of providing the substrate for linear binary classification. More specifically, given a collection of N input examples $\{\mathbf{x}^{(1)}, \mathbf{x}^{(2)}, \ldots, \mathbf{x}^{(N)}\}$ (the sources can be anywhere in the network) the system gives the N nodal values $\{\mathbf{y}^{(1)}, \mathbf{y}^{(2)}, \ldots, \mathbf{y}^{(N)}\}$ (Even though to fix the ideas we think of network with a purely feed forward architecture, the same ideas could be applied to networks with feedbacks provided that we refer to their steady-state values).

Definition: The *unbiased binary capacity of the map* is the fraction C_b of the 2^N possible dichotomies of N points $\{\mathbf{y}^{(1)}, \mathbf{y}^{(2)}, \ldots, \mathbf{y}^{(N)}\}$ in the n-dimensional Euclidean space that is obtainable with a linear operator.

If we allow the transformation from $\mathbf{y}$ to be affine (with the inclusion of a bias term in the tapping filter) we can give a similar definition (*biased binary capacity*).

The *unbiased binary capacity* of the map can be computed [1] to be:

$$C_b(N, n) = (\frac{1}{2})^{N-1} \sum_{k=0}^{n-1} \binom{N-1}{k}, \tag{1}$$

provided that every subset of n, or fewer vectors, is linearly independent. This is usually referred to as a set of points in general position. C_b is simply obtained as the ratio of all possible dichotomies induced by the linear combination and 2^N, which is the total number of possible dichotomies. $C_b(N, n)$ satisfies also the following recursion:

$$C_b(N, n) = \frac{1}{2} C_b(N-1, n) + \frac{1}{2} C_b(N-1, n-1), \tag{2}$$

with the boundary conditions

$$C_b(1, n) = 1 \ \forall n \geq 1; \ \ C_b(N, 0) = 0 \ \forall N \geq 1 \ .$$

The *biased binary capacity* of the map is $C_b(N, n+1) = C_b(N, n) + (\frac{1}{2})^{N-1}\binom{N-1}{n}$, provided that no k-flat (with $k \leq n - 1$) contains $k + 2$ points of the set $\{\mathbf{y}^{(1)}, \mathbf{y}^{(2)}, \ldots, \mathbf{y}^{(N)}\}$ in R^n, that is, there are no three points in a line, four points in a plane, etc.

The conditions for the application of the binary capacity expression are the key to understand the role of the neural network. In practice the points in Y have to be sufficiently scattered and not belong to degenerate subspaces, to be classifiable. This is an important aspect of the computational map where the sigmoidal functions embedded in the network play a key role in removing possible linear dependencies among the data points.

3 The Ordering Capacity of the Map

The question that arises at this point is: is it possible to generalize the concept of capacity to a nonbinary case? In other words, if the purpose of the map is to provide the substrate for multidimensional approximation, how do we generalize the dichotomy counting argument?

The approach we take in this paper is based on the following observations. Consider a multidimensional function $f(\mathbf{x})$ applied to a number of points $\{\mathbf{x}^{(1)}, \mathbf{x}^{(2)}, \ldots, \mathbf{x}^{(N)}\}$. The corresponding values are $\{f(\mathbf{x}^{(1)}), f(\mathbf{x}^{(2)}), \ldots, f(\mathbf{x}^{(N)})\}$. Figure 1(b) shows an example of corresponding N values ordered on the real line. If we desire to associate to those points a set of "desired values" $\{d^{(1)}, d^{(2)}, \ldots, d^{(N)}\}$, we can always find a strictly increasing function ϕ such that $\phi(f(\mathbf{x}^{(i)})) = d^{(i)}$ for every $i \in \{1, \ldots, N\}$. This suggests that the inherent multidimensional operation of f is related to its capability of rearranging the points on the real line. In other words, *two nonlinear functions*

f_1 and f_2 can be considered equivalent, with respect to a set of points, if they induce the same ordering. This is because it is always possible to find a strictly increasing mapping to equate f_1 and f_2 on these points. We can adopt the following definition: *the number of functions realizable by a class f is equal to the number of orderings induced by them.* Given N points $\{\mathbf{y}^{(1)}, \mathbf{y}^{(2)}, \ldots, \mathbf{y}^{(N)}\}$ in an n-dimensional space, *the unbiased ordering capacity $C_o(N, n)$* of a linear combiner is defined as the fraction between the number of possible orderings induced by a linear combiner and the number $N!$ of possible permutations of the N points. This observation was originally made by Cover [2] that showed that $C_o(N, n)$ satisfies the the following recursion:

$$C_o(N, n) = \frac{1}{N} C_o(N - 1, n) + \frac{N - 1}{N} C_o(N - 1, n - 1), \tag{3}$$

with the boundary conditions $C_o(2, n) = 1 \ \forall n \geq 1; C_o(N, 1) = \frac{2}{N!} \ \forall N \geq 2$. A closed-form solution is also computable. Figure 2 shows a comparison of C_0 and C_b. Note how we have full ordering capacity for $N \leq n$ and that the behavior is very sharp when $N \geq n$. The result with the ordering capacity confirms that the capability of a linear combiner of producing arbitrary mappings cannot go much beyond its number of degrees of freedom n, which is equal to the number of coefficients available for learning. This must be seen in contrast to the binary capacity that extends almost to $\frac{N}{n} \simeq 2$. This is expected since to implement arbitrary mappings must be much harder than provide dichotomies. The *biased ordering capacity* is easily computed analogously.

4 Generalization for Binary Functions

In designing a neural system, we have to face the problem of how well the system generalizes on inputs that were not included in the original data set. Capacity and generalization are two conflicting requirements. The role of the designer is to provide a good compromise between these two opposite needs. Consider the N output vectors of the computational map $\{\mathbf{y}^{(1)}, \mathbf{y}^{(2)}, \ldots, \mathbf{y}^{(N)}\}$ with a fixed map structure corresponding to the source vectors $\{\mathbf{x}^{(1)}, \mathbf{x}^{(2)}, \ldots, \mathbf{x}^{(N)}\}$. We know that the final unbiased perceptron can realize $2^N C_b(N, n)$ dichotomies of the output vectors set. This means that we can think of the space of the unbiased perceptron weights as divided into $2^N C_b(N, n)$ sets H_i ($i = 1, \ldots, 2^N C_b(N, n)$). Consider now a new source vector $\mathbf{x}^{(N+1)}$ and its relative output vector $\mathbf{y}^{(N+1)}$. If $sign(\mathbf{y}^{(N+1)T} \mathbf{w})$ is constant for every $w \in H_i$, we say that the i-th dichotomy is not ambiguous respect to $\mathbf{y}^{(N+1)}$; otherwise, we say that the i-th dichotomy is ambiguous [1]. The unbiased binary ambiguity A_b is the fraction of the $2^N C_b(N, n)$ homogeneously linearly separable dichotomies that are ambiguous. The *unbiased binary ambiguity* can be computed as $A_b(N, n) = \dfrac{C_b(N, n - 1)}{C_b(N, n)}$ provided that the $N + 1$ points satisfy the conditions for which (1) holds. The result provides a very compact way of considering the issue of generalization. The behavior of the generalization capacity, $1 - A_b(N, n)$, is clearly complementary to that of the network capacity as shown in Figure 2 for $n = 60$: increasing the value of $\frac{N}{n}$, the ambiguity of

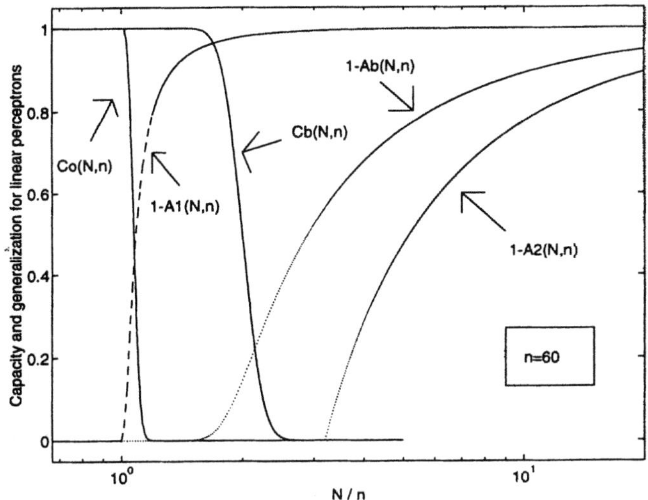

Figure 2: The binary capacity $C_b(N, n)$, the ordering capacity $C_o(N, n)$, the generalization abilities for the binary case expressed as $1 - A_b(N, n)$ and the upper and lower bound on the generalization abilities in the case of orderings expressed, respectively as $1 - A_1(N, n)$ and $1 - A_2(N, n)$, versus the ratio $\frac{N}{n}$, for the case $n = 60$.

the network is reduced together with its capacity. The biased ambiguity can be obtained analogously.

5 Generalization for Orderings

The unbiased ordering ambiguity A_o, as the fraction of the homogeneously linearly inducible orderings that are ambiguous can be considered in reference to the continuous mapping problem. An ordering is ambiguous with respect to a new vector if it is possible to change the ranking of the new vector without changing the ordering of the first vectors. A closed-form formula appears hard to compute without specific assumptions on the space Y. However, the following theorem provides an upper and lower bound on the behaviour of the unbiased ordering ambiguity.

Theorem: The unbiased ordering ambiguity $A_o(N, n)$ can be bounded as follows:

$$A_1(N, n) = \frac{C_o(N, n - 1)}{C_o(N, n)} \leq A_o(N, n) \leq \frac{N C_o(N, n - 1)}{C_o(N, n)} = A_2(N, n). \quad (4)$$

Proof: Consider the N output vectors of the computational map $\{\mathbf{y}^{(1)}, \mathbf{y}^{(2)}, \ldots, \mathbf{y}^{(N)}\}$ with a fixed map structure corresponding to the source vectors $\{\mathbf{x}^{(1)}, \mathbf{x}^{(2)}, \ldots, \mathbf{x}^{(N)}\}$. Consider now a new source vector $\mathbf{x}^{(N+1)}$ and its relative output vector $\mathbf{y}^{(N+1)}$. If $(\mathbf{y}^{(N+1)} - \mathbf{y}^{(j)})\mathbf{w} \neq 0$ for all j and for all $w \in H_i$ that create an certain ordering, we cannot change the ranking of the new vector without changing this certain ordering of the first N vectors.

The number of orderings (i.e. the number of sets H_i) that is ambiguous because $(\mathbf{y}^{(N+1)} - \mathbf{y}^{(j)})\mathbf{w} = 0$ is equal, accordingly to Cover [2], to $N! \cdot C_o(N, n-1)$. Because a value w and so an ordering cannot be ambiguous for two different value of j and because $j \in \{1, \ldots, N\}$, the number of ambiguous orderings cannot be larger than $N \cdot N! \cdot C_o(N, n-1)$ and lesser than $N! \cdot C_o(N, n-1)$. Dividing the number of ambiguous orderings by the number $N! \cdot C_o(N, n)$ of linearly inducible orderings, we obtain the bound of Eq. 4.

Figure 2 shows the bounds for the complement to unbiased ordering ambiguity $(1 - A_o(N, n))$; we see that the lower bound for A_o (that correspond to the upper bound on the complement) is not very useful because it saturates very quickly. Instead, the other bound shows how, by increasing the value of $\frac{N}{n}$ the ordering ambiguity of the network is reduced and can be used as a design criterion. Note that the ambiguity is a parameter that characterizes the neural network structure and does not depend upon the target ordering and the probability distribution of the source vectors. We can bound similarly the biased ambiguity.

References

[1] T.M. Cover, Geometrical and Statistical Properties of Systems of Linear Inequalities with Applications in Pattern Recognition, *IEEE Trans. Elect. Comp.*, Vol.14, pp. 326-334, 1964.

[2] T.M. Cover, The Number of Linearly Inducible Orderings, *SIAM Journal of Applied Math.*, Vol. 15, N. 2, March 1967.

[3] T. Kohonen, *Self-Organization and Associative Memory*, 2nd ed., Springer Verlag, 1988.

[4] F. Palmieri, Linear Self-Association for Universal Memory and Approximation, *Proc. of World Congress on Neural Networks*, Portland, OR, Vol. 2, pp. 339-343, 1993.

[5] V. Vapnik, *The Nature of Statistical Learning Theory*, 2nd ed., Springer, 1995.

Classification of Structures by Recurrent Neural Networks

Alessandro Sperduti, Antonina Starita

Dipartimento di Informatica, Università di Pisa

Corso Italia 40 - 56125 - Pisa, Italy

Abstract

We introduce neural networks for the classification of structures. LRAAM-based networks, Cascade-Correlation Networks, and Neural Trees for structures are presented. We show how to compute the gradient for these networks. Moreover, we report some results on the computational power of these models.

1 Introduction

The ability to recognize and to classify structured patterns is fundamental for several applications, such as medical and technical diagnosis (discovery and manipulation of structured dependencies, constraints, explanations), molecular biology and chemistry (classification of chemical structures, DNA analysis, quantitative structure-property relationship (QSPR), quantitative structure-activity relationship (QSAR)), automated reasoning (robust matching, manipulation of logical terms, proof plans, search space reduction). Up to now, neural networks have been only used for classification of unstructured patterns and sequences, since they are usually believed to be inadequate for structured patterns because of their feature-based approach. On the contrary, recent works [6, 7, 8, 11, 9, 10] have shown that neural networks can represent and also classify structured patterns.

In this paper we show that using the so called "generalized recursive neuron" basically all the supervised networks developed for the classification of sequences can be generalized to structures. Specifically, we discuss Simple Recurrent Networks, (obtained as special cases of LRAAM-based networks), Recurrent Cascade Correlation networks, and Neural Trees. Results on the computational power of these models are also given.

2 Preliminaries

We consider finite directed node labeled graphs without multiple edges. For a set of labels Σ, a *graph* X (over Σ) is specified by a finite set V_X of *nodes*, a set E_X of ordered couples of $V_X \times V_X$ (called the set of *edges*) and a function ϕ_X from V_X to Σ (called the *labeling function*). For a finite set V, $\#V$ denotes its cardinality. Given a graph X and any node $x \in V_X$, the function $out\_degree_X(x)$ returns the number of edges leaving from x, i.e., $out\_degree_X(x) = \#\{(x,z) \mid (x,z) \in E_X \ \wedge \ z \in V_X\}$. Given a total order on the edges leaving from x, the node $y = out_X(x,j)$ in V_X is the node pointed by the *jth* pointer leaving from x. The *valence* of a graph X is defined as $\max_{x \in V_X} \{out\_degree_X(x)\}$. A *labeled directed acyclic graph (labeled DAG)* is a graph, as defined above, without loops. A node $s \in V_X$ is called a *supersource* for X if every node in X can be reached by a path starting from s. The root of a tree (which is a special case of directed graph) is always the (unique) *supersource* of the tree. In the following, we deal with DAGs.

3 The First Order Generalized Recursive Neuron

The neural networks recently proposed for the processing of structures are based on *generalized recursive neurons*. A generalized recursive neuron is an extension of the recurrent neuron where instead of just reusing the output of the unit on the previous time step, the outputs of the unit for all the nodes which are pointed by the current input node are considered. Then the output $o^{(g)}(x)$ of the generalized recursive neuron to a node x of a graph X is defined as

$$o^{(g)}(x) = f(\sum_{i=1}^{N_L} w_i l_i + \sum_{j=1}^{out\_degree_X(x)} \hat{w}_j o^{(g)}(out_X(x,j))), \tag{1}$$

where N_L is the number of units encoding the label $l = \phi_X(x)$ attached to the current input x, and $\hat{w}_j$ are the weights on the recursive connections. Note that, if the valence of the considered domain is n, then the generalized recursive neuron will have n recursive connections, even if not all of them will be used for computing the output of a node x with $out\_degree_X(x) < n$.

When considering k interconnected generalized recursive neurons, eq. (1) becomes

$$o^{(g)}(x) = F(W l_x + \sum_{j=1}^{out\_degree_X(x)} \widehat{W}_j o^{(g)}(out_X(x,j))) \tag{2}$$

where $F_i(v) = f(v_i)$, $l_x \in \Re^{N_L}$, $W \in \Re^{k \times N_L}$, $o^{(g)}(x) \in \Re^k$, $o^{(g)}(out_X(x,j)) \in \Re^k$, $\widehat{W}_j \in \Re^{k \times k}$. In the following, we will refer to the output of a generalized neuron dropping the upper index.

To understand how generalized recursive neurons can generate representations for DAGs, let us consider a single generalized recursive neuron u and a single DAG X. The following conditions must hold: *i)* the generalized recursive neuron u must have as many recursive connections as the valence of the graph X; *ii)* the graph X must have a reference *supersource*. Note that, if the graph X does not have a supersource, then it is always possible to define a convention for adding to the graph X an extra node s (with a minimal number of outgoing edges) such that s is a supersource for the new graph.

If the above conditions are satisfied, we can adopt the convention that the graph X is represented by $o(s)$, i.e., the output of u to s. Consequently, due to the recursive nature of eq. (1), it follows that the neural representation for a DAG is computed by a feedforward network (*encoding network*) obtained by replicating the same generalized recursive neuron u and connecting these

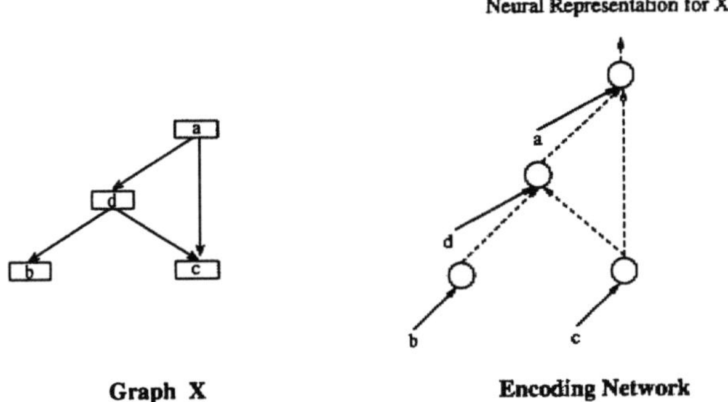

Figure 1: A labeled graph X and the associated encoding network.

copies according to the topology of the structure (see Figure 1). The encoding network fully describes how the representation for the structure is computed.

4 Neural Networks for Structures

Recently, several neural networks for the encoding of structures have been proposed. In this paper, we present some supervised neural networks for the classification of structures. Specifically, we discuss LRAAM-based networks [11, 10], Cascade-Correlation networks for structures [9, 10], and Neural Trees for structures [10].

An **LRAAM-based network** is constituted by two parts: the first part of the network is an LRAAM whose task is to devise a compressed representation for each structure. This compressed representation is usually obtained by using the standard learning algorithm for LRAAM [7]. The output of the network is then computed by the second part of the network through a multi-layer feedforward network with none or several hidden layers. In [11], it was noted that this kind of network constitutes a generalization of a Simple Recurrent Network (SRN) [1] to structures. This correspondence is obtained by removing the set of connections from the hidden to the input layer (which does not contribute

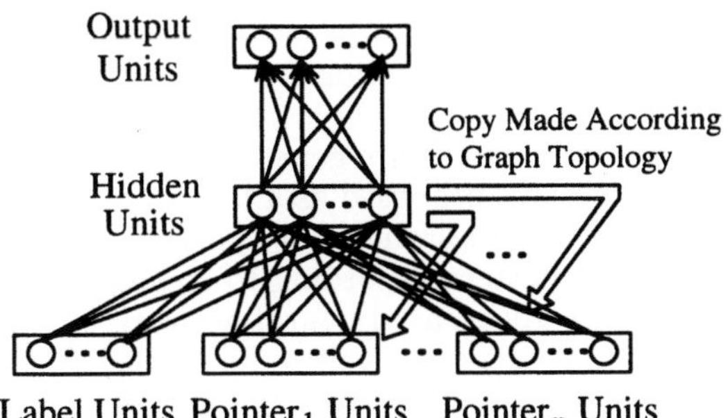

Figure 2: The LRAAM-based network for classification of structures.

to the computation of the output, while is used for learning purposes). In this paper, we reduces to this model, as shown in Figure 2. Essentially, the network has an hidden layer of generalized units and an output layer of standard units. In Figure 2, the pointer fields are represented explicitly. The label field is subdivided in two parts: one part is used to represent the label, while the second one encodes the *pointer condition bits*, i.e., there is a bit for each pointer field, and whenever the pointer field is void (all pointer units are set to 0), the corresponding bit is set to 1. Otherwise the corresponding bit is set to 0.

The **Cascade-Correlation Networks** are devised using the Cascade-Correlation algorithm. The Cascade-Correlation algorithm [3] grows a standard neural network using an incremental approach for classification of unstructured patterns. The starting network $\mathcal{N}_0$ is a network without hidden nodes trained with a Least Mean Square algorithm; if network $\mathcal{N}_0$ is not able to solve the problem, a hidden unit u_1 is added such that the *correlation* between the output of the unit and the residual error of network $\mathcal{N}_0$ is maximised[1]. The weights of u_1 are frozen and the remaining weights are retrained. If the obtained network $\mathcal{N}_1$ cannot solve the problem, the network is further grown, adding new hidden units which are connected (with frozen weights) with all the inputs and previously installed hidden units. The resulting network is a *cascade* of nodes. Fahlman extended the algorithm to classification of sequences, obtaining good results [2]. Cascade-Correlation can further be extended to structures by using generalized recursive neurons. The output of the kth hidden unit can be computed as

$$o^{(k)}(x) = f(u_k), \tag{3}$$

$$u_k = \sum_{i=1}^{N_L} w_i^{(k)} l_i + \sum_{v=1}^{k} \sum_{j=1}^{out\_degree_X(x)} \hat{w}_{(v,j)}^{(k)} o^{(v)}(out_X(x,j)) + \sum_{q=1}^{k-1} \bar{w}_q^{(k)} o^{(q)}(x)), \tag{4}$$

where $w_{(v,j)}^{(k)}$ is the weight of the kth hidden unit associated to the output of the vth hidden unit computed on the jth component pointed by x, and $\bar{w}_q^{(k)}$ is the weight of the connection from the qth (frozen) hidden unit, $q < k$, and the kth hidden unit. The output of the output neuron $u^{(out)}$ is computed as

$$o^{(out)}(x) = f(\sum_{i=1}^{k} \tilde{w}_i o^{(i)}(x)), \tag{5}$$

where $\tilde{w}_i$ is the weight on the connection from the ith (frozen) hidden unit to the output unit.

Neural Trees for Structures are generalizations of Neural Trees. Neural Trees (NT) have been recently proposed as a fast learning method in classification tasks. They are decision trees where the splitting of the data for each node, i.e., the classification of a pattern according to some features, is performed by a perceptron [5] or a more generalized neural network [4]. After learning, each node at every level of the tree corresponds to an exclusive subset of the training data and the leaf nodes of the tree completely partition the training set. In the

---

[1]Since the maximization of the correlation is obtained using a gradient ascent technique on a surface with several maxima, a pool of hidden units is trained and the best one selected.

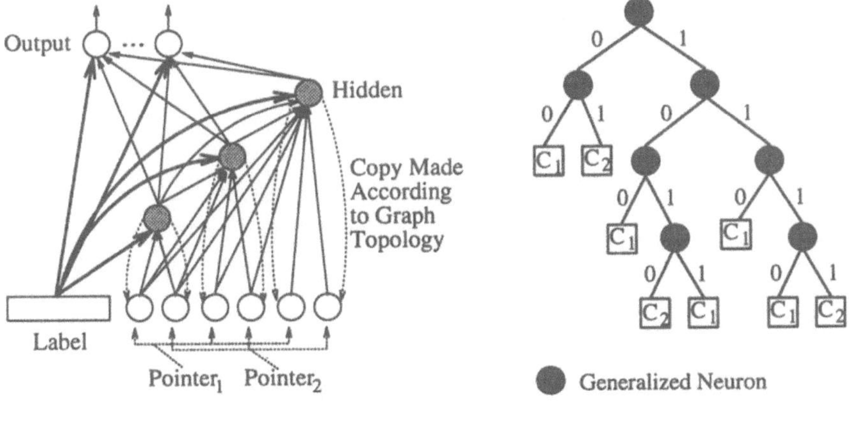

Figure 3: Cascade-Correlation Network and Neural Tree for classification of structures.

operative mode, the internal nodes route the input pattern to the appropriate leaf node which represents the class of it. The extension of these algorithms to structures is straightforward: the standard discriminator associated to each node of the tree is replaced by a generalized recursive discriminator (with step function). Thus, instead of evaluating an unstructured pattern, each node of the tree evaluates a structure (see Figure 3).

5 Computing the Gradient

In this section we show how to compute the gradient for a set of generalized neurons. Using this derivation, it is not difficult to reconstruct the learning algorithms for the models described above.

Let N_g be the number of generalized neurons, s the *supersource* of X, and

$$y = o(s) = F(Wl_s + \sum_{j=1}^{out\_degree_X(s)} \widehat{W}_j o(out_X(s,j))) \quad (6)$$

be the representation of X according to the encoding network, where n is the valence of the domain. The derivatives of y with respect to W and $\widehat{W}_i$ ($i \in [1, \ldots, n]$) can be computed from eq. (6):

$$\frac{\partial y_t}{\partial W_{tk}} = o'_t(s)((l_s)_k + \sum_{j=1}^{out\_degree_X(s)} (\widehat{W}_j)_t \frac{\partial o(out_X(s,j))}{\partial W_{tk}}), \quad (7)$$

where $t = 1, \ldots, N_g$, $k = 0, \ldots, N_L$ and $(\widehat{W}_j)_t$ is the tth row of $\widehat{W}_j$;

$$\frac{\partial y_t}{\partial (\widehat{W}_i)_{tq}} = o'_t(s)(o_q(out_X(s,i)) + \sum_{j=1}^{out\_degree_X(s)} (\widehat{W}_j)_t \frac{\partial o(out_X(s,j))}{\partial (\widehat{W}_i)_{tq}}), \quad (8)$$

where $t = 1, \ldots, N_g$, and $q = 1, \ldots, N_g$.

These equations are recursive on the structure X and can be computed by noting that if v is such that $out\_degree_X(v) = 0$, then

$$\frac{\partial o_t(v)}{\partial W_{tk}} = (l_v)_k, \text{ and } \frac{\partial o_t(v)}{\partial (\widehat{\boldsymbol{W}}_i)_{tq}} = 0. \tag{9}$$

This allows the computation of the derivatives alongside with the computation of the neural representations for the graphs.

6 Computational Power

The computational power of recurrent networks is usually characterized by resorting to *finite state automata (fsa)*. A recurrent network is considered to be powerful if it can simulate any fsa. In our case, instead, we refer to *tree automata* [12], i.e., generalizations of fsa to trees.

Concerning the computational power of the above models for the classification of structures, the following theorems can be proved:

Theorem 6.1 *An LRAAM-based network can simulate any Tree Automaton.*

Theorem 6.2 *A Cascade-Correlation network for structures cannot simulate any Tree Automaton.*

Theorem 6.3 *A Neural Tree for structures cannot simulate any Tree Automaton.*

Theorem 6.4 *Any Neural Tree for structures can be simulated by a Cascade-Correlation network for structures.*

Note that from the previous two theorems, and from the fact that Cascade-Correlation networks and Neural Trees for structures are generalizations of Recurrent Cascade-Correlation networks and Neural Trees for sequences, respectively, we have

Corollary 1 *A Neural Tree for sequences cannot simulate any Finite State Machine.*

Corollary 2 *Any Neural Tree for sequences can be simulated by a Recurrent Cascade-Correlation network.*

7 Conclusion

We have proposed a generalization of the standard neuron, namely the generalized recursive neuron, for extending the computational capability of neural networks to processing of structures. On the basis of the generalized recursive neuron, we have shown how several of the learning algorithms defined for standard neurons, can be adapted to deal with structures. The ability of such networks to perform classification tasks involving logic terms was experimentally demonstrated in [10]. Moreover, we have reported results on the computational power of the proposed models. Specifically, LRAAM-based networks can implement any Tree Automaton, while neither Cascade-Correlation networks, nor Neural Trees can. As a special case of the latter result, it is obtained that Neural Trees for sequences cannot implement any Finite State Machine. These limitations of Cascade-Correlation networks, and Neural Trees for structures should be considered when designing solutions for specific applications.

References

[1] J. L. Elman. Finding structure in time. *Cognitive Science*, 14:179–211, 1990.

[2] S. E. Fahlman. The recurrent cascade-correlation architecture. Technical Report CMU-CS-91-100, Carnegie Mellon, 1991.

[3] S. E. Fahlman and C. Lebiere. The cascade-correlation learning architecture. In D. S. Touretzky, editor, *Advances in Neural Information Processing Systems 2*, pages 524–532. San Mateo, CA: Morgan Kaufmann, 1990.

[4] A. Sankar and R. Mammone. *Neural Tree Networks*, pages 281–302. Neural Networks: Theory and Applications. Academic Press, 1991.

[5] J. A. Sirat and J-P. Nadal. Neural trees: a new tool for classification. *Network*, 1:423–438, 1990.

[6] A. Sperduti. Encoding of Labeled Graphs by Labeling RAAM. In J. D. Cowan, G. Tesauro, and J. Alspector, editors, *Advances in Neural Information Processing Systems 6*, pages 1125–1132. San Mateo, CA: Morgan Kaufmann, 1994.

[7] A. Sperduti. Labeling RAAM. *Connection Science*, 6(4):429–459, 1994.

[8] A. Sperduti. Stability properties of labeling recursive auto-associative memory. *IEEE Transactions on Neural Networks*, 6(6):1452–1460, 1995.

[9] A. Sperduti, D. Majidi, and A. Starita. Extended cascade-correlation for syntactic and structural pattern recognition. In *International Workshop on Syntactic and Structural Pattern Recognition*, 1996. To appear.

[10] A. Sperduti and A. Starita. Supervised neural networks for the classification of structures. Technical Report TR-16/95, Dipartimento di Informatica, Università di Pisa, 1995. Submitted to IEEE Transactions on Neural Networks.

[11] A. Sperduti, A. Starita, and C. Goller. Learning distributed representations for the classification of terms. In *Proceedings of the International Joint Conference on Artificial Intelligence*, pages 509–515, 1995.

[12] J. W. Thatcher. Tree automata: An informal survey. In A. V. Aho, editor, *Currents in the Theory of Computing*. Prentice-Hall, Englewood Cliffs, NJ, 1973.

SECTION 5

PATTERN RECOGNITION
AND ROBOTICS

Neuro-Fuzzy Processing of Remote Sensed Data

P. Blonda, A. Bennardo, G. Satalino
Istituto Elaborazione Segnali ed Immagini I.E.S.I.-C.N.R.
Via Amendola, 166/5 - 70126 - Bari - ITALY
Fax: + 39 80 5484311, Phone: +39 80 5421612
E-Mail: blonda@iesi.ba.cnr.it

Abstract

In this work the effectiveness of the Fuzzy Kohonen Clustering
Network (FKCN) has been explored in two classification experiments of remote sensed data. The FKCN has been introduced
in a multi-modular neural classification system for feature extraction before labelling. The unsupervised module is connected in
cascade with the next supervised module based on the backpropagation learning rule. The performance of the FKCN has been
evaluated in comparison with those of a conventional Kohonen
Self Organizing Map (SOM) neural network. Experimental results
have proved that the fuzzy clustering network can be used for complex data pre-processing, but further investigation on FKCN stop
criterion is required.

Keywords: neuro-fuzzy, Kohonen networks, remote sensed data,
modular neural system.

1   Introduction

In the last few years, great interest has been shown in the classification of remote sensed data using both neural networks (NN) and fuzzy
logic (FL) technologies [1]   [2]   [3] . More recent Artificial Intelligence
research has explored the merging of fuzzy logic and NN concepts and
methodologies in order not only to benefit from the advantages of the
two technologies but also to overcome possible disadvantages. Some authors have focused upon methods for the fuzzification of conventional
NN architectures. The Fuzzy Kohonen Clustering Network (FKCN)
was proposed by Bezdek [4] for merging the fuzzy c-Means model with
the Kohonen clustering network [5]. It was tested by the authors on

the well known Anderson's Iris data set. FKCN uses fuzzy membership values extracted directly from the training data as learning rates, it also controls automatically the learning rate distribution by means of the fuzziness parameter m, which represents the weighting exponent of each membership function. In addition, the Kohonen concept of neighborhood size and updating are embedded in the learning rate strategy. FKCN is non-sequential, therefore it is independent from the sequence of feeding input data, moreover it terminates when successive estimates of the network weights become close, independently of the learning iteration limit. In our work the FKCN has been applied to analyze remote sensed data.

In a previous paper [6] , we demonstrated that the learning task of the supervised Multi Layer Perceptron (MLP) classifier could be made more efficient by preprocessing the input with unsupervised neural modules for feature discovery. The results confirmed that modular learning can perform better than nonmodular learning with respect to learning speed without loss in classification accuracy. Input data with low (statistical) separability were preprocessed by the sequential Kohonen Self Organizing Map (SOM) network to reduce computation time. The SOM network has many advantages, but some input parameters such as the learning rate, the size of update neighborhood and the strategy to alter these two parameters during learning must be defined by the user. These parameters and the sequence of feeding input data can influence SOM performance when a low number of iterations is used for network training. The FKCN module seems not to be hampered by such limitations. Therefore it has been chosen to carry out feature extraction before classification of two data sets, one characterized by low, the other by high statistical separability, in order to optimize the classification system with respect to both learning speed and parameter selection. The results obtained will serve to evaluate and compare the performance of the two unsupervised networks.

2   The Neural Architectures

The architectures applied in our system involve unsupervised and supervised learning rule. The unsupervised module is the Fuzzy Kohonen Clustering Network (FKCN), which is well described in the paper of Bezdek [4] and synthesized in this section.

The supervised module of the classification system is a feed-forward network based on the Back Propagation (BP) learning rule. The well known BP algorithm is described in more details by Rumelhart et al. [7]. In this work a Single Layer Perceptron (SLP), i.e. a Multilayer Perceptron (MLP) architecture with no hidden neurons, has been used for data relabelling.

The FKCN consists of an input layer with p neurons and an output layer with c neurons.

Let $X = \{x_1, x_2, ..., x_n\} \subset R^p$ be the input vectors, and c be the number of clusters in the competitive layer.

Initialization: *Fix: c, $\epsilon > 0$ some small positive constant and a distance D.*

Step 1) *Initialize network weight vectors $v_0 = (v_{1,0}, v_{2,0}, ..., v_{c,0}) \in R^{cp}$.*

Select both the iteration limit for the algorithm , t_{max}, and some positive constant greater than one, m_0.

Set: $\Delta m = (m_0 - 1)/t_{max}$.

Step 2) *FOR $t = 1, 2, ..., t_{max}$:*

a) $m_t = (m_0 - t\Delta m)$;

b) FOR $i = 1, 2, ..., c$ and $k = 1, 2, ..., n$:

b.1) $D_{ik,t} = \parallel x_k - v_{i,t-1} \parallel^2$;

b.2) IF $D_{ik,t} = 0$

$$THEN \quad u_{ik,t} = 1 \text{ and } \alpha_{ik,t} = 1;$$

$$ELSE \quad u_{ik,t} = \frac{\left(\dfrac{1}{D_{ik,t}}\right)^{1/(m_t-1)}}{\sum_{j=1}^{c} \left(\dfrac{1}{D_{jk,t}}\right)^{1/(m_t-1)}}$$

$$and \quad \alpha_{ik,t} = (u_{ik,t})^{m_t},$$

where $\alpha_{ik,t}$ represents the learning rate.

Step 3) *FOR $i = 1, 2, ..., c$ the weights updating rule is expressed by:*

$$v_{i,t} = v_{i,t-1} + \frac{\left[\sum_{k=1}^{n} \alpha_{ik,t}(x_k - v_{i,t-1})\right]}{\sum_{s=1}^{n} \alpha_{is,t}}.$$

Step 4) *Compute $E_t = \|v_t - v_{t-1}\|^2 = \sum_i \|v_{i,t} - v_{i,t-1}\|^2$.*

Step 5) *IF $E_t \leq \epsilon$ stop; ELSE next t.*

3 Data Description

The data used for training and testing the modular system have been selected from three Landsat thematic mapper (TM) images acquired in April, July, and October 1986, respectively. In the first experiment, only two bands of the July image were used, whereas in the second experiment the multitemporal images were first coregistered and then analyzed. The ground truth of 25 fields, consisting of 7662 pixel, was determined by means of visual interpretation of aerial photographs and by local inspection. The classes are: 1) *Bare soil*; 2) *Urban areas*; 3) *Pasture*; 4) *Coniferous reafforestation*; 5) *Olives groves*; 6) *Vineyards*; 7) *Cropland*. In both experiments the 50% of the labelled pixels were randomly selected for training the networks, the remaining 50% were used for testing. All the data were normalized to have real values between 0.0 and 1.0 .

In this work, each pixel is represented by two vectors: 1) the *N - dimensional* vector representing the radiance values recorded in the N bands of the images; 2) the *7 - dimensional* vector, coded as $\hat{V}_i^\mu = 1$ and $\hat{V}_k^\mu = 0$ ($k \neq i, k = 1 \cdots 7$) for a pattern μ associated with class i.

4 Data Analysis

In the first experiment the focus was on the case of a low dimensional data set, i.e. on statistically *poor* remote sensed data, where the differentiation between similar classes can be invalidated not only by the management of imprecise data, but also by the lack of complementary information, such as multitemporal evolution. Since the original six bands data set of the July image provide a high class separability [8] , only two bands of the July image were considered in order to provide such a data set and to easily monitor the weights distribution in a bidimensional space. The Jeffries-Matusita (JM) distance used to provide a measure of data statistical separability is equal to 1.181.

In the second experiment, the 18-bands multitemporal data set was used as input to the modular classification system. In this case, the data are characterized by a high separability, i.e. the JM value is equal to 1.413.

In the two experiments the FKCN module was used to perform the decomposition of each data set in clusters. For each class, training data have neither equally populous samples, nor comparable variances. For this reason and in order to obtain, through the multimodular classification system, high classification performance, as the ones produced by a single MLP classifier [8], the data were oversegmented before labelling. Therefore, during the classification experiment several FKCN architectures, characterized by a different number of output nodes, were trained. In each simulation, the number of nodes in the input FKCN layer was equal to the one of input spectral bands, whereas the number of nodes in the output layer was equal to the number of demanded clusters. This paper will focus only on the results obtained with 30 nodes in the FKCN output layer. Once the FKCN had been trained, for each input training pattern both the FKCN output vector and the target vector associated to the same pattern were directly fed into the next SLP recognition module. The output layer of the last SLP module consisted of seven nodes.

The stability of the FKCN module has been investigated with respect to both m_0 and Δ_m parameters. The results are expressed in terms of percentage of overall classification accuracy obtained by the recognition module after FKCN preprocessing (FKCN+SLP). For training and testing data, the percentage values are reported in Table 1 a) and Table 1 b) at fixed $m_0 = 2$, variable Δ_m, first and second experiment, respectively. In these tables m_f means the final value of the m parameter, which is decreased during learning.
The results of the reverse study are illustrated by means of the error rate evolution with time, in Figure 1 and Figure 2, for low and high dimensional data, respectively.

In both Figures, the number of iterations is referred to the FKCN training, whereas the number of mistakes is computed, at each FKCN iteration, by one SLP iteration. The results shown in the previous Tables and Figures evidence the FKCN sensitivity to the m_0 parameter, both for low and high dimensional data. In addition, the number of iterations required to satisfy the FKCN stop criterion seems to depend on the initial m_0 value. Changes in the Δ_m parameter influence the FKCN stability in the case of high dimensional data. Moreover, the selection of the ϵ value seems very delicate, since changes in this value produced different performance. Certainly, the results suggest the need of further investigation on the stop criterion.

	$\epsilon = 10^{-6}$			
Δ_m	m_f	t_{out}	Train	Test
0.05	1.05	19	82.28%	82.01%
0.02	1.02	49	84.42%	83.08%
0.001	1.94	56	82.23%	82.16%
0.00001	1.99	63	81.74%	81.43%
a)				

	$\epsilon = 10^{-5}$			
Δ_m	m_f	t_{out}	Train	Test
0.05	1.05	19	96.43%	95.08%
0.02	1.02	49	97.41%	95.48%
0.001	1.93	65	88.48%	87.79%
0.00001	1.99	57	88.66%	87.74%
b)				

Table 1: Classification performance obtained trough the modular (FKCN+SLP) neural classifier. The FKCN was used with $m_0 = 2$, variable Δ_m: a) experiment 1, b) experiment 2.

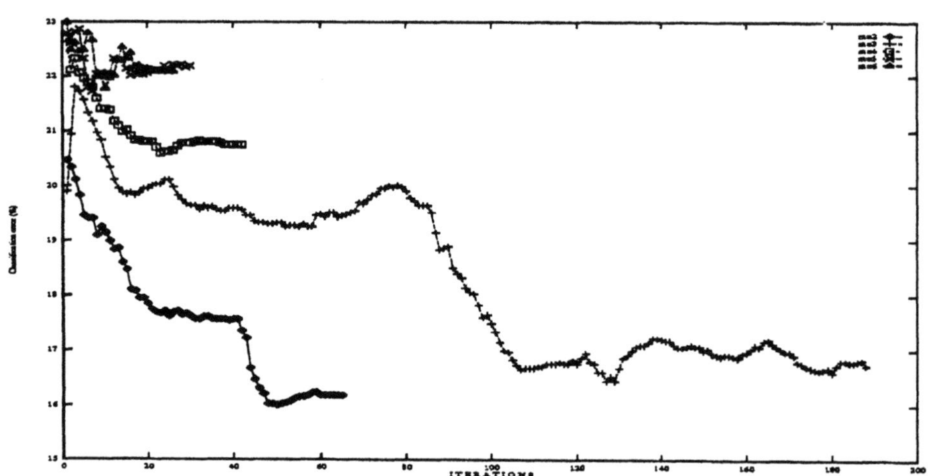

Figure 1: Experiment 1: FKCN error rate as a function of time with $\Delta_m = 0.01$, variable m_0, $\epsilon = 10^{-6}$.

5 Discussion

On the base of the results illustrated in the previous sections about the sensitivity of the FKCN with respect to changes in m_0 and Δ_m, the fuzzy network was then trained for all the T_{max} iterations. First, both the low and high dimensional data sets were preprocessed with $m_0 = 2$, different T_{max} values, i.e. variable Δ_m. As a result, both with the low and high dimensional data, the performance did not significantly change when increasing the T_{max} value, after about the first few (fifteenth-twenty) iterations.

The FKCN was also trained with fixed $\Delta_m = 0.05$ for different m_0 values, for all the T_{max} required iterations. The percentages of the overall accuracy are shown in Table 2 a) and in Table 2 b), for low and high dimensional data, respectively. The tables seem to evidence that comparable performance can be obtained when m_f values are made very close to 1. Therefore, if the algorithm is forced to stop before this condition different results are obtained.

6 Comparison with a SOM Module

Both low and high dimensional data sets were preprocessed by a conventional SOM module, for comparison purpose. As well known, one problem in dealing with the SOM neural network is the selection of the network parameters, i.e. the neighborhood size, the learning rate, the strategy to alter these two parameters and the number of iterations to run. In this study several simulations were carried out with different neighborhood sizes and for each neighborhood size by changing the number of learning iterations. In addition, the previous simulations were repeated with a new order of presentation of the training data, in order to valuate how this factor could influence the final results. The topology of the SOM module consisted of 30 nodes in the output layer and of N nodes in the input layer, where N is the band number. For each neighborhood size, the percentage of the overall classification accuracy obtained by the modular system (SOM+SLP) for the training and testing data are illustrated in Table 3 and in Table 4 for the low and high dimensional data set, respectively. The values of the neighborhood size and the iteration number used for SOM training are indicated in these two Tables, first line and first column, respectively. The learning rate was $\alpha = 0.2$.

The results illustrated both in Table 3 and in Table 4 evidence the difficulty related to the choice of a good set of SOM parameters, when

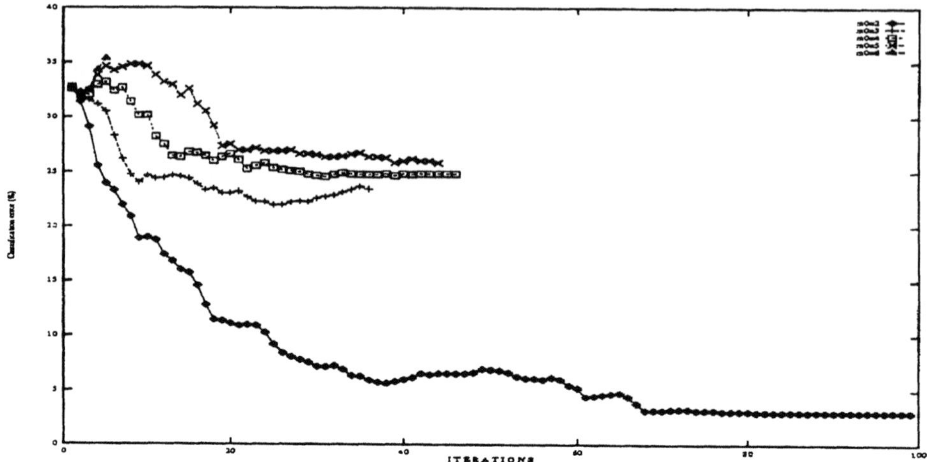

Figure 2: Experiment 2: FKCN error rate as a function of time with $\Delta_m = 0.01$, variable m_0, $\epsilon = 10^{-5}$.

m_0	t_{Max}	Train	Test
2	20	82.28%	82.01%
3	40	82.44%	82.06%
4	60	82.13%	81.67%
5	80	82.39%	82.35%
6	100	82.44%	82.63%

a)

m_0	t_{Max}	Train	Test
2	20	96.43%	95.15%
3	40	96.79%	94.97%
4	60	95.45%	94.68%
5	80	96.43%	94.67%
6	100	96.25%	94.15%

b)

Table 2: Classification performance obtained by the modular system in: a) experiment 1; b) experiment 2. FKCN was trained with fixed Δ_m, variable m_0

Neigh.	5		10		15	
T_{max}	Train	Test	Train	Test	Train	Test
5	78.46%	77.74%	76.40%	75.54%	74.30%	73.42%
10	82.36%	81.38%	78.64%	77.89%	79.37%	78.92%
15	83.12%	83.05%	83.38%	82.82%	78.49%	78.18%
20	81.35%	81.80%	81.56%	81.77%	82.44%	81.88%
25	84.05%	83.42%	82.96%	82.87%	81.56%	82.06%
30	83.53%	83.08%	82.93%	82.82%	81.01%	81.06%
40	83.17%	82.74%	84.18%	83.73%	81.56%	80.62%

Table 3: *Experiment 1: classification performance obtained trough the modular (SOM+SLP) neural classifier with the low dimensional data set.*

Neigh.	5		10		15	
T_{max}	Train	Test	Train	Test	Train	Test
5	94.20%	92.40%	91.61%	89.97%	90.00%	88.46%
10	95.27%	94.25%	92.77%	91.42%	93.48%	92.89%
15	95.71%	94.11%	93.75%	91.88%	94.64%	93.23%
20	96.16%	94.51%	93.48%	92.25%	93.66%	91.93%
25	95.45%	94.22%	93.21%	92.79%	93.66%	92.65%
30	96.16%	94.94%	93.30%	92.54%	93.93%	93.09%

Table 4: *Experiment 2: classification performance obtained trough the modular (SOM+SLP) neural classifier with the high dimensional data set*

working with a low number of iterations (here, one iteration means one complete presentation of the training set to the network). The best system performance obtained by using the SOM network, as preprocessing module in the system, are comparable with those obtained by using the FKCN, once FKCN has been trained until the m parameter become very close to 1. Therefore, when preprocessing labelled data to speed up the neural classification task, the FKCN can be used instead of the SOM network for the reasons in the following. The FKCN is a batch algorithm, i.e. it is independent of the input sequence. Moreover, FKCN does not require explicitly the definition of neighborhood size, learning rates and strategy to update these parameters. However, in order to obtain stable results with respect to changes in both m_0 and Δ_m parameters, all the iterations have to be run, since the classification performance depends on the choice ϵ values.

7 Conclusions

In this work an application of the fuzzy version of the Kohonen self-organizing map neural network has been considered. The FKCN has been used to preprocessing both low and high dimensional remote sensed data before classification, in order to reduce the amount of computation time required by the learning phase of a single MLP classifier. The FKCN algorithm uses fuzzy membership values as learning rates, automatically extracted during learning from the training data. The concept of the neighborhood size and neighborhood updating are embedded in the FKCN learning procedure. Therefore, it could provide an useful tool for data preprocessing. The results obtained with data characterized both by low and high statistical separability reveal the need of further study on the stop criterion.

References

[1] E. Binaghi and A. Rampini. Fuzzy decision making in classification of multisource remote sensing data, from numerical to symbolic image processing: Systems and Applications. Optical Engineering, Special issue (G. Vernazza eds.), 1993; 3:1193-1204.

[2] P. Blonda, G. Pasquariello, S. Losito, A. Mori, F. Posa, D. Ragno. An experiment for the interpretation of multitemporal remotely

sensed images based on a fuzzy logic approach. Int. J. Remote Sensing, 1991; 12:463-476.

[3] F. Wang. Fuzzy supervised classification of remote sensing images. IEEE Trans. Geosci. Remote Sensing, 1990; 28:194-201.

[4] E. C. Tsao, J. C. Bezdek and N. R. Pal. Fuzzy Kohonen clustering network. Pattern Recognition, 1994; 27:757-764.

[5] T. Kohonen. Self-organization and Associative Memory, 2nd Edition. Springer Verlag, Berlin, 1987.

[6] P. Blonda, V. la Forgia, G. Pasquariello and G. Satalino. Feature extraction and pattern classification of remote sensed data by a modular neural system. Optical Engineering, 1996; 35:536-541.

[7] D. E. Rumelhart, G. E. Hinton, and R. J. Williams. Learning Internal Representation by Error Propagation. In Parallel Distributed Processing, vol I, D. E. Rumelhart, J. L. McClelland. Eds., Cambridge: MIT Press., 1986.

[8] G. Pasquariello and P. Blonda. Multitemporal remote sensing data classification using neural networks. In: International Archives of Photogrammetry and Remote Sensing, XXIX, L. W. Fritz, J. R. Lucas Eds., Committee of the XVII Int. Congress for Photogramm. and Remote Sensing, 1992, pp 922-929.

The use of neural networks for the automatic detection and classification of weak photometric sub-components in early-type galaxies

G. Di Sciascio, G. Longo

Osservatorio Astronomico di Capodimonte

80131 Napoli, Italy

R. Tagliaferri

Dipartimento di Informatica ed Applicazioni, Università di Salerno and INFM

84081 Baronissi (Sa), Italy

M. Capaccioli

Osservatorio Astronomico di Capodimonte

80131 Napoli, Italy

G. Richter

Astrophysikalisches Institut Potsdam

D-14482 Potsdam, Germany

Abstract

Weak photometric subcomponents offer an important tool to understand the present structure and past evolution of early-type galaxies. The detection of such structures can be performed either by detailed modeling of the 2-d light distribution or by means of expecially tailored filtering techniques such as, for instance, the Adaptive Laplacian Algorithm. In this paper we present vive application of a Multi Layer Perceptron and a Self-organizing neural nets to the detection and classification of some Laplacian morphologies. In particular we discuss the construction of a training set of images of artificial galaxies and some preliminary results.

1 Introduction

Early type galaxies are gas-poor stellar systems, i.e. objects with a very smooth light distribution which, at first glance, appears not perturbed by features which are typical of later morphological types such as spiral or irregular galaxies (such as spiral arms, inner disks, dust lanes and patches). In the late eighties it became evident that such simplicity was only ficticious and that early-type galaxies were instead rather complex multicomponent objects harbouring a whole variety of inner structures such as polar rings, bars, stellar disks, etc.. Each of these components was shown to be a powerful tool to investigate both the present structure and the past history of the individual objects. Many questions remain, however, to be answered: are these component genetic – i.e. did they form together with the host galaxy – or rather they are evolutionary varieties which result from cannibalism or merging phenomena ? In order to detect and study these components we applied the Adaptive Filtering Algoritm described in [1] to the detection of faint variations (down to 5%) of the local value of the radius of curvature in the galaxy light distribution. The application of this algoritm to a sample of 28 early-type galaxies reveals an impressive variety of structures in the Laplacian images [2]. The true nature of these structures, however, is often difficult to understend. This is the case, for instance, of polar rings and stellar bars, which in spite of having very different properties, present more or less identical Laplacian morphologies.

In order to attempt an objective classification of these structures, we have explored the possibility to train neural nets to recognize and classify Laplacian morphologies. In what follows we shall shortly outline the construction of the training set (Section 2 and 3); the characteristic of the nets used (Section 4) and the results (Section 5).

2   The construction of the training set

The use of neural nets can be split into three different phases: training, testing and application. The training and testing are usually performed using sets of data for which the result is *a priori* known. Since in our case such set was not available we had to recur to sets of simulated multi-component galaxy images. In what follows we shall describe in some detail the hyphotheses behind the simulated images, the results of the training and test phases and some applications. In order to obtain realistic galaxy images we had to design:

– routines to simulate the performances of a typical instrumental setup (telescope + filters + CCD) and,

– routines producing realistic models of multicomponent early type galaxies.

2.1   The instrument simulation package

To model a realistic data aquisition chain several parameters need to be taken into account:
– size of the primary mirror and obstruction factor introduced by the secondary mirror;
– diffuse light and luminosity loss introduced by the optics and by the filters;
– instrumental point spread function (PSF);
– CCD quantum efficiency, sampling frequency and readout noise;

The overall effect of atmospheric and instrumental blurring is described as the convolution of the 'intrinsic' galaxy image with the following PSF:

$$G(R, \sigma_1, \sigma_2, \sigma_3) = K \cdot \left[\frac{k_1}{\sigma_1^2} \cdot e^{-R^2/\sigma_1^2} + \frac{k_2}{\sigma_2^2} \cdot e^{-R^2/\sigma_2^2} + \frac{k_3}{\sigma_3^2} \cdot e^{-R^2/\sigma_3^2} \right]$$

where $k_1 = 1$, $k_2 = k_3 = 0.08$, $\sigma_3 = 3\sigma_2 = 9\sigma_1$ (σ_1 varying from 0.55" to 1")
[3]. This formula has one degree of freedom only: the value of the dispersion
σ_1 of the principal gaussian component (the so-called 'seeing').

Once the radial trend of the PSF has been established, we may proceed to
convolve our two-dimensional model with the sum of gaussians. The convolution spreads the central part of the galaxy image and redistributes the light
over a larger area.

2.2 Simulation of the background

Objects appear projected onto a luminous background which is the result of the
overposition of several different components: light scattered by the atmospheric
dust, light scattered by the Zodiacal dust and light emitted by faint background
unresolved sources. In order to simulate a real observation we applied a Poisson
noise to each pixel of the convolved images.

2.3 Construction of two-component artificial galaxies

As a first approximation, we assumed the galaxy to consist of two components
only: bulge and disk.

– The Bulge was modeled as a spheroid with luminosity profile modeled accordingly to the so-called de Vaucouleurs law [4]

$$I_b(R) = I_e \cdot 10^{-3.33[(R/R_e)^{1/4}-1]},$$

where $R = \sqrt{x^2 + \frac{y^2}{(b/a)_b}}$, $(b/a)_b$ is the bulge axial ratio and I_e is the surface
intensity at R_e, the radius that contains one-half of the total light of the model.
The coordinates (x, y) are the apparent distances from the galactic center along
the major and minor axis (a and b) respectively.

This model corresponds to an axisymmetric spheroid with the projected
surface density distribution constant on similar ellipses.

– The Disk was instead modeled as a flattened spheroid with an exponential
luminosity profile

$$I_d(R) = I_0 \cdot e^{-R/R_h},$$

where $R = \sqrt{x^2 + \frac{y^2}{(b/a)_d}}$, $(b/a)_d$ is the disk axial ratio, R_h is the scale length
and I_0 is the central surface intensity.

– Twisting. The simulation also takes into account the so-called *twisting profile* (id est, the variation of the position angle α of the isophotal major axis with the radius). The equation of the bulge isophotes, for a galaxy with twisting, can be written as

$$\frac{(x \cdot cos\alpha - y \cdot sin\alpha)^2}{a^2} + \frac{(x \cdot sin\alpha + y \cdot cos\alpha)^2}{b^2} = 1,$$

where the angle α is a function $f(a)$ of the major axis a. We reproduce the twisting by computing in every pixel (x, y) the major axis values a which satisfy the previous equation.

3 The simulations

In order to produce the sets of simulated galaxy images we first fixed some basic instrumental parameters (i.e. the size of the mirrors and the characteristics of the CCD). The remaining degrees of freedom are: the exposure time, the sky background level, the relative luminosity of the bulge and disk, the bulge axial ratio, the disk axial ratio, the bulge and the disk size, the amount of twisting. These parameters define the parameter space which needs to be explored by the models.

By setting realistic limits to the variations of these parameters, we produced what we call *experiments*: i.e. grids of models obtained by fixing the remaining instrumental parameters and the bulge/disk luminosity ratio to some values and changing the other parameters at constant steps. These experiments allowed us to have a complete sampling of the parameters space.

For the present work we have performed 5 experiments, varying the bulge/disk luminosity ratio varying from 0.5 to 0.9. Assuming, for the other parameters the following values: bulge flattening $(b/a)_b = 0.3, 0.4, 0.5, 0.6, 0.7, 0.8, 0.9, 1.0$, disk flattening $(b/a)_d = 0.1$, $R_e = 15"$, $R_h = 10"$. Galaxies are affected by 0^0, 30^0 and 60^0 twisting (see, for instance Fig. 1).

This adds up to a total of 120 galaxies. Each image was then processed through the Adaptive Laplacian Algoritm in order to produce the Laplacian images (see, for instance Fig. 2) to be used as training and test sets for the neural nets.

4 The neural networks

We used the above experiments to train both a Supervised Multi-Layer Perceptron (MLP) [5] and a Self Organizing (SOM) Neural Network [6] to recognize and classify the Laplacian morphologies. After the preprocessing, the images were segmented [7] in 3 classes (background, disk, bulge) by using SOM neural nets. The input is a 3 x 3 window of pixels feature vector and the output is a 3 x 1 neuron grid. After the segmentation a supervised classifier was used to identify the 3 image components we wanted to analyze (spheroid, disk, twisting). In this case an MLP with back-propagation learning algorithm was used to solve the problem. The input was a 16 values feature vector made as follows: there were 4 cuts of the images (vertical, horizontal and the 2 diagonal lines) and for each of these it was calculated the number of pixels belonging to each of the classes obtained in the segmentation step.

5   Experimental results

After several experiments, we obtained the best results in segmentation by using two steps in the learning process of the SOM net: in the first one, the net had a learning rate $\eta(t)$ varying from 0.05 to 0 and random initialized weights with low values; in the second one, $\eta(t)$ was decreased from 0.02 to 0.

We obtained the best performance in the morphology classification by using an MLP composed by 4 input neurons (the i-th input element being the sum of the pixels of the above mentioned feature vector belonging to the i-th segmentation class, with $i = 1, \ldots, 4$), by 1 hidden layer of 32 neurons and by 3 output neurons. The net parameters used in this net were the learning rate $\eta = 0.5$, $\alpha = 0.8$ in the "momentum" term, while the weights were initialized with random values between -0.5 and $+0.5$.

The training set was made of 120 images and the test set of 60 ones, following the above mentioned techniques. The system faulted only in the classification of 5 images, obtaining a classification error of 8% in the test set.

Acknowledgments

One of us (R. T.) has been supported in part by IIASS and Murst 40% *"Metodologie e tecniche di progettazione per sistemi eterogenei basati su architetture neurali"*. This work has been partially sponsored by the Italian Space Agency.

References

[1] Lorenz H. et al. 1994, AA, 277, 321.

[2] Capaccioli et al., 1996, in Proceed. of VI Italo-Korean Meeting, Word Scientific Press (Singapore), in press

[3] Capaccioli M and de Vaucoulers G., 1983 ApJ S., 52, 465

[4] de Vaucouleurs G, 1948, Ann. Ap., 11, 247

[5] D.E. Rumelhart, J.L. McClelland, "Parallel Distributed Processing", MIT Press, Cambridge, MA (1986).

[6] T. Kohonen "The Self Organizing Map", Proceedings of the IEEE, vol 78, n.9 (1990).

[7] N.R. Pal and S.K. Pal " A review on image segmentation techniques", Pattern Recognition, vol. 9, 1277 (1994).

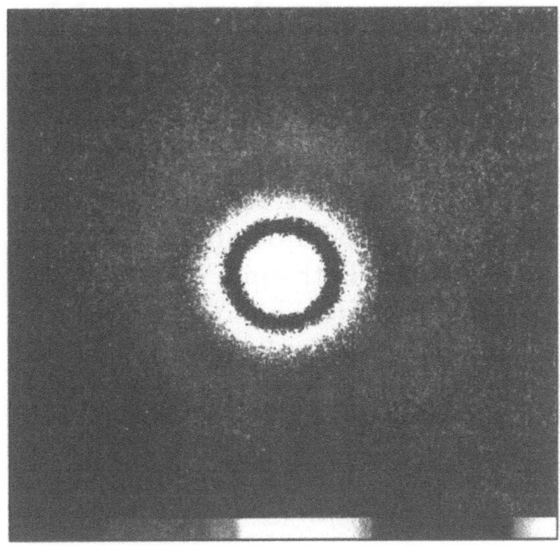

Figure 1: Simulated galaxy image affected by 60^0 twisting

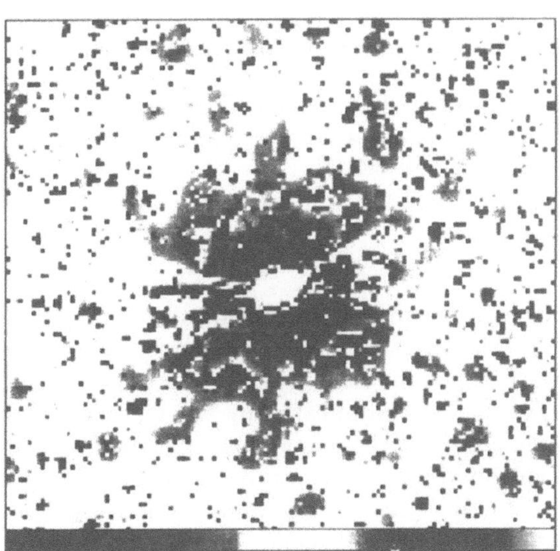

Figure 2: Laplacian image of the galaxy shown in Fig. 1

A Generalized Regularization Network for Remote Sensing Data Classification

Michele Ceccarelli
CPS-CNR, ITALY
ceccarel@mathp0.dma.unina.it
Alfredo Petrosino
IRSIP-CNR, ITALY
alfredo@irsip.na.cnr.it

July 25, 1996

Abstract

Real-time classification of remote sensed data is a challenging application in pattern recognition area, expecially when the spatial and temporal resolution increase. Several recent studies proved that the neural network approach is an interesting alternative to conventional statistical methodologies. In this paper we focus on the RBF networks applied to Sinthetic Aperture Radar data classification. RBF networks are used as a classification module of the textural features extracted by a log-Gabor pyramidal analysis of the original image. The problem of fast and reliable estimation of RBF parameters is addressed and a data-dependent approach is proposed. In particular, spectral properties of the areas to be recognised are used for the choice of the feature extraction parameters, whereas the statistical properties of the features are used for a suitable design of a Generalized Regularization Network [1].

1 Introduction

Terrain classification is one of the primary applications of high resolution satellite imagery such as Sinthetic Aperture Radar (SAR); it consists into assigning each pixel of the image to a terrain class (*e.g.*, urban area, agriculture, park, lake, river, etc.). The problem is complicated by the fact that each pixel of a SAR image contains the measure of the energy backscattered by a small area in the observed environment providing information about its surface state. Such information turns out to be not sufficient to exactly classify the scene content in radar images due to the presence of speckle deriving from the coherency of the image formation process, and to the presence of texture depending on the structure of the illuminated surface and on the illumination angle.

Recently, neural classifiers have been applied to remotely sensed data to overcome the limits of classical classifiers such as a priori hypothesis on the data distribution, sequentiality, etc. Decatur [2] used three layered back–propagation networks to classify SAR data comparing his results with maximum likelihood classifiers. Lee *et al.* [3] used a four layered back–propagation network for cloud classification of Landsat MSS data comparing the results with various statistical classifiers. Benediktsson *et al.* [4] made an empirical comparison of neural networks and statistical techniques in the classification of a multisource remote sensing and geographic data. Bishof *et al.* [5] reported the application of a three layered back-propagation network for classification of Landsat TM data on a pixel-by-pixel basis, with the integration of contextual information. Heermann *et al.* [6] explored a back-propagation neural network for classification of multispectral image data comparing the results with supervised and unsupervised classification algorithms. Azimi-Sadjadi *et al.* [7] proposed to train a neural classifier on salient features extracted by the principal component analysis scheme applying the system on SAR images.

In the above approaches the effect of speckle is efficiently mitigated, whereas no specific attention is dedicated to the structure (more or less complex) of shadows, specular reflection, edges and so on. An attempt to the solution of this problem has been reported in [8, 9], where the textural information, *i.e.* the spatial variability of neighbouring pixel values within the image, captured by multichannel Gabor filtering [10], has been combined with the radiometric information provided by the pixel tone variation. Back-propagation and LVQ networks trained on such features were applied to classify SAR images. In this paper we improve the proposed system by using a *data dependent approach*. In particular, the feature extraction phase is tailored on the spectral characteristics of the specific areas to be recognised. Indeed, if a library of reference areas (urban, agricoltural, etc.) is provided, a remote-sensing image interpreter can be tailored to recognize such areas. Following the original idea of Bovik *et al.* [10], we choose the central frequencies of the Gabor filters as the peaks of the spectrum of the reference areas. Moreover, the statistical properties of the extracted features in the training phase are used to shape the activation function of the Radial Basis nodes in the neural classifier. The system is composed by a multichannel log-Gabor filtering phase and a neural network based classification phase as described in the next two sections.

2 Textural Feature Extraction Phase

Orientation and frequency responces are extracted from local areas of the input image to form a representative feature vector. In this work we use the log-Gabor pyramid [11], or equivalentely the Gabor wavelet decomposition, to define a finite set of filters. The use of Gabor filters in texture analysis has been recently taken in consideration because texture segmentation requires simultaneous measurements in both the spatial and frequency domains. An important property of such filters is that they optimally and uniquely achieve simultaneous localization in both domains.

The original image I is first convolved with a bank of orientationally tuned filters and then fed into a Gaussian pyramid:

$$G_0^\alpha = [g_\sigma \cdot e^{(i\vec{k}_\alpha \cdot \vec{r})}] * I \tag{1}$$

where $\vec{r} = x\vec{i} + y\vec{j}$, x and y being the spatial coordinates of the filter, and $\vec{k}_\alpha = \cos\theta_\alpha \vec{i} + \sin\theta_\alpha \vec{j}$. θ_α is the preferred orientation of the α-th filters and g_σ a Gaussian function centered in the origin with variance σ.

A recursive procedure allows the creation of each of Gaussian pyramid through the images labelled $G_1^\alpha, \ldots, G_n^\alpha$ as follows

$$G_{i+\frac{1}{2}}^\alpha = H * G_i^\alpha \qquad G_{i+1}^\alpha = \text{subsampled} \quad G_{i+\frac{1}{2}}^\alpha \tag{2}$$

where the mask H is a bidimensional separable filter, Gaussian in shape, with values $(1/20, 1/4, 2/5, 1/4, 1/20)$.

The selection of the bank parameters θ_α is crucial to design an efficient image classification system. In order to have a reduced number of Gabor filters, we choose the frequencies on the basis of the intrinsic spectral properties of the areas to be recognized. Specifically, the preferred orientations are selected as the coordinates of the peaks in the spectrum of the reference areas. This turns out to produce a number of filters which is a multiple of the number of training areas.

3 RBF Classifiers

The main idea behind RBF networks is the construction of complex decision regions by superposition of simple kernel functions. The centers and widths of the kernel functions are the main parameters to be estimated during the learning phase. RBF networks are 3-layered systems where each output unit computes the following function

$$f_i(\mathbf{x}) = \sum_{j=1}^{M} c_{ij}\phi(||\mathbf{x} - \mathbf{m}_j||) \quad 1 \leq i \leq C \tag{3}$$

where C is the number of output units and the ϕ are M kernel functions computed by the hidden units. The vector $\mathbf{m}_j$ represents the input stimulus to which the unit j is maximally receptive. A Gaussian function is typically used for the kernel functions:

$$\phi(r) = \exp\left(-\frac{r^2}{2\tau}\right).$$

In general, when the number of basis functions (hidden units) is significantly larger than the number of input units, the hidden layer perform a pseudo-orthogonalisation of the sample points through a mapping from the feature space onto a higher dimensional space [12]. A complex classification problem in the original space can be considerably simpler in this higher dimensional space.

The points is how to choose the number and location of kernel centers. In [13] a study concerning the influence of the number and the distribution of centers on the RBF network approximation capabilities is reported, and the kernels are centered on the points of maximum curvature of the approximating function. In [14] a kernel function is placed on each training pattern and the c_{ij} weights are fixed by solving a linear system. In general, the number of kernel centers can be reduced by clustering the training data and the selection of kernel centers as the cluster code vectors. The hidden–to–output weights c_{ij} can be then estimated by LMS. Another global learning scheme for RBF networks is the use of gradient descent method for simultaneous supervised learning of center locations $\mathbf{m}_j$ and weighting coefficients c_{ij}.

Since RBF nodes model receptive fields, it is reasonable that they differ each other or, at least, depend on the data to be analysed. Indeed, it is possible that some components of the input pattern are more relevant than others, for example a component could be a linear combination of others, or could be too affected by noise to be significant, etc. This problem can be approached by shaping the kernel functions ϕ. This point is accurately analysed in [1] where the *Generalized Regularization Network* is proposed in order to approximate the functions $F_i(W\mathbf{x})$ instead of $f_i(\mathbf{x})$. W is an $d \times d$ which operates a linear transformation of the original set of variables $\mathbf{x}$, accounting the relevance of a component with respect to the others. This turns out to be equivalent to adopt a weighted norm instead of the Euclidean one, *i.e.* the equation (3) is replaced by

$$f_i(\mathbf{x}) = \sum_{j=1}^{M} c_{ij}\phi(||\mathbf{x} - \mathbf{m}_j||_W) \quad 1 \leq i \leq C \tag{4}$$

where $||\mathbf{x}||_W^2 = \mathbf{x}^T W^T W \mathbf{x}$. The choice of the matrix W is a rather complex problem (see for example [15]). An efficient approach for solving this problem is the use of the Karhunen-Loeve Transform (KLT) [16]. The method consists into two steps: the first step is a substitution of the coordinated axes in such a way the components of the input vectors become uncorrelated; the second step is the calculation of a set of weights which accounts for the information carried out by each of the new components.

Let H be the matrix of the eigenvectors of the covariance matrix of the data, arranged in decreasing order of the corresponding eigenvalues. The linear transform

$$\mathbf{x} \to \tilde{\mathbf{x}} = H^T \mathbf{x} \tag{5}$$

has the property of making uncorrelated the components of the random vector

$\tilde{\mathbf{x}}$. Moreover, the eigenvalues $\lambda_1, \ldots, \lambda_d$ of the covariance matrix give a measure of the statistical relevance of each component of the vector $\tilde{\mathbf{x}}$. A *relative measure* of the statistical relevance of the i-th component of $\tilde{\mathbf{x}}$ can be defined as

$$\tilde{\lambda}_i = \frac{\lambda_i}{\sum_{k=1}^{d} \lambda_k}. \tag{6}$$

The values $\tilde{\lambda}_i$ can therefore be used to adopt a weighted norm with the required properties, provided that the input vectors are first transformed according to (5). The proposed approximation scheme consists of using the equation (4) with

$$W = \tilde{\Lambda}^{1/2} H^T \tag{7}$$
$$\tilde{\Lambda} = \text{diag}(\tilde{\lambda}_1, \ldots, \tilde{\lambda}_N) \tag{8}$$

or explicitely

$$\phi(\|\mathbf{x} - \mathbf{m}_j\|_W) = \exp\left[-\sum_{i=1}^{N} \tilde{\lambda}_i(\tilde{x}_i - m_{ij})^2\right] \tag{9}$$

where we have retained the same notation for the $\mathbf{m}_j$'s as they are free parameters in both equations (4) and (9). This scheme can be also adopted for the dimensionality reduction of data and the consequent reduction of the number of classifier parameters (less computational costs). Indeed, since the value $\tilde{\lambda}_i$ measures the statistical relevance of the i-th input component, the components with a relevance value less than a given threshold can be neglected without a substantial loss of classification performance.

More specifically, the previous approach relies on a four layered feedforward network where:

- The input layer with d input nodes corresponding to d features.

- The first hidden layer with d linear nodes

$$\tilde{x}_i = \mathbf{w}_i^T \mathbf{x}$$

- The second hidden layer with M RBF nodes

$$y_j = \exp\left[-\sum_{i=1}^{d} \tilde{\lambda}_i(\tilde{x}_i - m_{ij})^2\right]$$

- The output layer with C sigmoidal nodes

$$z_k = \frac{1}{1 + exp\left[-\sum_{j=1}^{M} y_j c_{jk}\right]}$$

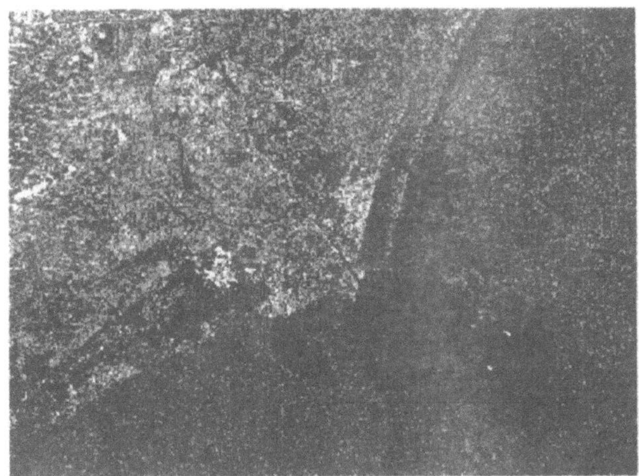

Figure 1: The ERS1 image *Fiumicino*.

	Class1	Class2	Class3	Class4
Class1	1.000	0.000	0.000	0.000
Class2	0.000	0.972	0.000	0.038
Class3	0.000	0.000	1.000	0.000
Class4	0.000	0.391	0.000	0.609

Table 1: Case 1: Confusion matrix for unshaped RBF

It is easy to show that if:

- $\mathbf{w}_i$ is set to the i-th eigenvector of the $d \times d$ covariance matrix Σ of the data components

- $\tilde{\lambda}_i = \dfrac{\lambda_i}{\sum_{k=1}^{N} \lambda_k}$.

the first layer behaves as a "withening" transform to uncorrelate the components of data, the scale factors $\tilde{\lambda}_i$ measure the statistical relevance (amount of variance in the i-th direction), providing an hyperellipsoidal shaping of the kernels.

4   Experimental Results

We applied the proposed image segmentation system to the ERS1 image *Fiumicino*, shown in figure 1, depicting the mouth of the Tiber, near the city of Rome and to the ESAR image of *Oberpfaffenhofen*, shown in figure 2, depicting the area around the airport of Oberpfaffenhofen in Germany.

	Class1	Class2	Class3	Class4
Class1	1.000	0.000	0.000	0.000
Class2	0.000	0.875	0.000	0.125
Class3	0.000	0.000	1.000	0.000
Class4	0.000	0.156	0.000	0.844

Table 2: Case 1: Confusion matrix for KLT–shaped RBF

	Class1	Class2	Class3
Class1	0.750	0.125	0.125
Class2	0.453	0.531	0.016
Class3	0.000	0.016	0.984

Table 3: Case 2: Confusion matrix for unshaped RBF

	Class1	Class2	Class3
Class1	0.703	0.172	0.125
Class2	0.156	0.844	0.000
Class3	0.000	0.031	0.969

Table 4: Case 2: Confusion matrix for KLT-shaped RBF

4.1 Case 1: ERS-1 image of *Fiumicino*

The aim was to partition the image in four mainly representative regions: urban area, agricultural area, sea, pollution.

A subimage of size 32×32 for training and a subimage of size 64×64 for testing are considered for each of the four classes. The training areas were used for the estimation of the first two preferred orientations of the Gabor filters as described in section 2, while the number of levels of the log-Gabor pyramid was set to 2. This produced a set of 1024 eight-valued feature vectors which were used to train a RBF neural classifier. Four outputs were used to indicate the four distinct classes with *"1 of C"* ($C = 4$) desired output coding, *i. e.* "1000" for the urban class; "0100" for the agriculture class; "0010" for the sea class and "0001" for pollution. In the testing phase, the decision rule used for the labelling of the test pixels was based on the output neuron with the largest activation. The RBF parameter estimation was performed by using the Conjugate Gradient Algorithm with restarts (Powell's method). Two kinds of experiments were performed on the data: one with the unshaped RBF network, equation (3); the other with the proposed KLT–shaped RBF network, equation (9). The number of hidden nodes was experimentally set to 16, and, for scaling purposes, at the output nodes the functions $f_i, i = 1, \ldots, 4$, are passed through a sigmoidal function. Tables 1 and 2 show the confusion matrices of both experiments. For the unshaped RBF we obtain a correct classification probability of 0.895, while for the KLT-shaped RBF a probability of 0.930 is obtained.

4.2 Case 2: ESAR image of *Oberpfaffenhofen*

The aim was to partition the image in three mainly representative regions: urban area, agricultural area, forest area.

As before, a subimage of size 32×32 for training and a subimage of size 64×64 for testing are considered for each of the three classes. The training set consists of 3072 patterns corresponding to image pixels extracted by 3 subimages of size 32×32, while the test set consists of 12288 patterns corresponding to image pixels extracted by 3 subimages of size 64×64. The RBF parameter estimation was performed by using the Conjugate Gradient Algorithm with restarts (Powell's method). The number of hidden nodes was experimentally set to 12. Tables 3 and 4 show the confusion matrices of both experiments. For the unshaped RBF we obtain a correct classification probability of 0.755, while for the KLT-shaped RBF a probability of 0.839 is obtained.

5 Conclusions

The Generalized Regularization Network has been tailored by using the KLT techique for finding appropriate widths of kernel functions. Tests on the recognition of terrain classes in SAR images have been conducted to verify the validity

Figure 2: The ESAR image *Oberpfaffenhofen*.

of the proposed approach. The main feature of the work is the use of a data-dependent approach for both the feature extraction phase and the architectural design of the classifier. The achieved results indicate the robustness of the system in coping with noisy real-world applications.

References

[1] F. Girosi, M. Jones and T. Poggio. Priors, stabilizers and basis functions: from regularization to radial, tensor and additive splines. MIT AI Lab. Memo, 1994.

[2] S. E. Decatur. Applications of neural networks to terrain classification. In Proceed. of IJCNN 1990, Washington DC, Laurence Erlbaum, 1990, pp. 283-288

[3] J. Lee, R. C. Weger, S. K. Sengupta and R. M. Welch. A neural network approach to cloud classification. IEEE Transactions on Geoscience and Remote Sensing 1990; 28:846-855.

[4] J. A. Benediktsson, P. H. Swain and O. K. Ersoy. Neural network approaches versus statistical methods in classification of multisource remote sensing data. IEEE Transactions on Geoscience and Remote Sensing 1990; 28:540-552.

[5] H. Bischof, W. Schneider and A. J. Pinz. Multispectral classification of Landsat images using neural networks. IEEE Transactions on Geoscience and Remote Sensing 1992; 30:482-490.

[6] P. D. Heermann and N. Khazenie. Classification of Multispectral remote sensing data using a back-propagation neural network. IEEE Transactions on Geoscience and Remote Sensing 1992; , 30:81-88.

[7] M. R. Azimi-Sadjadi, S. Ghaloum and R. Zoughi. Terrain classification in SAR images using principal component analysis and neural networks. IEEE Transactions on Geoscience and Remote Sensing 1993; 31:511-515.

[8] M. Ceccarelli, A. Farina, A. Petrosino, R. Vaccaro and F. Vinelli. SAR image segmentation using textural information and neural classifiers. L' Onde Electrique 1994; 74:24-28.

[9] M. Ceccarelli, A. Petrosino. Multi-Feature Adaptive classifiers for SAR image segmentation, accepted for publication on Neurocomputing: An International Journal, 1995.

[10] A. C. Bovik, M. Clark and W. Geisler. Multichannel texture analysis using localized spatial filters. IEEE Transactions on PAMI 1990; 12:55-73.

[11] D. H. Field. Relations between the statistics of natural images and the response properties of cortical cells. Jour. Optical Society of America A 1987; 4: 2379-2394.

[12] C. Bishop. Improving the generalization properties of radial basis function neural networks. Neural Computation 1991; 3:579-588.

[13] V. D. Sanchez. On the number and the distribution of RBF centers. Neurocomputing: An International Journal 1995; 7:197-202.

[14] D. F. Spect. Probabilistic neural networks. Neural Networks 1990; 3:109-118.

[15] M. T. Musavi, W. Ahmed, K. H. Chan, K. B. Faris and D. M. Hummels. On the training of radial basis function classifiers. Neural Networks 1992; 5:595-603.

[16] R. Gnanadesikan. Methods for Statistical Data Analysis of Multivariate Observations. Wiley, NY, 1977.

A Hybrid Neural Network Architecture for Dynamic Scene Understanding

Antonio Chella, Salvatore Gaglio

Dip. di Ingegneria Elettrica, Università di Palermo

Palermo, Italy

Marcello Frixione

IIASS - Istituto Internazionale per gli Alti Studi Scientifici

Vietri sul Mare (SA), Italy

Abstract

A hyprdid (neural and symbolic) architecture allowing for a deep understanding of moving scenes is described. The architecture is based on a working and effective integration among three levels of representation of data coming out from external sensors.

1   Introduction

The working environment of an autonomous agent is populated by moving and interacting objects, as moving and oscillating mechanisms, other moving agents, walking people and so on. The agent, operating in this dynamic and evolving environment must be able not only to perceive the shapes of the surrounding objects, but also to track and anticipate their trajectories according to the physical constraints. In order to accomplish these tasks, the agent needs simple reasoning capabilities about the physical behaviour of the external word.

We describe a hybrid architecture based on neural networks in which object perception and basic physical reasoning are two aspects of the same process of interpretation of sensor inputs. In the proposed architecture, basic reasoning processes about object perception and their physical behaviour are not located at the symbolic level, as in the qualitative physics systems, but at an intermediate level between the sensory level and the symbolic level.

The proposed architecture has been previously successfully employed to the description of static scenes acquired by a camera (see [3, 2]).

2   The three representation levels

We hypothesise three representation levels as the basis of our architecture: the *subsymbolic* level, in which the information directly comes from the sensory data; the *linguistic* level, in which information is expressed by a symbolic language; and an intermediate, prelinguistic *conceptual* level.

At this level the information is represented by a metric space defined by a certain number of cognitive dimensions, independent from any specific lan-

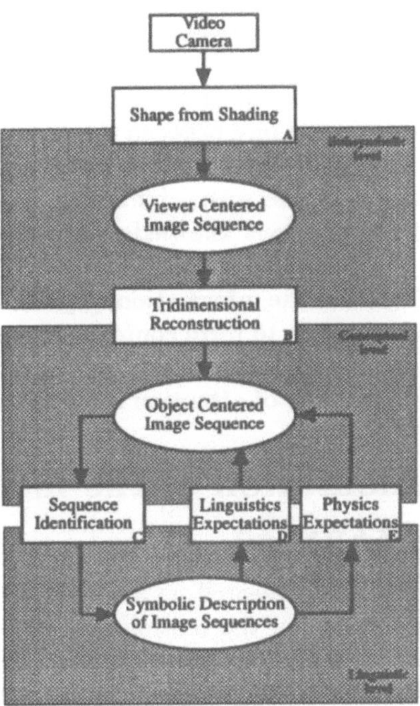

Figure 1: Block representation of the proposed architecture.

guage [4]. This level generates the very internal representation of the agent's external environment. A point in the conceptual space is called *knoxel* and it is our epistemological primitive element at the considered level of analysis. In the proposed architecture the conceptual level describes the 3-D characteristics of the acquired objects along with their motion status.

Our architecture employes *superquadrics* [7] as building blocks of object description; superquadrics are versatily primitive well suited to describe both natural and man-made forms. Existing reconstruction algorithms are robusts. These 3-D primitives allows for a concise and compact description of an object. By employing superquadrics as 3-D primitives, the description of dynamic scenes at the conceptual level is performed in terms of moving superquadrics; a superquadric with its motion status is our knoxel. Therefore the description of moving scenes is prior to any linguistic level inferences.

According to these premises, a suitable combination of similar moving superquadrics describes a basic physical process, e.g. an inertial motion of a cylinder-like object. A set of these sets describes a complex physical process, e.g. a moving cylinder approaching a quiet box. The role of the linguistic level is the symbolic categorisation and interpretation of acquired moving superquadrics at the conceptual level.

3 The attentive processes

A correlated aspect of the proposed architecture is the main role of attentive and expectation processes in the link between the linguistic and the conceptual level. We hypothesise a sequential attentive mechanism, that suitably scans the internal representation of the scene dynamics. The focus of attention is driven on the basis of the previous knowledge in order to detect and predict the relevant events in the perceived scene dynamics. Hence, it is a task of the higher level components to use the information acquired through the perceptual system to create expectations or to form contexts in which hypotheses may be verified and, if it is necessary, adjusted. A perception act is therefore a sequence of knoxels obtained by the focus of attention movements scanning the scene.

4 The implementation of the mapping routines by neural networks

A simple motion process is represented as a convex set of knoxel in the conceptual space. A simple model of the motion process is a *point attractor* of a suitable energy function defined over the conceptual space. A generic motion process, defined as a suitable combination of convex sets, may be described by an energy function containing several point attractors. The motion process may be therefore learned by an *attractor neural network* [5], [1], [6]. The general expression of the energy function of an attractor neural network for a motion process is:

$$E_1(t) = -\sum_{i=1}^{m} \sum_{j=1, j \neq i}^{m} T_{ij} \kappa_i(t) \kappa_j(t) \tag{1}$$

where m is the number of units of the network, $\mathbf{T}$ is the connection matrix storing the attractors describing the simple process component of the process P, $\kappa(t)$ is the knoxel representing the current state of the network. In order to describe a perception act of the motion process representing the scanning of the focus of attention, a sequential operation in the corresponding attractor neural network is implemented. The implementation is based on time–delayed connections among units. These connections store the order of sequence of knoxels of the process perception act [6]. The resulting energy term is:

$$E_2(t) = -\sum_{d=1}^{s} \sum_{i=1}^{m} \sum_{j=1, j \neq i}^{m} D_{ij}^d \kappa_i(t) \kappa_j(t - d\tau) \tag{2}$$

where τ is the time delay unit among two subsequent knoxels in the process perception act, s is the amplitude of the time window of interest, $\mathbf{D^d}$ is the delayed synapses connection matrix storing the order of knoxels in the perception act by time delay $d\tau$, $\kappa(\mathbf{t})$ and $\kappa(\mathbf{t} - \mathbf{d}\tau)$ are respectively the current and the past $d\tau$-th knoxel in the sequence.

The global external input to the network is modelled by the energy term:

$$E_3(t) = -\sum_{i=1}^{m} \sum_{j=1, j \neq i}^{m} F_{ij} \kappa_i(t) I_j(t) \tag{3}$$

where $\mathbf{F}$ is the input connection matrix, $\mathbf{I(t)}$ is the current input of the network.

The global energy function of the time delayed synapses attractor neural network is the sum of eqs. 1, 2,3:

$$E(t) = E_1(t) + \lambda E_2(t) + \varepsilon E_3(t) \tag{4}$$

where λ and ε are the weighting parameters respectively of the time delayed synapses and the external input synapses. Referring to eq. 4, an attractor is stable for a time period significantly long due to the E_1 term, so that the output knoxel is easily observed. After several $d\tau$ times, the term E_2 is able to destabilize the attractor and to carry the state of the network toward the successive attractor of the sequence representing the successive knoxel of the stored perception act. The neural network therefore outputs in a sequence all the knoxels of the stored perception act, thus proposing the knoxels to be searched in the conceptual space.

It should be noted that in eq. 4 the role of parameter λ is to tune the contribution of $E_2(t)$ with respect to $E_1(t)$. In fact, if $\lambda < 1$, $E_2(t)$ alone is not able to make the network state escape from a point attractor. To make the transition happen in this case, it is necessary also the contribution of the energy due to the external input $(E_3(t))$; therefore $\varepsilon > 0$. In this case the network is driven from the external input.

When the sequence of transitions among knoxels presented as input to the network by $E_3(t)$ coincide with the stored sequence in $E_2(t)$, the network *resonates*. The neural network in this case is therefore able to recognise an input process perception act when it resonates with one of the stored process perception acts.

5 Experimental Results

In this section the experimental results obtained by the implementation of the described architecture are presented. The experimental framework consists in dynamic scenes made up by moving blocks of woods; all the objects are situated on a black table. Sensory data are 2-D sequences of images acquired by a CCD video camera (two dimensional array of pixels) representing a sequence of views of the dynamic scene. Fig. 2 shows a sample sequence representing the bump of a moving cilinder with a quiet one. The fig. represents one frame out of five.

The operation of the block A of fig. 2 is the extraction of the depth map sequence from the acquired image sequence. While several techniques have been proposed in the computer vision literature for this problem, we employed a simple shape from shading algorithm.

184

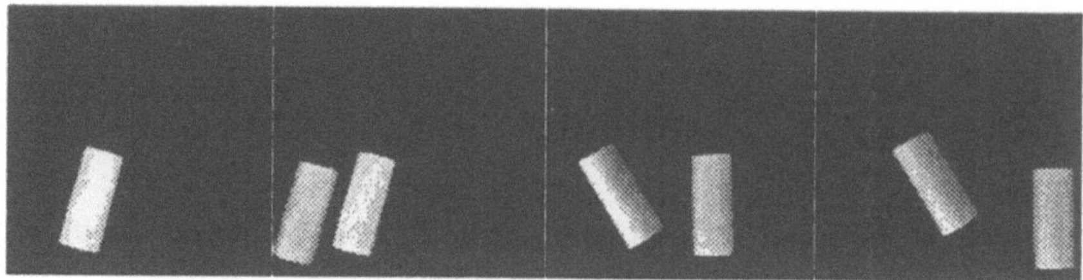

Figure 2: A sample sequence representing the bump of a moving cilinder with a quiet one. One frame out of five is represented.

To extract the knoxel representing the superquadrics from the acquired dynamic scene, a segmentation step is performed by contour extraction and region growing processes on the recovered surface. Each region found will give rise to a superquadric. The segmented depth map sequence are fed is input to block B of fig. 2. The first operation of this block is the volumetric representation of the depth map sequence by a spatial array sequence. The result is a discrete representation of the spatial occupancy of the objects present in the dynamic scene in term of *voxels*. Each voxels matrix extracted from the depth map sequence is the input to a best fitting operation in order to find the superquadrics best describing the acquired scene.

The expectation functions suitably drive the focus of attention in order to find the relevant process perception acts in the acquired image sequence. The *linguistic* expectation function in particular generates hypothesis by inferences at the linguistic level. With reference to the previous sequence, the description at the linguistic level of the *Approach* process states that an Approach process is made up by a moving knoxel filling the role of *mover* and a quiet knoxel filling the role of *target*. Therefore the architecture hypotheses the presence of a moving knoxel and a target knoxel: the linguistic expectation block (bloch C of fig. 2) generates the process perception acts hypotheses for the filler of the role *mover* and the filler of the role *target*. The operation of the block D of the architecture is based on the neural networks described in the previous section. This block generates the expected process perception acts for the mover filler and the target filler. When some of the expected knoxels are satisfied by some corresponding knoxels in the image sequence, the process perception act made up by the so found knoxels is sent to the sequence identification block. The relevant generated assertions at the linguistic level describing this operation are the following:

```
Time-window(t2)
Knoxel(k1)
Knoxel(k2)
Approach(s2)
during(s2,t2)
has-participant(s2,k1)
has-participant(s2,k2)
```

```
mover(s2,k1)
target(s2,k2)
```

The architecture is therefore able to recognize that an approach process is in act; that this process has the two knoxels as participants and it is able to find the roles for the two knoxels.

6   Conclusions

The proposed architecture is strongly based on a working and effective integration among three levels of data representation of data coming out from external sensors. This integration allows the linguistic level of the architecture to describe dynamic data coming out from external sensors.

The proposed architecture is not limited to the analysis of data coming out from artificial vision algorithms, but it can be also applyed to the description of a vaste range of dynamic data, as those coming out from mathematical simulations of differential equations, of the description of the dynamics coming out in the interaction between an agent and its environment. Further development will concern the generalisation of the proposed methodology to describe these more general kinds of dynamics.

References

[1] D. Amit, H. Gutfreund, and H. Sompolinsky. Spin-glass models of neural networks. *Physical Review A*, 32:1007–1018, 1985.

[2] E. Ardizzone, A. Chella, and S. Gaglio. A hybrid architecture for shape reconstruction and object recognition. *International Journal of Intelligent Systems*, 1996. (in press).

[3] A. Chella, M. Frixione, and S. Gaglio. Conceptual spaces and attentive mechanisms for scene analysis. In M. Marinaro and R. Tagliaferri, editors, *Neural Nets, WIRN Vietri 95*, pages 93–98, Singapore, 1995. World Scientific.

[4] P. Gärdenfors. Three levels of inductive inference. In D. Prawitz, B. Skyrms, and D. Westsrståhl, editors, *Logic, Methodology, and Philosophy of Science IX*. Elsevier Science, Amsterdam, The Netherlands, 1994.

[5] J.J. Hopfield. Neural networks and physical systems with emergent collective computational abilities. *Proc. Nat. Acad. Sci. USA*, 79:2554–2558, 1982.

[6] D. Kleinfeld. Sequential state generation by model neural networks. *Proc. Nat. Acad. Sci. USA*, 83:9469–9473, 1986.

[7] A.P. Pentland. Perceptual organization and the representation of natural form. *Artif. Intell.*, 28:293–331, 1986.

Proposal of a Darwin-Neural Network for a Robot Implementation

Carlotta Domeniconi
Università degli Studi di Milano
Ospedale di Desio-Servizio Universitario di Patologia Clinica
Piazza Benefattori, 1 20033 DESIO (Milano) Italy
email: domenico@dsi.unimi.it

Abstract

The objective of this work is the proposal of a Darwin-neural network that simulates an automaton with an adaptive behavior. We describe the environmental framework within the automaton can move, the areas the automaton is made of, the network dynamic, the transfer function that characterizes the state transition of neurons, the learning algorithm and the overall behavior of the network.

1. Introduction

A Darwin-neural network is a neural network model based on structural paradigms and learning processes introduced by the neural-Darwinism theory [1,2]. A Darwin-neural network learns specific tasks through interaction with an unknown environment; the behavior the network develops depends on the experience acquired through interaction with the environment [3].

2. Definition of the Problem and Automaton Presentation

The environmental framework is a bi-dimensional plane, within the automaton can move towards right or left. We suppose that objects of different shapes (rounds and squares) descend perpendicularly to the automaton movement direction (fig. 1).

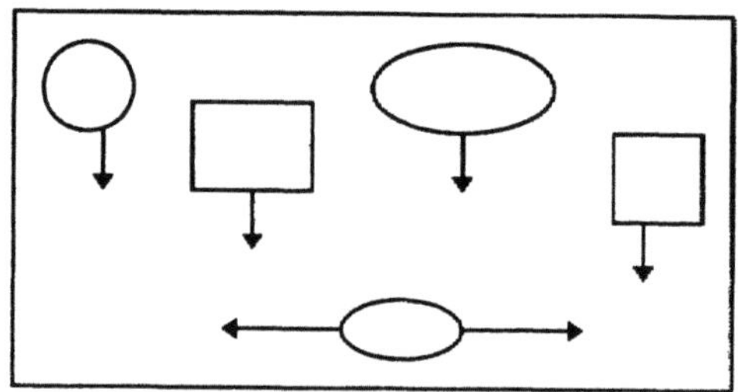

Fig. 1: Bi-dimensional environment for automaton operation

The squared objects represent a danger for the automaton; the rounded objects can be thought as energy or food sources. The automaton has to learn to come closer to rounded objects (energy sources) and to go away from squared ones (hot objects, for example) [4,5,6].

The automaton we define is made of five areas (fig. 2), each area corresponding to a neural net.

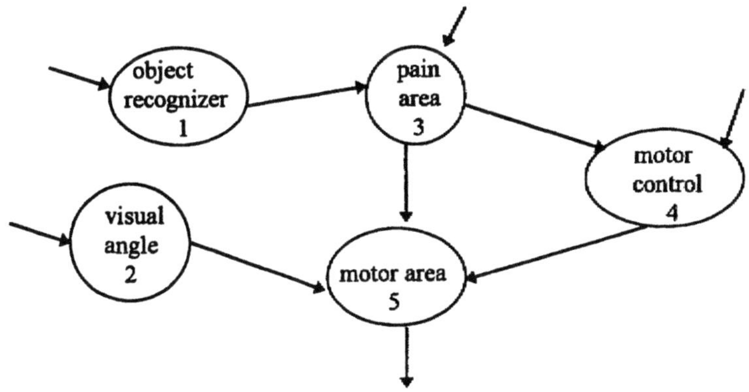

Fig. 2: Automata scheme

2.1. The Object Recognizer

Block 1 (fig.2) has the task to supply an internal description of the object-stimulus. This area is structured as a classification couple [1] that performs a parallel sampling of the external environment. Block 1 has to be able to detect the characteristics that allow the network to classify the object-stimulus as a member of a specific category. The classification couple realizes this task correlating the object properties through the reentrant connections between the maps of abstractions of the sampling logic modalities. Actually, block 1 receives in input numerical values, which represent the object distinctive characteristics. These features will be invariant respect to the object location in space and time.

2.2. The Visual Angle

The visual angle objective is to send a stimulus which shows the movement to perform to the motor area, once the automaton has acquired enough experience. The configurations the neuronal groups of this area assume have to discriminate the automaton relative position with respect to the object-stimulus. The input for block 2 is represented by a numerical list. This input is a function of the automaton initial position and it evolves as the automaton moves in the environment. It is reasonable to build an input list as a function of the ratio between the automaton distance from the intersection point of the perpendicular to the automaton direction of movement, passing through the object-stimulus, and the distance of the object from this intersection point. The object moves at discrete times with an established velocity, whereas the automaton moves at discrete times of a certain distance which is function of the motor area output value.

2.3. The Pain Area

The pain area produces the memory of the object that caused pain so that the automaton can learn to go away from the situation classified as dangerous [7]. At the beginning the automaton, having no experience of the external environment, will establish contact with dangerous objects. The pain area will receive from the environment a signal with high value which will allow the activation of its neuronal groups. Connections between blocks 1 and 3 will organize in order to produce, next time a dangerous object will be presented as stimulus, the memory of the pain that kind of object causes. The organization of connections between blocks 3 and 4 will cause the inhibition of the behavior that brings the automaton closer to objects (inhibition of genetic curiosity). Connections between blocks 3 and 5 will develop a motor answer that allows the automaton to go away from dangerous objects.

2.4. The Motor Control

The motor control realizes a simplification of feedback mechanisms which allow the individual to control his movements, in relation with himself and with the surrounding environment [8]. Since the organization of the several mechanisms that contribute to the development of this consciousness is very complex, we suppose the automaton knows when it is coming closer or going away from a specific object. This awareness is implemented as an error signal given by the environment as an input to the motor control area. The network will organize the connections between blocks 4 and 5 to give the proper motor answer to the perceived error.

2.5. The Motor Area

The motor area elaborates its inputs to give in output a signal that represents the movement the automaton decided to perform. The motor area is formed by two sub-areas:

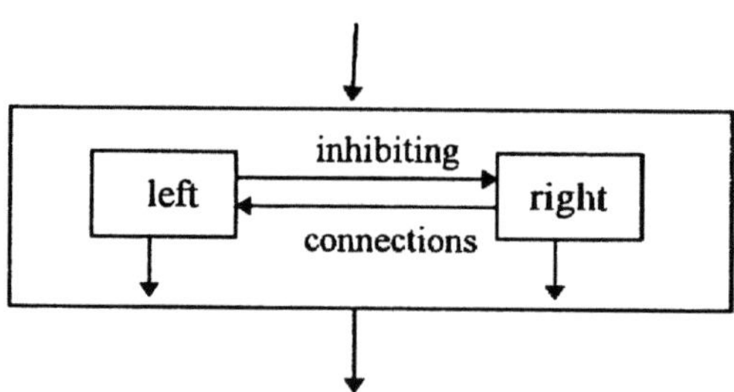

Fig. 3: Internal structure of the motor area

They correspond, depending on which one is active, to a movement towards left or towards right. The two sub-areas have inhibiting lateral connections, to avoid a situation in which both are active. In general, however, there will be groups active in both areas: we consider active the area with the greater concentration of groups of active neurons. Therefore, the motor answer will be a weighted average of the two outputs coming from these areas, and it will give both the direction and the length of the movement [9].

3. The Network Dynamic

The Cartesian axis of the plane representing the environment are given by the lines passing respectively through the automaton and the object, along the correspondent directions of movement. The dynamic of the network that formalizes the automaton evolves according to the following sequence:

1. The object starts from a given position and moves at regular intervals.
2. The movement of the object produces an input for each area of the network (except for the motor area).
3. Each area has a given time interval to elaborate the external input signal and give an answer as output. The activation state of the motor area represents the last movement of the automaton.
4. During the next time interval the areas interact between themselves. The connections weights will be strengthened or weakened, depending on the pre- and post-synaptic groups activities and on the learning rules established between areas.
5. The system produces a motor output.
6. The object moves again and the sequence goes back to point 2.

4. The State Transfer Function

The state transfer function we consider is the following.
Given

$$I_{tot} = \sum_j c_{ij} HS\left(S_j - \theta_E\right) S_j - \sum_k \beta HS\left(S_k - \theta_I\right) S_k \qquad 1$$

where $HS(x)$ is the Heavyside function:

$$HS(x) = 1 \quad \text{if} \quad x \geq 0$$
$$HS(x) = 0 \quad \text{if} \quad x < 0 \qquad 2$$

we have,

$$S_i(t) = \omega S_i(t-1) + N \qquad \text{if} \qquad \theta_N \leq I_{tot} \leq \theta_P$$
$$S_i(t) = I_{tot} + N \qquad \text{if} \qquad I_{tot} > \theta_P$$
$$S_i(t) = N \qquad \text{if} \qquad I_{tot} < \theta_N \qquad 3$$

In Eq. 1 and Eq. 3 $S_i(t)$ represents the state of group i at time t; c_{ij} is the connection weight of input j to group i; S_j is the state of group j, in other words of the group connected to the j-th input of group i; θ_E is the threshold of the exciting inputs: only inputs $S_j \geq \theta_E$ are considered; β is a given coefficient of inhibition;

S_k is the state of group k that belongs to an inhibiting neighborhood of group i; θ_l is the threshold of inhibiting inputs: only inputs $S_k \geq \theta_l$ are considered; N is a normally distributed noise; $\omega = e^{-1/\tau}$ is a persistence parameter, with τ a characteristic time constant. ω denotes time decadence, according to an exponential law, of groups activities when global inputs range between θ_N and θ_P. In other words, when the input weighted sum belongs to an interval of 0 (exciting and inhibiting inputs cancel one another), it does not affect the activation state of post-synaptic groups. When $I_{tot} > \theta_P$, the global input is excitatory and it causes the post-synaptic group activation. When $I_{tot} < \theta_N$, the global input effect is inhibiting and it causes the annulment of post-synaptic activity. The transfer function thresholds (especially θ_E and θ_I) are averaged on values the inputs can have.

5. The Learning Algorithm

The learning rule of connections weights between blocks is

$$c_{ij}(t+1) = \frac{c_{ij}(t) + \delta(S_i - \theta_i)(S_j - \theta_j)}{NOR} \qquad 4$$

where S_i and S_j are respectively the post-synaptic and the pre-synaptic groups activation states. θ_i and θ_j represent the post-synaptic and the pre-synaptic groups amplifying thresholds. δ $(-1 \leq \delta \leq 1)$ is an amplifying parameter, and NOR is a normalization factor (it keeps the weights absolute values below the unity). The kind of learning the automaton is submitted to depends on values of the thresholds θ_i and θ_j, and on the sign of δ.

5.1. The Learning Process Between the Visual Angle and the Motor Area (2-5)

The connections between the visual angle and the motor area organize in order to bring the automaton closer to objects. Specifically, connections between active groups in both areas ($S_2 - \theta_2 > 0$ and $S_5 - \theta_5 > 0$) are strengthened, whereas connections between active groups in the visual angle and inactive groups in the motor area ($S_2 - \theta_2 > 0$ and $S_5 - \theta_5 < 0$) are weakened. In these two cases, δ is set to a positive value; in the others two no learning takes place, and δ is set to 0.

The active area within the motor area represents the "right" direction of movement. It follows the strengthening of connections between active groups that correspond to an internal representation of the input received by block 2 and active groups within the motor area. The motor control reverses the current direction of movement when it receives an error input from the environment. Typically at the beginning the environmental input will cause wrong movements in regard to the given task (due to random initial weights values). The motor control error perception allows, through organization of connections 4-5, the correction of movement's direction. As a consequence, connections 2-5 can organize in order to strengthen the correct action. As the automaton interacts with the environment, the motor control becomes active less often (the automaton is learning to move correctly), and the visual angle learns the correct direction for coming closer to objects to show the motor area.

5.2. The Learning Process Between the Motor Control and the Motor Area (4-5)

The connections between the motor control and the motor area have the task to correct the automaton movement in presence of error perception. If the environmental input for the motor control is 0, also the motor control activation state is 0 (except for the residual activity which has a fast decadence). In this case block 4 won't affect the activation state of block 5. Connections weights 4-5 are subject to changes only when the motor control is active ($S_4 - \theta_4 > 0$). Specifically, connections between active groups in both areas ($S_4 - \theta_4 > 0$ and $S_5 - \theta_5 > 0$) are weakened; connections between active groups in the motor control and inactive groups in the motor area ($S_4 - \theta_4 > 0$ and $S_5 - \theta_5 < 0$) are strengthened. In these two cases, δ is set to a negative value; in the others two no learning takes place, and δ is set to 0.

At the beginning the motor control will be frequently active. There will be an intensive learning activity on connections 4-5. The absolute value of δ is initially high so that the learning process on connections 4-5 can be as fast as possible allowing a proper organization of connections 2-5 depending on the motor area activation state. As learning proceeds, the absolute value of δ rapidly decreases; it can again be set to a high value when a new error perception occurs.

5.3. The Learning Process Between the Object Recognizer and the Pain Area (1-3)

Connections between blocks 1 and 3 organize in order to remind the automaton the pain a dangerous object produces. In this way the automaton can learn to go away from objects classified as dangerous. Connections between active groups within the object recognizer ($S_1 - \theta_1 > 0$) and currently active groups within the pain area ($S_3 - \theta_3 > 0$) are strengthened. To obtain this δ is set to a positive value; otherwise δ is set to 0.

This learning process performs an association between the internal representation of an object-stimulus and the currently active groups within the pain area. This association allows to classify an object as dangerous. In fact, when a dangerous object is presented more than once as stimulus to the automaton, its internal representation within the object recognizer becomes connected to the pain area via connections with high valued weights. The signals traveling on these connections are amplified causing the activation of post-synaptic groups within the pain area. This activation realizes the reminding of the pain caused by the present object-stimulus, without establishing contact with it. The threshold values of the excitatory inputs of neurons of block 3 are high. It is therefore necessary at the beginning an external input to make active neuronal groups within the pain area. It follows that initially only the visual angle and the motor control organize the automaton movement. At this stage the automaton develops a curiosity that brings it closer to objects. Once the automaton has made contact with dangerous objects, it develops a "mistrust" regard these objects and it learns to go away from them. The amplification threshold of groups of block 3 has a low value to allow a fast recognition of dangerous objects.

5.4. The Learning Process Between the Pain Area and the Motor Control (3-4)

The connections 3-4 organize to inhibit the contingent activity within the motor control, in case the pain area is active. In this way, the pain area becomes able, through connections 3-5, to show the motor area the withdrawal direction from dangerous objects. Specifically, connections between active groups in both areas ($S_3 - \theta_3 > 0$ and $S_4 - \theta_4 > 0$) and connections between active groups in the pain area ($S_3 - \theta_3 > 0$) and inactive groups in the motor control ($S_4 - \theta_4 < 0$) are weakened. To obtain this, in the first case δ is set to a negative value, whereas in the second case it's set to a positive value; otherwise δ is set to 0.

The task pursued applying this rule is to establish inhibiting connections between the pain area and the motor control. The value of δ decreases with time, reaching a 0 value when connections 3-4 have weights adequately inhibiting. At this stage of learning process, the activity within the pain area is due only to the interaction with the object recognizer (no more contacts take place with dangerous objects). The groups activity within the pain area will cause the inhibition of the motor control contingent activity.

5.5. The Learning Process Between the Pain Area and the Motor Area (3-5)

The connections 3-5 have to show the motor area the withdrawal direction from dangerous objects. This direction can be thought as fixed: in presence of danger the automaton moves always towards the same direction. Under this hypothesis, connections 3-5 are fixed. In alternative, the withdrawal direction can be interpreted as the direction opposite to the ones of movement when the automaton perceives the danger. Under this second hypothesis the connections 3-5 undergoes the same learning process applied to connections 4-5. Specifically, connections between active groups in both areas are weakened ($S_3 - \theta_3 > 0$ and $S_5 - \theta_5 > 0$), whereas connections between active groups within the pain area ($S_3 - \theta_3 > 0$) and inactive groups within the motor area ($S_5 - \theta_5 < 0$) are strengthened. In both cases δ is set to a negative value. Also for this learning process, the absolute value of δ rapidly decreases with time; it can again be set to a high value if the automaton gets in contact with a misclassified dangerous object.

6. Conclusions

The work presented is based on the idea that the use of a neural-Darwin network is an interesting and useful approach to solve robotics problems. The neural network defined represents a coherent and stable model. The automaton develops an adaptive behavior which depends on the experience acquired through interaction with the environment. If the automaton has experience of many non-dangerous objects it develops a strong curiosity and a tendency to come closer to all objects, including the dangerous ones. If the automaton has experience of many dangerous objects it develops a tendency to go away from all objects, on account of strengthening of connections between the visual angle and the motor sub-area that represents the withdrawal movement direction (frequently active under the hypothesis made).

Acknowledgments

This work is the result of fruitful exchange of ideas with Dario Russi, to whom I am especially grateful. I want to thank Alberto Bertoni, Paola Campadelli, and Marco Dorigo for helpful discussions. A special thanks goes to Anna Esposito, for the encouragement given in writing the paper.

References

1. G. M. Edelman, Neural Darwinism: The theory of Neuronal Group selection, Basic Book, 1987.

2. G. M. Edelman, Bright Air, Brilliant Fire On the matter of the Mind, Basic Book, 1992.

3. B. Manderick, Selectionism as a Basis of Categorization and Adaptive Behavior, Ph.D.Thesis, AI Lab VUB Brussels, 1991.

4. R. O. Duda, P. E. Hart, Pattern Classification and Scene Analysis, New York: Wiley,1973.

5. J. H. Holland, Adaptation in natural and artificial systems, Ann Arbor: The University Michigan Press, 1975.

6. R. von Mises, Mathematical Theory of Probability and Statistics, Academic, New York, 1964.

7. L. B. Booker, D. E. Goldberg and J. H. Holland, Classifier Systems and Genetic Algorithms, Artificial Intelligence 40, 1989.

8. N. Bernstein, The coordination and Regulation of Movements, Oxford: Pergamon, 1967.

9. M. Dorigo, U. Schnepf, A bootstrapping approach to robot intelligence: first results, Politecnico di Milano, 1990.

A Novel Hypothesis on Cortical Map: Topological Continuity

F. Frisone   V. Sanguineti   P. Morasso

DIST - Department of Informatics, Systems and Telecommunication

University of Genova

Via Opera Pia 13, 16145 Genova, Italy

e-mail:friso@dist.dist.unige.it

Tel.: +39-10-353-2801/+39-10-353-2749

Fax: +39-10-353-2154

Abstract

The paper proposes that some cortical maps are indeed topologically continuous, distributed representations, i.e. representations that also contain and make use of topological information. It is shown that the hypothesis is consistent with experimental observations and might suggest how the cortex can carry out computations that involve spatial or topological reasoning, like trajectory generation, visuomotor transformations.

Keywords: Cortical maps, topological continuity, lateral connections, modular organization.

1   Introduction

The observed modular organization of brain cortex into *columns*, communicating by means of a dense network of *lateral* connections (i.e. parallel to cortical surface) suggests that it can be seen as a massively interconnected set of elementary *processing elements*, i.e. as a *computational map* [5].

The structure of lateral connections is not genetically determined but, indeed, depends mostly on electrical activity during development. More precisely, they have been observed to grow exuberantly after birth, and to reach their full extent within a short period; during the subsequent development, a pruning process takes place so that the mature cortex is characterized by a well defined pattern of connectivity.

Observations on the primary visual cortex have shown that the final connectivity is closely related to that of afferent connections, and both self-organize simultaneously and synergetically, driven by the external visual input.

Lateral connections are always reciprocal and, although they are significantly weaker than that of cortical afferences, are believed to play a crucial role in mediating cooperation and competition between columns.

Most models of brain cortex assume that lateral connections are local and radially symmetric, whose strength varies according a *mexican hat* profile. Such a pattern of connectivity has been shown [1] to be crucial for supporting a competitive dynamics, responsible for the observed phenomenon of *peristimulus inhibition* (i.e. the observed localization of a region of activation in response to an external stimulus), and for the spontaneous emergence (or self-organization) of topographic representations of the input stimuli. Moreover dynamic behaviors like *diffusion, dispersion, convection*, useful in processing of visual inputs, can be efficiently supported by local or quasi-local connections.

However, the assumption of locality of lateral connections implies that, from the point of view of circuitry, the cortex is basically a 2-dimensional lattice; such a conceptual model is more or less implied in several artificial neural network models inspired by the visual cortex, like the *self-organizing feature maps* [6, 7] or the *elastic net* model [2].

The observed patterns of lateral connectivity are much more complicated: in particular, the existence of lateral connections that are not local does suggest an higher dimensional topology, thus ruling out all the models implying a 2-dimensional circuitry.

This paper proposes a new hypothesis on the functional role of lateral connections, accounting for their observed complexity, and suggesting how the cortical architecture can carry out computations like trajectory planning (with or without obstacle avoidance), perceptual reversal, visuomotor transformations. In this hypothesis the rule of not local lateral connections in the cortex is to represent higher dimensional stimuli.

It is known that long range connections are very significant in representing complex high-dimensional input spaces (e.g. they are present in cortical areas where each column is sensitive to multimodal stimuli). Thus, while lower-level sensory representation in the brain seems to be organized in 2-dimensional maps (such as *retinotopic maps*), the representation of higher dimensional quantities may need long-range connections to represent more complex similarity relationships [9]. Therefore, although some maps may appear to be *non-topographic* they may be *topologically continous*.

2 Topological continuity

We suggest a novel hypothesis for interpreting quasi-topographic maps:

> *lateral connections are used to implement higher dimensional lattices, so that maps are topological representations of the stimulus space and although some maps may appear to be non-topographic or quasi-topographic, they may actually be topologically continuous.*

This hypothesis is also motivated by the fact that lateral connections in self-organized topology representing maps reflect the dimensionality of the repre-

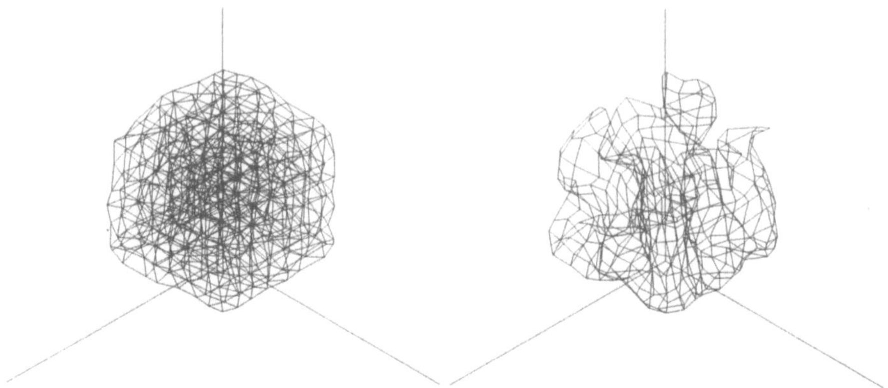

Figure 1: Homeomorphic (left) and non-homeomorphic (right) map

sented domains [3].

We give the following formal definition of *topological continuity*:

Definition *(topological continuity): Given two topological spaces X and Y and their associated distance measurements $d_x(\cdot)$, $d_y(\cdot)$, a function $y = y(x) : X \mapsto Y$ is said to be* topologically continuous *in the point $x_o \in X$ if and only if $\forall \varepsilon > 0 \; \exists \delta > 0$ so that $d_y(y(x), y_o) \leq \varepsilon$ implies $d_x(x, x_o) \leq \delta$ (where $y_o = y(x_o)$).*

A function is simply said to be topologically continuous if the above condition is satisfied for each $x_o \in X$.

In the case of Euclidean spaces, topological continuity is coincident with the more familiar notion of continuity. More formally, we can give the following

Definition *(topology-preserving map): A map is said to be* topology-preserving *or* homeomorphic *if the mapping $f = f(\mathbf{x})$ and its inverse are topologically continuous.*

In the case of lattices, it is easy to show that

Proposition *(topology-preserving lattice): A* lattice *is topology-preserving if and only if its order equals the dimension N of the input space.*

Fig. 1 shows the same convex 3 dimensional domain, covered by a lattice of order 3 (left) and by a lattice of order 2 (right); in the first case the map is homeomorphic whereas in the second case it is non-homeomorphic.

An important property of homeomorphic maps is that a continuous trajectory $\mathbf{x}(t)$ in the input domain X is translated into a sequence of adjacent winning units; in the case of Fig. 1(right), continuous paths in the domain may correspond to discontinuous sequences of winning units.

Point out that topological continuity is a not a property of maps but one of representations: in other words, topological continuity involves the semantic aspects of maps and more, relates them with structure.

Perceptual distortions put into evidence that internal representations should necessarily possess a *topologic* aspect, which is essential for representing knowledge of *spatial/metric* type, like the concepts of *continuity* and *straightness* [10].

The hypothesis of topological continuity implies that the average number of lateral connections is rapidly increasing with the dimension n of the represented *space*. However, since lateral connections grow very rapidly with n, there must be physical constraints on the maximum dimensionality n of a topologically continuous map that could be represented by the cortex. On the other hand, there are physical boundaries in the sensitivity of single columns to multi-modal stimuli [11]: upper bound in dimensionality of topologically continuous maps has been estimated to be 3 to 5.

3 Conclusions

Many regions in the brain are known to be topographic representations of stimuli (e.g. sensory surfaces). We showed how the topologic information, supported by long-range connections in associative areas, can represent *spatial* or *metric* knowledge (*continuity, straightness* etc.)

The *necessity* of representations of geometric/topologic type is even more clear in the case of coordinate transformations, forward as well as inverse, which are necessary for controlling complex musculoskeletal systems [4, 8].

References

[1] S. Amari. Dynamics of pattern formation in lateral-inhibition type neural fields. *Biological Cybernetics*, 27:77–87, 1977.

[2] R. Durbin and G. Mitchison. A dimension reduction framework for understanding cortical maps. *Nature*, 343:644–647, 1990.

[3] F. Frisone, F. Firenze, P. Morasso, and L. Ricciardiello. Application of Topology-Representing Networks to the Estimation of the Intrinsic Dimensionality of Data. In *ICANN95- Int. Conf. on Artificial Neural Networks*, volume 1, pages 323–327, Paris, October 1995.

[4] A.P. Georgopoulos, M. Taira, and A. Lukashin. Cognitive neurophysiology of the motor cortex. *Science*, 260:47–51, 1993.

[5] E. I. Knudsen, S. du Lac, and S.D. Esterly. Computational maps in the brain. *Annual Review of Neuroscience*, 10:41–65, 1987.

[6] T. Kohonen. Self organizing formation of topologically correct feature maps. *Biological Cybernetics*, 43:59–69, 1982.

[7] T. Kohonen. The self organizing map. *Proceedings of the IEEE*, 78:1464–1480, 1990.

[8] P. Morasso and V. Sanguineti. Self-organizing body-schema for motor planning. *Journal of Motor Behavior*, 26:131–148, 1993.

[9] C. Von der Malsburg and W. Singer. Principles of cortical network organization. In P. Rakic and W. Singer, editors, *Neurobiology of Neocortex*, pages 69–99, New York, 1988. Wiley.

[10] D. M. Wolpert, Z. Ghahramani, and M. I. Jordan. Are arm trajectories planned in kinematic or dynamic coordinates? an adaptation study. *Experimental Brain Research*, 1995. in press.

[11] E. Zohary. Population coding of visual stimuli by cortical neurons tuned to more than one dimension. *Biological Cybernetics*, 66:265–272, 1992.

A Mlp-Based Digit And Uppercase Characters Recognition System

G. A. M. Gioiello[1,2], E. Martire[1], F. Sorbello[1], G. Vassallo[2]

[1]DIE Dipartimento di Ingegneria Elettrica, Università di Palermo
Viale delle Scienze, 90128 Palermo, Italy

[2]CRES Centro per la Ricerca Elettronica in Sicilia
Via Regione Siciliana 49, 90046 Monreale, Italy

Abstract

A simple software solution for digit and uppercase handwritten characters recognition is presented. The proposed solution is based on a two-layer Multi Layer Perceptron (MLP) trained by a conjugate gradient descent (CGD) optimization algorithm. This neural network is embedded in a software tool for automatic processing of forms achieved using a scanner. The chosen solutions allow us to obtain good results both in terms of recognition rate and speed. In the paper are fully described design details and experimental results.

1 Introduction

Significant research efforts are currently aimed at designing fast and affordable devices to automatically solve the handwritten characters' recognition problem.

We implemented a complete software system to be used in conjunction with a computer and a scanner. The software code is written in ANSI C++ and is thus fully portable. The proposed solutions show good performances in terms of speed and recognition rate.

The handwritten character's database supplied by the National Institute for Standard and Technology (NIST) has been used for testing and training phases.

2 Remarks on preprocessing and neural net performances

The use of an apt pre-processing algorithm is necessary to obtain a good performance of the whole recognition system [1, 2]. In what follows we indicate with ①, ②, ③, the pre-processing primitives and describe the related implementation algorithms.

Primitive ①: matrix dimension reduction from m*m to n*n;
Primitive ②: matrix dimension reduction from 24*24 to 8*8;
Primitive ③: character rotation.

In our tries we adopted the following pre-processing algorithms: 'A' type reduces the original 128*128 binary matrix to a 16*16 binary matrix by using primitive ①. 'B' type reduces the original 128*128 binary matrix to a 8*8 integer matrix the elements of which range in [0,9] by using primitive ① and ②. 'C' type reduces the slant of the character starting from the original 128*128 binary matrix. Furthermore, it reduces the new 128*128 binary matrix to a 8*8 integer matrix the elements of which range in [0,9] by using primitive ①, ② and ③.

In all cases the Conjugate Gradient Descent optimization algorithm proposed by Powell [3] have been used for training the nets.

Table 1 reports the comparative performances obtained using the above algorithms. A more detailed discussion on theoretical background and experimental setup may be found in [1, 2].

Table 1

Net	Activation function of hidden/output layers	Preprocessing type applied to trained set	Preprocessing type applied to test set	% Rec. of test set
UPPERCASE				
256-158-26	log/log	A	A	91.4
64-220-26*	sin/exp	B+C	B	94.5
64-220-26*	sin/lin	B+C	B+C	**96.2**
DIGITS				
256-90-10	log/log	A	A	94.9
64-128-10**	slog/lin	B+C	B	95.8
64-128-10**	sin/lin	B+C	B+C	**97.0**

Legend: ***** and ****** - indicates that the same net was used in the different test as reported; **lin** - linear function; **log** - asymmetric logistic function; **slog** - symmetric logistic function.

3   The proposed application

We implemented a complete software system to be used in conjunction with a computer and a scanner. The software code is written in ANSI C++. The whole system is available under Windows 3.1, under MS-DOS, under Mac-OS and under Unix. The sheet are achieved using a scanner, and saved on a file using the TIFF format, using one of the following compression algorithms:

- Packbits;
- CCITT Group III 1D/2D;
- CCITT Group IV;
- No compression.

Sheets are to be pre-printed (forms), i.e. the characters are to be found on the sheet in specific locations.

In order to deal with little rotation of the paper sheet during the scan phase, the same paper sheet must have two marks, aligned horizontally: no matter the shape of the marks, these latter must be invariant with respect to rotations, as circles or rings.

Each form is described using a simple language that states where the characters are to be found: a brief but exhaustive description of the language is reported in table 2. Notice that in the table is reported just a single record, but there is no limit on the number of records that can be included in a single sheet description.

A real world example for the sheet showed in figure 1 is reported in table 3.

Table 2

```
TEMPLATEDOCUMENTNAME <name of the document>
RANGEDOCUMENT <max column> <max rows>
LEFTFIXEDSYMBOL <x coordinate> <y coordinate>
/* Represent the center of the left and right mark to deal with
little paper rotations while scanning is on */
RIGHTFIXEDSYMBOL <x coordinate> <y coordinate>
/* Individuates the area where the marks are to be found */
RANGEFIXEDPOINTS <y-min value> <y-max value>
/* Refers to mark size */
RADIUSFIXEDSYMBOL <radius>
RESOLUTIONX <horizontal resolution>
RESOLUTIONY <vertical resolution>
/* Minimum size area to be considered */
DIMFIXEDSYMBOL <width> <lenght>
RECORD <name of the record>
FIELD <name of the field> <number of cells> <field type>
CELL <x coord.> <y coord.> <width> <lenght> <cell type> [<class>]
...
ENDFIELD
...
ENDRECORD
```

Table 3

```
TEMPLATEDOCUMENTNAME AACM.TPL
RANGEDOCUMENT 2472 3494
LEFTFIXEDSYMBOL 177 142
RIGHTFIXEDSYMBOL 2293 142
RANGEFIXEDPOINTS 0 400
RADIUSFIXEDSYMBOL 30
RESOLUTIONX 299
RESOLUTIONY 299
DIMFIXEDSYMBOL 50 50
RECORD DatiStudente
FIELD Matricola 7 1
CELL 118 462 83 86 1
CELL 212 462 83 86 1
CELL 306 462 83 86 1
CELL 400 462 83 86 1
CELL 496 462 83 86 1
CELL 586 462 83 86 1
CELL 684 462 83 86 1
ENDFIELD
FIELD Nome 11 0
CELL 118 697 83 86 0
...
ENDFIELD
...
ENDRECORD
...
```

The Windows 3.1 demo version and the Mac-OS demo version also allows the user to write a characters within a box using the mouse and to submit the same to the system for recognition. For evaluation purposes it is also possible to change the thickness of the drawing.

Fig. 1: An example of a real world form sheet.

The output is in plain ASCII text: we plan to make this output available directly to a user dependent application. Recognition speed varies accordingly to the power of the platform, to the resolution used to scan the character, and to the number of presentations of the same character to the net.

In table 4 are summarized, in characters per second, the speed performances on different platforms.

The values refers to the following environmental choices:

- the paper sheet is a A4 format sheet;

- the paper sheet was scanned using a 300 dpi resolution;

- each character was sent to the net three times: the first one the character was simply rescaled as stated in chapter 2; the second and the third time the previously rescaled character is slightly rotated on the left and on the right side, as more accurately described in [4].

Figures 2-9 shown a typical work session while employing our software system under Windows 3.1. Figures 6, 7, 8, 9 refers to the use of the dialog box for

Table 4

Platform and operating system	Rec. speed (char/sec)
486DX 33 Mhz // Windows 3.1	2
486DX 33 Mhz // MS-DOS + Dos Extender (32 bits code)	8
Pentium 60 Mhz // Windows 3.1	8
Pentium 60 Mhz // MS-DOS + Dos Extender (32 bits code)	25
Digital Alpha AXP3000 // Unix	35

interactive testing of the system: this is ad additional feature of the demo versions under Windows 3.1 and Mac-OS.

4 Conclusions

We introduced a simple software solution for digits and uppercases handwritten characters' recognition. The proposed solution is based on a two-layer Multi Layer Perceptron (MLP) trained by a conjugate gradient descent (CGD) optimization algorithm embedded in a software tool for automatic handwritten character recognition. The software tool, to be used in conjunction with a scanner. is written in ANSI C++ and is portable, thus several platforms can be used.

The whole system is available under Windows 3.1, under MS-DOS, under Mac-OS and under Unix. The whole output is made available in plain ASCII: we plan to make this output available directly to a user dependent application, as a DBMS, for example. Recognition rate is platform independent: recognition speed varies accordingly to the power of the platform.

References

[1] G. A. M. Gioiello, F. Sorbello, A. Tarantino, G. Vassallo, *Simple Techniques for an Efficient Recognition of Handwritten Characters using a MLP*, submitted to International Journal of Intelligence Systems.

[2] G. A. M. Gioiello, F. Sorbello, A. Tarantino, G. Vassallo, *A neural solution for the hand-written character recognition task using a MLP digital architecture*, Twelfth IASTED International Conference February 20-23 1995, Innsbruck, Austria.

[3] M. J. D. Powell, *Restart Procedures for the Conjugate Gradient Method*, Mathematical Programming, Vol. 12, pp. 241-254.

[4] G. Vassallo, G. A. M. Gioiello, C. Condemi, F. Sorbello, *A MLP-Based character recognition system*, Proc. of the International Conference On Artificial Neural Networks, ICANN'94, Sorrento, Italy, 26-29 May, 1994. Springer Verlag eds.

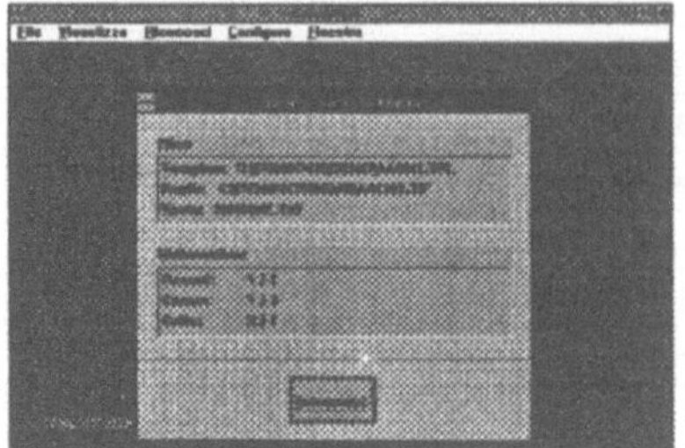

Fig. 2: The system while running.

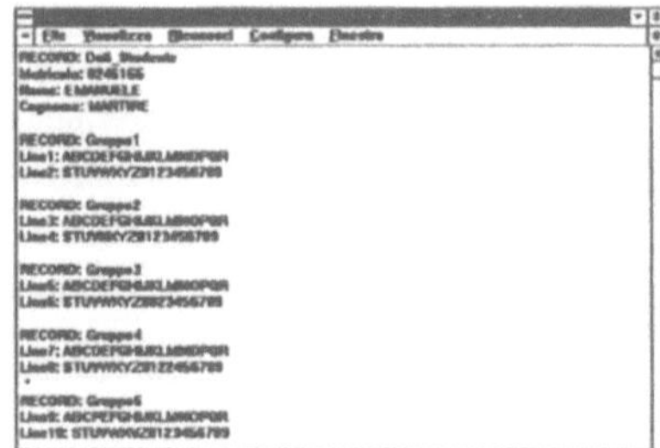

Fig. 3: The plain ASCII output.

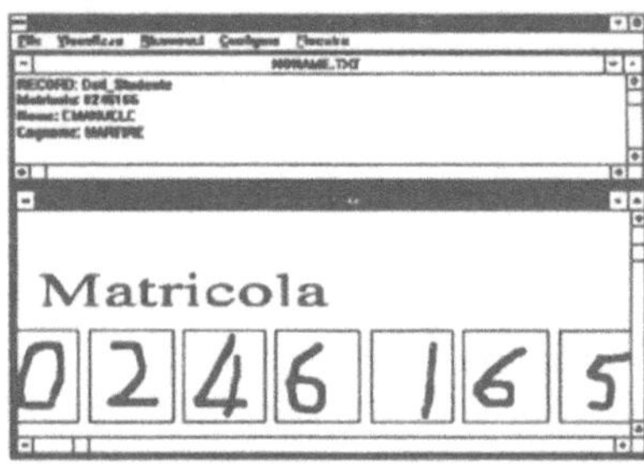

Fig. 4: The original ones and the recognized characters.

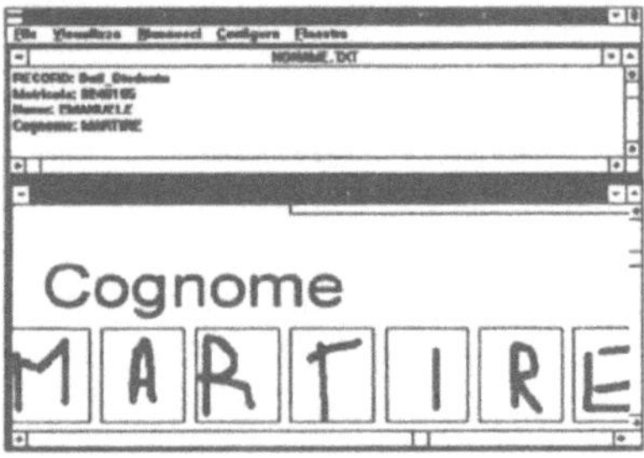

Fig. 5: A second example.

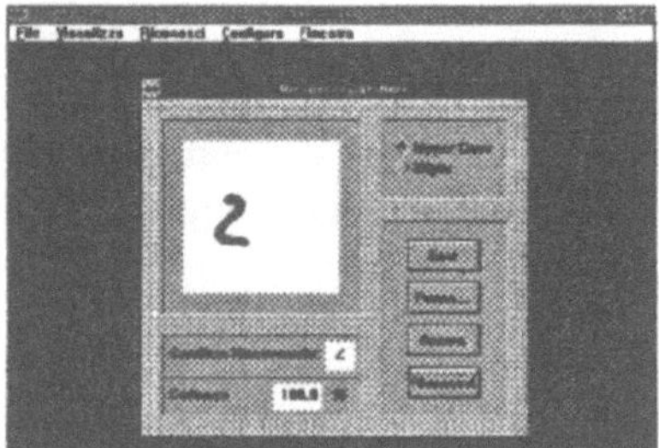

Fig. 6: An example of a **2** digit recognition. Confiance level is 100%.

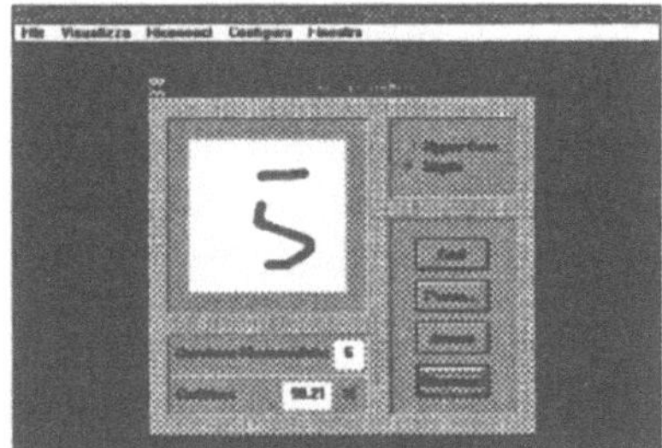

Fig. 7: An example of a **5** digit recognition. Confiance level is 98.21%.

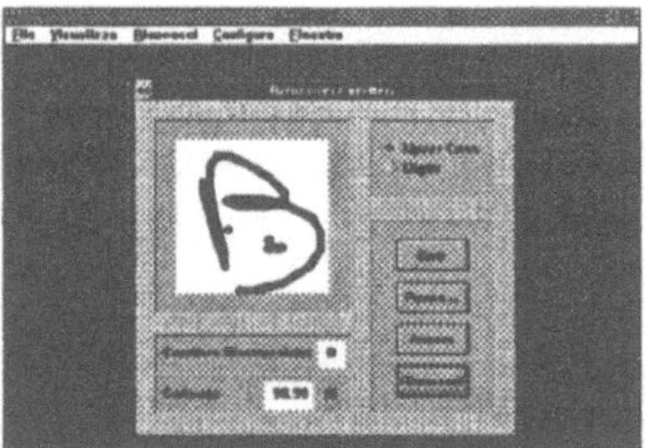

Fig. 8: An example of a **B** character. Confiance level is 98.90%.

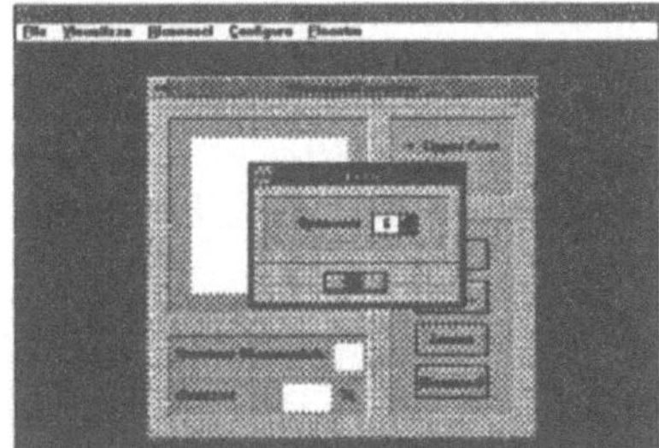

Fig. 9: The *choose drawing thickness* dialog box.

Are Multilayer Perceptrons Adequate for Pattern Recognition and Verification?

Marco Gori

Facoltà di Ingegneria, Università di Siena
v. Roma, n. 56, Siena, Italy

Franco Scarselli

Dipartimento di Sistemi e Informatica, Università di Firenze
v. S. Marta, n. 3, Firenze, Italy

Abstract

This paper discusses the ability of multilayer perceptrons (MLP) to model the probability distributions of the inputs in typical pattern recognition problems. It is shown that multilayer perceptrons may be unable to model patterns distributed in typical clusters, since in most practical cases these networks draw open separation surfaces in the pattern space. Unlike multilayer perceptrons, autoassociators and radial basis function networks (RBF) create closed separation surfaces. This make them more suitable especially for pattern verification, but also for dealing with large pattern recognition problems with many classes, where modular structures are typically used.

1 Introduction

In the last few years multilayer perceptrons have been used massively in a number of different problems and, especially, in the area of pattern recognition. The experimental results have been impressive in some applications involving a limited number of classes, whereas serious troubles have been reported when dealing with a large number of classes.

In this paper we give a theoretical interpretation of this problem by focusing attention on the capability of MLPs to draw close separation surfaces in the pattern space. This study is motivated by a simple analysis on common classifier desiderata. Classifiers must certainly exhibit high discrimination capability, but they must also capture effectively the probability distribution of the examples. Multilayer perceptrons are very well-suited for discriminating tasks, since the separation surface they create must meet a sort optimization criterion involving the classification error. Similar criteria, however, do not necessarily give the network the ability to reject patterns that do not belong to the probability distribution of the training set. From a geometrical point of view, this phenomenon is associated with the separation surfaces created during the learning phase. In order to capture the probability distribution of typical patterns, these surfaces must obviously be closed.

The main conclusion of this paper is that, in the case in which the network has more inputs than hidden units, these surfaces are open. In most practical pattern recognition problems this assumption on the number of units is certainly met, since networks currently used have a pyramidal architecture with a large number of inputs [4]. This results explains the limitation currently experimented when using MLP for problems of pattern verification [3] and for classification tasks with many classes. Any problem of pattern verification requires in fact to discriminate a given class with respect to any possible negative examples and, therefore, the separation surfaces must be closed. Similarly, when increasing the number of classes the network has to discriminate each class with respect to all the remaining examples.

When using more hidden units than inputs we show that MLP can indeed draw closed separation surfaces but, however, can also lead to open surface. We also show that autoassociators and a subclass of radial basis function networks are not plagued by this problem since they inherently create closed separation surfaces in the pattern space.

2 MLP for Pattern Recognition: Notation

Application of pattern recognition and verification that are based on MLPs are quite common. In similar applications, the patterns are represented by vectors in $\mathcal{R}^m$ and the classes by subsets of $\mathcal{R}^m$. A pattern recognition problem consists of partitioning $\mathcal{R}^m$ in different classes $S = \{C : C \subset \mathcal{R}^m\}$. Basically, given a pattern $\mathbf{y}$, the task of the network is to say whether $\mathbf{y}$ belongs to C for each C in S. A verification problem is a special version in which one has to check whether or not $\mathbf{y} \in C$. In practice, one or more neural networks are fed with pattern $\mathbf{y}$ and the network outputs give the answer according to a certain coding.

Let us consider the case in which each class C is modeled by a multilayer network with a single output neuron. In that case, C is defined as the set of patterns for which the output of the network is greater than a fixed threshold, that is

$$C = \{\mathbf{x} : f(\mathbf{x}) \geq L, \mathbf{x} \in \mathcal{R}^m\}, \tag{1}$$

where $f : \mathcal{R}^m \to \mathcal{R}$ is realized by an MLP network and L is a fixed real threshold. In general, p classes $C_1, \ldots, C_p$ can be modeled by a network with p output nodes. In this case the output of networks i-th defines class C_i.

Another possible way of implementing a classifier is that of using autoassociators, i.e. multilayer networks with as many inputs as outputs. Each autoassociator must model a class. During the learning phase, the autoassociator is feeded with the patterns of the class and it is asked to perform the copy of the inputs to the outputs. During the test, in order to establish whether a pattern is in the class modeled the autoassociator, the output is compared with the input in terms of geometrical distance. The following criterion is typically used

$$C = \{\mathbf{x} : \|f(\mathbf{x}) - \mathbf{x}\|_\infty / \|f(\mathbf{x})\|_\infty \leq L\} \tag{2}$$

where $\| \cdot \|_\infty$ denotes the superior norm.

Pattern classification can be implemented simply by considering as many autoassociators as classes and by feeding any incoming pattern to all the autoassociators. The class can be established by looking for the autoassociator for which the input/output distance is minimum. Whereas, problems of pattern verification can be approached by inspection of the input/output distance of the network when feeding it with incoming patterns. They turns out to be verified provided that such distance is below a threshold defined in advance.

3 Separation Surfaces

Let us consider networks with three layers, one input, one hidden, and one output layer, and assume that the activation function $\sigma(\cdot)$ is monotone. Such a network realizes functions in the form of $f(\mathbf{x}) = \sigma\left(\sum_{i=1}^{n} w_i \sigma(\mathbf{v}_i^T \mathbf{x} + b_i) + c\right)$, where $\mathbf{v}_i \in \mathcal{R}^m$, $w_i, b_i, c \in \mathcal{R}$, $1 \leq i \leq n$, are the parameters of the network and $\mathbf{x}$ is the vector of the input signals. Finally, suppose to deal with pattern recognition problems where the patterns of a given class belong to C (1).

Recent results on approximation theory prove that multilayer networks can model any class on a compact subset of $\mathcal{R}^m$ up to any degree of precision, provided that the perceptron includes a sufficient number of hidden layer neurons [6, 5]. On the other hand, in practical applications, the number of hidden units needed to fulfill the purpose is exaggerated and would lead to overtraining phenomena with a consequent scarce generalization to new examples. The following theorem gives some insights on the case in which the input units are more then the hidden neurons. This is commonly verified in many practical cases where the dimension of the input space is very large.

Theorem 1 *Let $\sigma(\cdot)$ be a monotone increasing function, and consider the examples belonging to $C \neq \emptyset$. If and $n \leq m$ then C is not limited.*

Proof: Since $n \leq m$ the vectors $\mathbf{v}_1, \ldots, \mathbf{v}_{n-1}$ are not a basis of $\mathcal{R}^m$. Consequently, there exists a vector $\bar{\mathbf{y}}$ such that $\mathbf{v}_i^T \bar{\mathbf{y}} = 0$, $1 \leq i \leq n-1$.

Let $\mathbf{y}_0 \in C$ be and $\alpha \in \mathcal{R}$, and consider $\mathbf{y} = \alpha \bar{\mathbf{y}} + \mathbf{y}_0$. Let us assume that $w_n \sigma(\mathbf{v}_n^T \mathbf{y} + b_n) = w_n \sigma(\mathbf{v}_n^T \mathbf{y}_0 + \alpha \mathbf{v}_n^T \bar{\mathbf{y}} + b_n) \geq w_n \sigma(\mathbf{v}_n^T \mathbf{y}_0 + b_n)$ holds. The last inequality can easily be fulfilled by choosing α such that $\alpha w_n \mathbf{v}_n \bar{\mathbf{y}} > 0$. Consequently, we get

$$
\begin{aligned}
f(\mathbf{y}) &= \sigma\left(\sum_{i=1}^{n} w_i \sigma(\mathbf{v}_i^T \mathbf{y}_0 + \alpha \mathbf{v}_i^T \bar{\mathbf{y}} + b_i) + c\right) \\
&= \sigma\left(\sum_{i=1}^{n-1} w_i \sigma(\mathbf{v}_i^T \mathbf{y}_0 + b_i) + w_n \sigma(\mathbf{v}_n^T \mathbf{y} + \alpha \mathbf{v}_n^T \bar{\mathbf{y}} + b_n) + c\right) \\
&\geq \sigma\left(\sum_{i=1}^{n-1} w_i \sigma(\mathbf{v}_i^T \mathbf{y}_0 + b_i) + w_n \sigma(\mathbf{v}_n^T \mathbf{y}_0 + b_n) + c\right) = f(\mathbf{y}_0) \geq L
\end{aligned}
$$

Finally we find that $\mathbf{y} \in C$ and, since $\mathbf{y}$ can be chosen arbitrarily large, we conclude that C is not limited. ∎

Notice that the condition given in this theorem is only necessary. There are indeed examples in which multilayer perceptrons with more hidden units than

208

inputs can be trained to model a class whose patterns lie in a limited domain, but the separation surfaces of the corresponding network are open.

This can be illustrated straightforwardly by the experiments shown in figures 1 and 2. A set of random patterns in R^2 with module 0.5 and 2 was used for producing the positive and negative examples, respectively. The positives patterns are represented by stars, and the negative examples by circles. It was assumed that a pattern belongs to the modeled class when the output of the network was larger than 0.8, whereas the pattern belongs to the complementary class when the output is smaller than 0.2. The continuous and the dotted line represent the contour with an output equal to 0.8 and the contour of the level 0.2, respectively. The employed network has 5 hidden neurons, and, although the number of hidden units is greater than the number of inputs, separation surfaces are not closed in the example of Fig. 1. In practice, closeness of separation surfaces depends also on the distribution and the number of the negative patterns. In fact, unlike what happens in Fig. 1, in Fig. 2 the number of the negative is large enough to to obtain a closed separation surface.

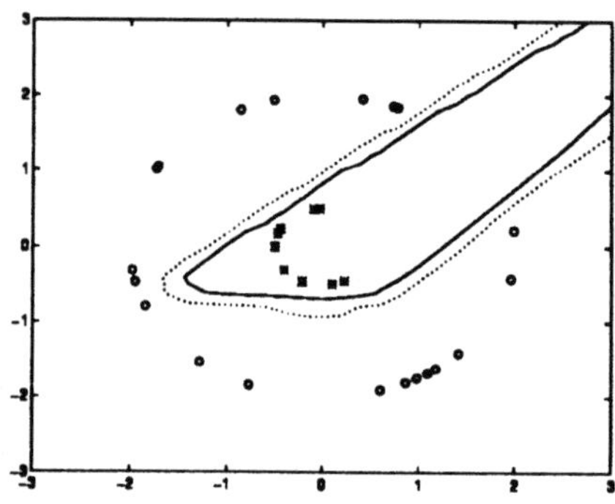

Figure 1: A class modeled by a multilayer perceptron.

Figure 2: The same class when using more examples.

4   Autoassociators

Recently, Bianchini *et al.* [1] have proven that if C is defined according to (2) and $f(\cdot)$ is the function realized by a multilayer perceptron with sigmoid activation function in the hidden layer and linear activation function in the output layer, then domain C is bounded. The result remains valid also if $\|f(\mathbf{x}) - \mathbf{x}\|_\infty / \|f(\mathbf{x})\|_\infty$ in (2) is replaced by $\|f(\mathbf{x}) - \mathbf{x}\|_\infty$[1] and if the superior norm is replaced by other norms.

It is interesting that the result does not depend on training, that is an autoassociator always produce bounded domains independently of the training. Fig. 3 gives an example of typical bounded domains created by autoassociators. For each patter, the corresponding error in the autoassociation process turns out to be $e(\mathbf{x}) = \|f(\mathbf{x}) - \mathbf{x}\|^2$, where $\|\cdot\|$ denotes the Euclidean norm and $\mathbf{x}$

---

[1] That is the distance between $f(\mathbf{x})$ and $\mathbf{x}$ is measured by the norm of the difference.

the input pattern. The network we used had two inputs, two output neurons and one hidden neuron.

The inherent property of drawing closed separation surfaces makes autoassociators very well suited for problems of pattern verification (see e.g. [3]). Autoassociators, however, are also of interest for pattern recognition, especially when dealing with many classes. They are in fact a truly modular behavior and can be effectively used as building blocks in large connectionist systems [2].

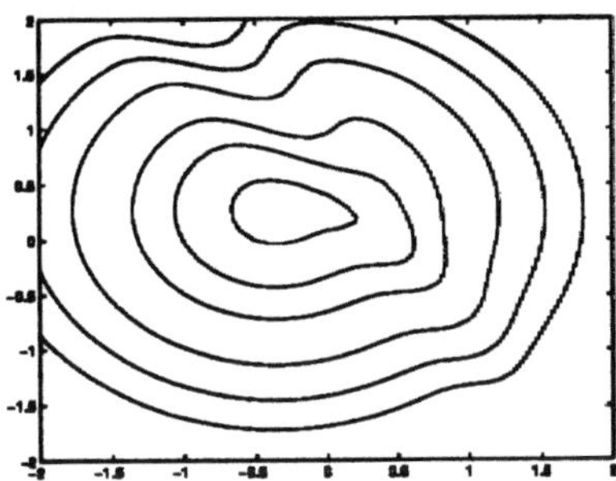

Figure 3: The contour lines of the function that, for each pattern, returns the autoassociator error.

5 Radial Basis Functions

Let us consider radial basis function networks (RBF) and suppose again that C is defined by equation (1). A similar network computes functions in the form of $f(x) = \sum_{i=1}^{n} w_i k \left(\|\mathbf{x} - \mathbf{v}_i\|/b_i \right) + c$, where $k(\cdot) : \mathcal{R} \to R$ is a symmetric function, decreasing in $[0, \infty)$, for which $lim_{x \to \infty} k(x) = 0$, $\| \cdot \|$ is the Euclidean norm, $\mathbf{v}_i \in \mathcal{R}^m$, and $b_i, w_i, c \in R$, $1 \leq i \leq n$ are the parameters of the network.

If we place no restriction on the networks parameters, the separation surfaces drawn by these networks can either be closed or open. However, when restricting the weights connecting the hidden and the output layer to be nonnegative, the separation surfaces drawn be RBF turn out to be closed.

Theorem 2 *Let C be a domain defined according to (1). Let $f(\cdot)$ be the function computed by a RBF network. Furthermore, assume that that parameters $w_1, \ldots, w_n$ are nonnegative and that $C \neq \mathcal{R}^m$. Then domain C is bounded.*

Proof: (Sketch). Note that, since $C \neq \mathcal{R}^m$, there exists $\mathbf{y} \in \mathcal{R}^m$ so that $f(\mathbf{y}) < L$. Consequently, from the definition of $f(\cdot)$ and the non-negativeness of $w_1, \ldots, w_n$, we get $c < L$. Now we can immediately see the reaction of the network as $\| \mathbf{x} \| \to \infty$:

$$\lim_{\|\mathbf{x}\| \to \infty} f(\mathbf{x}) = \lim_{\|\mathbf{x}\| \to \infty} \sum_{i=1}^{n} w_i k(\|\mathbf{x} - \mathbf{v}_i\|/b_i) + c = c < L.$$

Hence, we conclude that "large" patterns are not in C. ∎

This theorem states that RBF networks share with autoassociators the property of generating bounded domains for pattern classification and verification.

It is worth mentioning that the non-negativeness assumption on the output weights does not limit the computational capabilities of general RBF networks. Theoretical results on the universal interpolation of RBF are indeed gained under that assumption (see e.g. [7]).

6   Conclusions

In this paper, we have discussed the ability of multilayer perceptrons to create bounded domains in classification and verification tasks. We have proved that three layered perceptrons with less hidden units than inputs cannot model bounded classes effectively, since the created separation surfaces are open. This limits severely the application of these networks to pattern verification and also to problems of pattern recognition involving many classes.

Multilayer networks acting as autoassociators do not suffer for this problem. Autoassociators, however, do not exhibit the typical high discrimination capability of classical MLP and, therefore, hybrid systems using properly both the architectures seem to be more promising for very large pattern recognition problems [2]. Moreover, it has been shown that under a simple non-restricting assumption on the output weights, RBF networks resemble autoassociators in their capability of drawing closed separation surfaces.

References

[1] M. Bianchini, P. Frasconi, and M. Gori. Learning in multilayered networks used as autoassociators. *IEEE Transactions on Neural Networks*, 6(2):512–515, March 1995.

[2] A. Frosini, M. Gori, and P. Priani. A neural network-based model for paper currency recognition and verification. *IEEE Transactions on Neural Networks*, 1996. To appear.

[3] M. Gori, L. Lastrucci, and G. Soda. Autoassociator-based models for speaker verification. *Pattern Recognition Letters*, 1996. To appear.

[4] M. Gori and A. Tesi. On the problem of local minima in backpropagation. *IEEE Transactions on Pattern Analysis and Machine Intelligence*, PAMI-14(1):76–86, January 1992.

[5] K. Hornik. Some results on neural network approximation. *Neural Networks*, 6:1069–1072, 1993.

[6] M. Leshno, V. Lin, A. Pinkus, and S. Shocken. Multilayer feedforward networks with a polynomial activation function can approximate any function. *Neural Networks*, 6:861–867, 1993.

[7] J. Park and I. W. Sandberg. Approximation and radial-basis-functions networks. *Neural Computation*, 5:305–316, 1993.

Computational Intelligence in Electromagnetics

F.C. Morabito
Dipartimento di Ingegneria Elettronica e Matematica Applicata
Università di Reggio Calabria
Reggio Calabria, Italy
E-mail: morabito@csiins.csii.unirc.it

Abstract

A combined use of the three techniques globally referred to as computational intelligence (i.e., Artificial Neural Networks, Fuzzy Logic and Evolutionary Computation) is proposed to cope with the solution of inverse problems. Electromagnetics is presented as an interesting application field of the methods. Firstly, it is shown how simple autoassociative ANN might successful play a role in fault tolerant feature extraction. Secondly, hybrid neuro-fuzzy systems are shown to have better reconstruction performance with respect to ANN and FL systems working alone. Finally, evolutionary techniques are used to improve the strategy of optimization of a neural processor in a simple identification problem.

1. Introduction

A lot of interesting as well as commonly encountered tasks in electromagnetics can be formulated as inverse problems. Relevant examples are the optimization and/or the synthesis of a device starting from the operative requirements and the identification of inaccessible systems from non invasive measurements taken outside the inspected body. From an analytical viewpoint, the solution of such problems implies the inversion of an integral expression in which the unknown function to be solved for is operated by a kernel which has a smoothing effect. The inversion of the operator is extremely sensitive to small changes or errors which reveals the ill-posedness of a task. In addition, there is an implicit loss of information related to the smoothing process. Several difficulties arise in numerically treating the inverse problem: the three-dimensional and non linear nature of typical electromagnetic problems further complicate the search for a solution, if any and unique. To refine the solution, iterative procedures are typically introduced which involve the minimization of some suitable criterion functions. Apart from the quantitative approach of searching for approximate solutions involving linearization, and of developing some knowledge on the corresponding direct problems, the approach to inverse problems is rather based on qualitative analysis. Instead of attempting to find the exact solution of a set of non linear equations, it seems often more informative to gain an approximate insight on the problem by investigating a bank of input-output relationship obtained from a suitable scheduled experimental campaign. Approximate solutions also imply lower computational costs as well as the possibility of easily introducing *a priori* knowledge on the solution which has the same effect of adding regularizing terms to the cost function.

In this respect, intelligent computation in electromagnetics can play a relevant role. However, the related procedures cannot be compared to the standard approaches in simple benchmark cases. When a standard approach is sufficient to determine a solution, computational intelligence cannot add nothing and its proposal can be suspiciously interpreted. Artificial Neural Networks (ANN), Fuzzy Logic (FL), and Evolutionary Computation (EC) are indeed currently investigated as powerful tools which are able to overcome some of the limitations of standard numerical techniques, particularly when properly combined.

In this paper, we consider some aspects of these strategies of computation which seem at the same time new and promising, as well as reliable enough to suggest their application in practical electromagnetic problems. ANNs have already been applied in electromagnetics particularly to solve inverse problems and to optimize design [1-4]. Fuzzy Logic (FL) has motivated a special interest, particularly in Japan, for engineering applications. The evolutionary thinking has influenced the researchers as a natural way to cope with optimization with constraints [5].

We present here a model reduction application of ANNs in electromagnetics as well as an example of how hybrid ANN-FL approaches can do better with respect to each one of them operating alone. In the final section of the paper, we show how evolutionary computation can also have an impact in optimal design.

2. ANN as a Fault Tolerant Principal Component Analyzer

One of the most important steps in multidimensional data analysis is related to the determination of the directions along which patterns of data show maximum variance. It is in fact impossible to discriminate among different vector inputs if they are simply equal with each other! This key problem in data manipulation is commonly referred to as *feature extraction* in pattern recognition theory. In real-time operating systems, we have to deal with lots of spatially correlated data coming from sensors and we would like to just process the most significant combinations of the patterns for the problem under study. Principal Component Analysis (PCA) basically allows to determine the optimal projection directions by operating like an unsupervised learning algorithm. PCA is classically a *batch* technique which is not suitable for applications where data come in the on-line way. Several self-organizing rules have been proposed in the recent literature which are also capable to cope with this problem [6, 7]. However, in this paper, we wish to point out a different aspect related to the *fault tolerance* of data reduction procedures commonly implemented via standard and neural PCA.

It is well known that one of the important features of PCA is that the set of determined projections are uncorrelated: this leads to a better conditioned optimization procedure when using gradient descent, like in backpropagation. However, the loss of one of the principal projections has the direct consequence of destroying the reconstruction. For this reason, it can be helpful to have a more distributed representation in which those projection directions are artificially duplicated during learning. It is possible to easily reach this goal by using a single hidden layer ANN which can discover the subspace spanned by the first k principal components without exactly determine the best projection directions. The cited ANN learns in supervised autoassociative mode: once fixed the number of the hidden nodes, the learning procedure aims to determine the reduced configuration which is optimal in terms of reconstruction error. The reduced representation is not an orthogonal one and, far from being a drawback, this can be viewed as a desired feature of the obtained compressed vector.

In the autoassociative ANN, the input layer is driven by the sensors. The output layer has the same number of nodes of the input layer. The single hidden layer acts as a feature extractor. Each neuron of the network is linear. Once trained with the standard backpropagation algorithm, using as desired response the same data used as inputs, each hidden node of the resulting linear ANN is related to a PC. In particular, by selecting an ANN with one hidden neuron, the training yields exactly the first PC. The PC's learning is a simple task because of the absence of non-linearities which implies the absence of local minima of the cost function to be minimized.

We have here used the above described ANN-PCA technique to reduce the number of inputs feeding an ANN which has the task of recovering the equilibrium parameters of a plasma column evolving in a tokamak reactor[3]. We have made reference, just as an example, to the ITER (International Thermonuclear Experimental Reactor) geometry and to a possible configuration of magnetic sensors (flux loops and field probes). A set of about 1000 patterns have been generated by using an equilibrium code. In the course of the simulated experiments, we have supposed of picking up 77 magnetic measurements from suitable locations along the boundary of the machine . The computed measurements are the inputs of the ANN which has a variable number of hidden nodes. By projecting the input pattern onto a subspace of 20 dimensions we obtain a very good reconstruction of the input pattern from its reduced representation. The RMS error is reported in Fig.1(a). The effect of introducing a greater number of PC's is made useless by the presence of the measurement noise which prevent to obtain a zero error and from the consideration that a standard eigenanalysis suggest a number of 10 as suitable enough for the data set here analyzed. The determination of the minimal number of hidden nodes is a matter of compromise between the need of having an enough good reconstruction error and the need of adequately guarantee the fault tolerance of the model. The achieved representation is such that a fault in whatever hidden node does not reduce the reconstruction accuracy in a measurable way. On the contrary, the exclusion of each one of the first three PC's basically destroy the reconstruction.

By introducing a sigmoidal activation function to the hidden nodes, one can add some extra robustness to the model, also extracting higher-order statistics. However, in this case we have to deal with the problem of local minima of the cost function and thus with the dependence from the weights initialization.

This complicates the training, particularly considering our task, where the input dimensionality is very high and the input measurements are spatially correlated. Fig.1(b) reports a typical RMS curve for this nonlinear PCA. It is clear that the training is getting stuck in a local minimum which prevents to reach a good reconstruction.

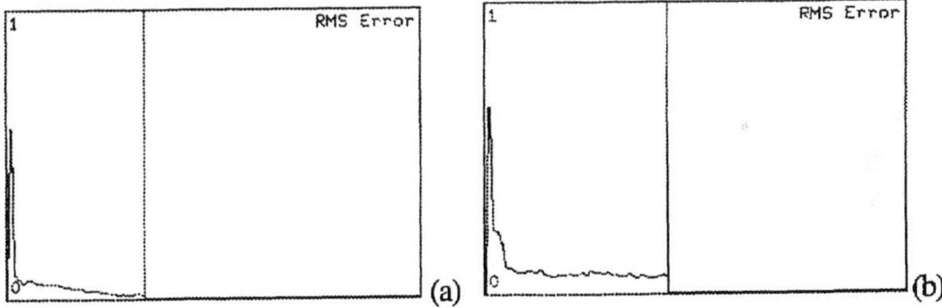

Figure 1: RMS error curve for a Linear and a Non-Linear ANN-PCA

3. A Fuzzy Approach to ANN Modeling: Identification of a Circular Hole by Using an Adaptive Hybrid Processor

FL and its engineering developments can be used to significantly reduce the computational burden of some iterative tasks, namely, the determination of the extrema of a cost function. Fuzzy systems can be represented as series expansions of fuzzy basis functions, which have been proved capable of universal approximation [10]. In such a way, linguistic IF-THEN rules, possibly obtained from human experts familiar with the system under study, can be used either to simplify a decision process or to reduce the computational complexity of a task, like the training of an ANN. Furthermore, by considering a sigmoidal ANN the fuzzy techniques can be exploited to introduce vague information about the solution into the model. This can be used in inverse problems to reduce the ill-posedness of the task.

In the example here presented, we consider the detection and characterization of a circular hole on an earthed plane based on the analysis of the deformation of the eddy current pattern obtained by exciting the configuration by means of a solenoidal coil. The identification problem consists of determining the location (x_h and y_h) and the size (a) of the hole starting from a set of magnetic field and flux measurements. While we have shown elsewhere [11] that it can be useful to introduce some rules into the expansion approximation, we suppose here of not possessing any *a priori* information to be expressed in terms of fuzzy rules from which to approximately assess the defect location and size. Nonetheless, a combined neuro-fuzzy processor can yet be used to improve the identification performance. Fig.2 depicts a hybrid model which includes both a block based on linguistic rules and an adaptive part composed of a fuzzy basis function and a sigmoidal section.

We then suppose that there are no rules from which synthesizing the fuzzy part of the processor. Consequently, we propose of adaptively generating the fuzzy section. Even if the information are extracted from the training data set which is also used to train the ANN part, the use of an adaptive fuzzy inference system is convenient in that it can explain how the decisions about the identification problem are taken. In detail, we assign to each input variable (differential magnetic flux measurement) a triple of fuzzy values within the expected range (which can be approximately estimated from the training data set). The output variables may assume five fuzzy values (namely, LARGE NEGATIVE, SMALL NEGATIVE, ZERO, SMALL POSITIVE and LARGE POSITIVE for x_h and y_h and VERY SMALL, SMALL, ZERO, LARGE, VERY LARGE for the hole size, a). The choice of the membership functions and their parameters is carried out by supposing a total absence of prior information. Thus, the range of each variable is preliminary subdivided into subregions of similar sizes.

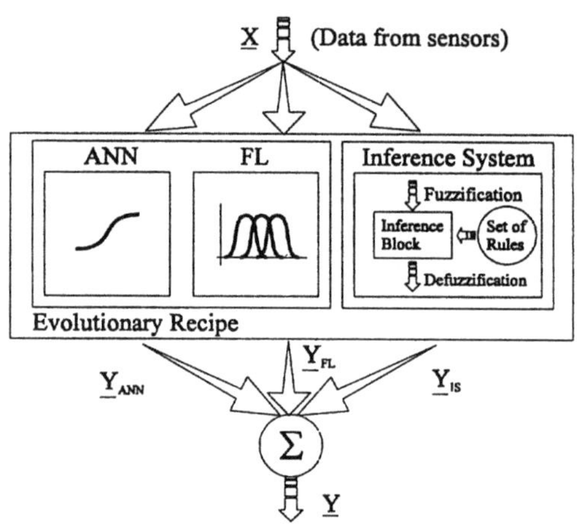

Figure 2: The hybrid numeric-linguistic network

The selected kind of MF is the generalized bell curve of equation: $f(x,a,b,c) = (1 + |(x-c)/a|^{2b})^{-1}$ where the parameter c locates the center of the curve, and $b>0$ controls the curvature. This MF guarantees to the final mapping satisfying smoothness properties. In addition, the selection of the ideal center is less constraining with respect to a Gaussian function. Once specified the MF's type and number, an optimization procedure is started which aims to determine an optimal configuration for all the free parameters of the model (i.e. the MF parameters). The procedure uses least-squares and backpropagation for linear and non-linear parameters, respectively. Fig. 3 shows the results of the optimization procedure for two input variables, the flux differences ($\psi_5-\psi_7$) and ($\psi_6-\psi_8$).

Basically, the final MFs adhere to the local features of the training data set. The adaptive technique also automatically generate a set of fuzzy rules whose consistency can be verified off-line. Once the fuzzy representation has been determined, the next step is again the tuning of the identification by means of the training of the sigmoidal part of the mapping. Table 1 reports the performance of the combined model compared to that obtained by using both a standard ANN and the fuzzy part of the processor.

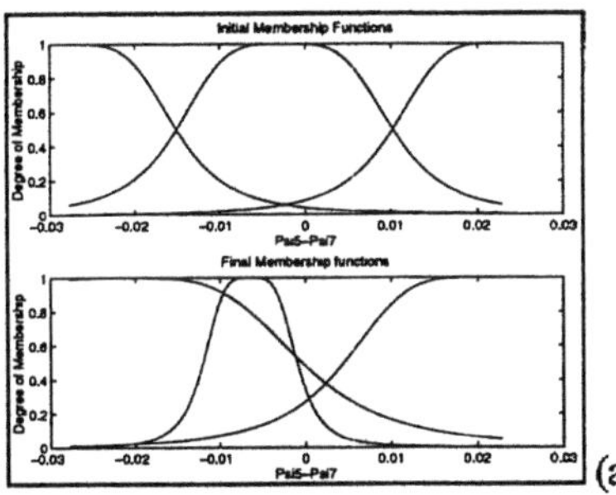
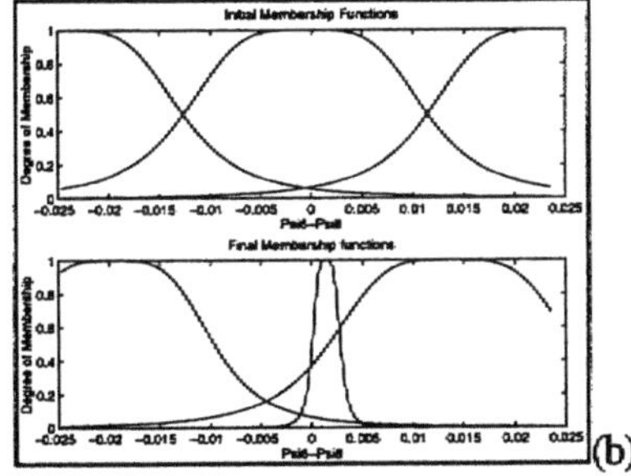

Figure 3: Membership functions before and after the optimization procedure for the fuzzy variables $(\psi_5 - \psi_7)$, (a), and $(\psi_6 - \psi_8)$, (b).

Output Variable	x_h	y_h	a
Range (mm)	$0 \div 250$	$0 \div 120$	$1 \div 5$
Standard ANN's	2.5	2.7	0.65
Fuzzy Block	3.4	3.9	0.85
Hybrid System	2.1	2.2	0.45

Table 1: Comparison of the hole identification performances from different techniques in terms of physical RMSE (mm).

4. Evolutionary Computation in Optimization.

The basic approach to optimization is to define a type of measurement, a cost function (also in presence of a regularizer), that expresses the index of performance and iteratively improve this performance by selecting from among possible alternatives. The typical strategy is to generate a deterministic sequence of trial solutions based on gradient descent (or higher-order statistics) of the cost function. This is also the technique of learning used by backpropagation. In this respect, evolutionary computation can play an important role [5]. Here, we will show by a simple example how simulated evolutionary optimization can improve the performance of a standard ANN model. The proposed example concerns the determination of the three state parameters describing a circular plasma evolving in the vacuum chamber of a tokamak reactor (namely, the plasma center R- and Z- co-ordinates, and the asymmetry factor, Λ). The training of standard ANNs on a suitably generated data set of patterns allows to find a local minimum of the cost function, and special topologies can favor an asymptotically convergence towards a good minimum [2, 3]. In principle, by using an evolutionary algorithm, we can achieve the global minimum, and also determine a solution when random perturbations are imposed on the cost function. In our example, we train an ANN until a satisfactory minimum of the LMS error function is obtained. Unfortunately, due to the nature of the gradient descent procedure and to the convoluted behavior of the error function related to the presence of nonlinearities, the optimal weight vector is difficult to achieve. Furthermore, the final result of the optimization procedure depends on the initial condition, i.e. the initial weights seeding. A major drawback of evolutionary computation, as well as other stochastic techniques, i.e. simulated annealing, is the computation time, often prohibitive for any engineering applications. However, the convergence toward an optimal weight vector can be simplified when starting the optimization procedure near a good minimum. We solve this problem by the following simple evolutionary recipe:

1) 20 different ANNs are initialised with different weights matrices into a proper range (-0.5, 0.5)

to generate an initial population of parent vectors;

2) An offspring matrix is created from each parent by adding a Gaussian random variable with zero mean and pre-selected standard deviation to each weight;

3) A selection criterion based on the evaluation of a fitness function determines which matrices survive to the next generation by comparing each parent matrix to its offspring;

4) The process is iterated until a sufficient solution for the initial weights matrix is reached.

Once the best ANNs were selected, a training procedure was started for each one of them. The evolutionary process is repeated in the vicinity of the convergence procedure stopping with a similar technique to determine the best solution.

We don't discuss here in more details the global optimization procedure, however, Table 2 reports the results obtained in our test case. Moreover, we wish to stress the point that an evolutionary computation strategy might be the only way to solve more difficult problems.

Normalized RMS error	%R	%Z	%Λ
ANN + in-out link	1.4	1.3	1.5
Evolutionary ANN	0.7	0.4	0.8

Table 2 : Performance of the evolutionary ANN model in terms of full-scale percent RMSE

5. Concluding Remarks

In this paper, we have shown that a combination of techniques related to the computational intelligence can greatly improve standard numerical processing of data in electromagnetics. The first candidate problem for exploiting the methods here presented is the inverse problem typically encountered in such areas. The paper just show some encouraging results in this respect, but the aim of current researches is to demonstrate, in practical problems like Non Destructive Testing, that the proposed methods can yield a satisfying even if qualitative approach when standard techniques fails to work.

References

1. E.Coccorese, R.Martone and C.Morabito, A Neural Network Approach for the Solution of Electric and Magnetic Inverse Problems, IEEE Trans. on Magnetics, 1994, 30: 2829-2839.
2. F.C.Morabito and M.Campolo, Advances in Neural Network Non Destructive Testing, Proc.2nd Intl.Conf.on Comp. in Electromagnetics, IEE, Nottingham, UK, 1994, 75-79.
3. E.Coccorese, R.Martone, F.C.Morabito, Identification of non-circular plasma equilibria using a neural network approach, Nuclear Fusion, 1994, 34: 1349-1363.
4. O.A.Mohammed, et al., Design Optimization of Electromagnetic Devices using Artificial Neural Networks, IEEE Trans.on Magn., 1993, 29: 1775-1778.
5. Special Issue on Evolutionary Computation, IEEE Trans. on Neural Networks, 1994, 5.
6. E.Oja, Principal Components, Minor Components, and Linear Neural Networks, Neural Networks, 1992, 5: 927-936.
7. S.Y.Kung, and K.I. Diamantaras, A Neural Network Learning Algorithm for Adaptive Principal Component Extraction (APEX), Proc. ICASSP, Albuquerque, NM, 1990: 861-864.
8. S.Haykin, Neural Networks: A Comprehensive Foundation, IEEE Press, 1993.
9. E.Coccorese, F.C.Morabito, A Fuzzy Modeling Approach for the solution of an inverse electrostatic problem, to be published on IEEE Trans. on Magnetics, 1996 issue.
10. IEEE Trans. on Neural Networks, Special Issue on Fuzzy Logic and Neural Networks, 1992, 3.
11. F.C.Morabito et al., Identification of Circular Holes in Thin Plates using a Neural Network Approach, paper presented at the ISEM Conference, Cardiff, UK, 1995.

M.A.I.A. Unsupervisioned Neural Network in the Compression and Decompression of Images.

Giuseppe Pappalardo, Domenico Rosaci, Giuseppe M.L. Sarnè
DIEMA - Facoltà di Ingegneria - Università di Reggio Calabria
Reggio Calabria, 89128 - Italia

Abstract

In this paper a new technique for compressing the images in a codified way and the subsequent explosion in a standard form is presented. Through the use of a neural network based on a new technique of information arrangement the problem of image processing has been well resolved. The results obtained are compared with those obtained through the use of the techniques used at present.

1 Introduction

All the algorithms for the compression and decompression of images have the common objective of reducing the image to the minimum number of information necessary for its reconstruction. Among the most interesting methods actually studied by the researchers there are those based on the mathematical theories of fractal compression (known as *Iterated Function Systems* and *Recurrent Iterated Function Systems* [1]); in particular, the fractal compression [4] (and its variants [6]), uses some transformations acting on blocks of images through a strategy of exhaustive search of redundant information. Recently, different techniques of image compression using artificial Neural Networks (NN) have been proposed (NNs search the relationship between data on the basis of a *learning* made on examples of pairs of data belonging to the sets to be related) [3], [2], [7], [5].

The Modulate Asynchronous Information Arrangement (MAIA) theory [10] is a technique for the data processing and the MAIA model can be well applied in a NN framework [9] [11], in particular in order to assess an effective learning strategy. The MAIA NN is particularly useful for the study of the image compression; in fact, by modelling the image through the arrangement a significative reduction of the image dimension can be obtained. Furthermore, the MAIA NN can be identified as a mathematical model of compression likewise the fractal compression techniques, but it can also be used in a neuronal framework, with a large economy of computing time. In this work a unsupervisioned MAIA NN has been used to compress a Black/White (B/W) image, whose dimensions are 128 pixels x 128 pixels, with 256 tonality of grey.

2 The Image Compression Generality

A B/W image can be represented through a matrix of points (pixels) on a rectangular background. A level of grey can be associated to each pixel with a value belonging to the range 0 (black)+255 (white) and a byte can be associated to each pixel. Each image can then be seen as a matrix of bytes $(\mathbf{I}_{m,n})$, whose dimensions are m and n (the same dimensions of the image). The objective of the compression is the transformation $T{:}\mathbf{I}_{m,n} \to \mathbf{I}_{d*}$ where $\mathbf{I}_{d*}$ is the vector having a dimension equal to $d < mxn$. Naturally, it is better that d is smaller than mxn as much as it is possible compatibility with the loss of information. The ratio $(mxn)/d$ denotes the obtained *compression degree*.

When the image has been codified in the vector $\mathbf{I}_{d*}$ it can be reconstructed through a suitable transformation of $\mathbf{I}_{d*}$. Naturally a check needs in order to establish if there was any loss of information that could worsen the quality of the image. The goodness of the reconstruction can be evaluated through some indicators: the *signal-to-noise ratio* (SNR) and the *peak value of* SNR itself. These two indicators evaluate the ratio signal/noise present in the compression.

In this paper a compression of an image-matrix nxn into an image-vector $n+\lambda n^2$ is presented (λ is a coefficient <1). The square dimension of the pixel matrix of the image does not affect the generality of the method proposed in the following. The searched transformation T is represented by a MAIA model, whose theoretical aspects are presented in the following as well as the resolution algorithm of the MAIA model for the two cases of compression and decompression of an image.

3 The MAIA Model

The problem relating pairs of data can be defined as follows: *"given two sets X and Y of atomic elements (respectively of cardinality equal to K and H), the unknown relationship R between the elements of X and the elements of Y has to be reached"*.

The relationship R is defined by its functional form (given in this case by the sequentially application of a finite number of elementary mathematical operations) and a suitable set of parameters. If in a first stage the functional form of R is supposed to be explicitly given, the previously problem can be expressed as follows: *"a set P of real numbers such that for each $y \in Y$ then $y=R^P(x_1,x_2,...,x_K)$ has to be reached, R^P being the functional form explicitly given and specified by the set of parameters P and the K elements $\{x_1,x_2, \ ,x_K\} \in X.$"*. An average error ε (*a priori* fixed) can be made on the whole set Y in the calculation of each y defined as above. The proposed functional form R relates the elements of X to the elements of Y through a weighted graph G defined as follows: *"G:$\forall x \in X$ a node $i \in G$ exists on which the value of x is defined and $\forall y \in Y$ a link $r \in G$ exists on which the values of y is defined"*. G is then formed at least by K nodes and H links and this condition is always possible. In fact, if a node is connected to all the others, $K(K-1)$ links are obtained, and if $K(K-1) \geq H$ the condition is verified, while in the other case K^* auxiliary nodes have to be introduced until $(K+K^*)(K+K^*-1) \geq H)$.

The Kolmogorov theorem [6] always assures the existence of at least an explicit functional form of R. This is undoubtedly of a great theoretical importance, but in the practice the solution can difficulty be obtained, and is strictly dependent on the relationship R that has to be reproduced.

In this note an heuristic solution of the problem is proposed. The R function is *ad hoc* constructed through the definition of some elementary mathematical functions that outline together the functional form of R. The problem is then simply converted in the estimation of the suitable parameters of these elementary relationships, i.e. those parameters that allow actually relating the sets X and Y through R. If H links $\in G$ are chosen, some $y \in Y$ element can be respectively associated to each of them, and an artificial transformation of the value of y is introduced; y is supposed to vary in the time through a cosinusoidal law defined by:

$y_h(t) = y_h \cos(\omega_i + \phi_h)$, ω_i and ϕ_h being suitable parameters respectively of the node and the link. Furthermore, for each node $i \in G$ this functional form is defined:

$$O_i(t) = T\left(x_i, y_{h_1^i}, y_{h_2^i}, \ldots y_{h_{M_i}^i}, t\right) = \tanh\left(a_i\left(x_i(t) + \sum_{j=1}^{M_i} y_{h_j^i}(t)\right) + b_i\right) \quad (1)$$

where T is a function of the values of x associated to the node i and of the M values y_{i_k} associated to the links connected to the node i, a_i e b_i being some suitable parameters of the hyperbolic tangent function. The hyperbolic tangent introduces a clipping in the sum of the periodical functions while the subsequent tuned circuits suitably mask the superior harmonic function. The transformation T has to be chosen such that: $O_i(t)=O_i cos(\omega_i t)$, i.e. such that $O_i(t)$ is periodic cosinusoidal and ω_i is the *oscillation* frequency that is the same for all links starting from the ith node. Finally, for each link the following condition has to be verified:

$$y_h = O_{i_h} / w_h \quad (2)$$

where O_i is the *harmonic oscillation modulus* defined at the beginning of the link h and w_h the weight of the link h. Then Eq. 1 and Eq. 2 define the functional form R.

The fundamental hypothesis is that of establishing a relationship between $y_h(t)$ and $O_i(t)$, such that at each time all the $y_h(t)$ are the sum of a part proportional to $O_i(t)$ (a direct effect of the transformation at the node), a part proportional to the variation in the time of $y_h(t)$ (an effect of the short time memory of the link), and a part proportional to the sum of all values previously assumed by $y_h(t)$ till the time t (an effect of the long time memory). These contributions, respectively called c_1, c_2 and c_3, can be expressed as function of a term that we call *link resistance*, α_h, for the hth link. c_1, c_2 and c_3 are inversely proportional to α_h, and they depend also on some other parameters as the following expression shows: $c_{1h} = 1/\alpha_h$; $c_{2h} = -\beta_h/\alpha_h$; $c_{3h} = -1/\gamma_h\alpha_h$. The MAIA law can then be expressed as:

$$y_h(t) = \left[O_i(t) - \beta_h(dy_h(t)/dt) - \int_0^t y_h(t) \Big/ \gamma_h\right] \Big/ \alpha_h \quad (3)$$

The Eq. 3 can be derived with respect to the time variable, and by organising the different terms the following differential equation is obtained:

$$\beta_h\left(d^2 y_h(t)/dt^2\right) + \alpha_h\left(dy_h(t)/dt\right) + y_h(t)/\gamma_h = dO_i(t)/dt \quad (4)$$

Through Eq. 4 the estimation of the parameters that define R are obtained. The Eq. 2 can also be expressed as:

$$\left(O_i(t)/w_h \cos(\omega_i t)\right) - y_h = 0 \tag{5}$$

The solution of the relationship R can then be obtained defining for each link $r \in G$ a system of the two equations (Eq. 4 and Eq. 5) in K variables.

4 MAIA Neural Network

The MAIA theory has been used to train a neural network; given the hypothesis made on $y_h(t)$ and $O_i(t)$ and the theory of the differential equations, the resolution of (Eq. 4) through a MAIA NN is true if the following conditions hold:

$$\phi_h = \tan^{-1}\left(\left(\beta_h \omega_i - (\gamma_h \omega_i)^{-1}\right)/\alpha_h\right) \tag{6}$$

$$y_h = O_i \Big/ \sqrt{\alpha_h^2 + \left[\beta_h \omega_i - \frac{1}{\gamma_h \omega_i}\right]^2} \tag{7}$$

Using the Eq. 6 the weight can be expressed as:

$$w_h = \left[\sqrt{\alpha_h^2 + \left[\beta_h \omega_i - (\gamma_h \omega_i)\right]^2}\right]^{-1} \tag{8}$$

and from the comparison between Eq. 5, Eq. 8 and Eq. 6, it can be written:

$$T\left(x_i, y_{h_1^i}, \ldots, y_{h_{M_i}^i}, \alpha_{h_1^i}, \ldots, \alpha_{h_{M_i}^1}, \beta_{h_1^i}, \ldots, \beta_{h_{M_i}^i}, \gamma_{h_1^i}, \ldots, \gamma_{h_{M_i}^i}, \omega_{h_1^i}, \ldots, \omega_{h_{M_i}^i}, t\right)\left[y_h \cos(\omega_i t)\right]^{-1} = w_h \tag{9}$$

All these equations do not need any differential equation, and at each iteration the following objective function has to be minimised.

$$\sum_{h=1}^{H}\left|T\left(x_i, y_{h_1^i}, \ldots, y_{h_{M_i}^i}, \alpha_{h_1^i}, \ldots, \alpha_{h_{M_i}^1}, \beta_{h_1^i}, \ldots, \beta_{h_{M_i}^i}, \gamma_{h_1^i}, \ldots, \gamma_{h_{M_i}^i}, \omega_{h_1^i}, \ldots, \omega_{h_{M_i}^i}, t\right)\left[y_h \cos(\omega_i t)\right]^{-1} - w_h\right| \tag{10}$$

This function evaluates the global error made at the tth iteration in the resolution of the system formed by the Eq.4 and Eq. 7. A further hypothesis on the relationship R is that of the preservation of the information at the node i, that is:

$$\sum_{j=1}^{M_i} y_j \cos\left(\omega_{i_j} t + \phi_j\right) = 0 \tag{11}$$

that can also be expressed as the equilibrium between all the input nd all the output quantities. In this condition the MAIA NN can work in a unsupervisioned way through a learning mechanism expressed by Eq.11, and an energy function expressed by Eq.10.

5 Compression/Decompression of Images with MAIA NN

If the image to be compressed has a $n \times n$ dimension the NN must have n neurones and each neurones must be connected to all the others, included it self. In the compression stage the bytes associated to each pixel of the image to be compressed will be stored over the $H = n^2$ links of the graph as weights, and beginning from an arbitrary configuration ω_h, the resolution of the problem (Eq.12) subject to (Eq.13) will give the $n+\lambda n^2$ values (n outputs x_n at neurones and λn^2 value at links) that represent the compression I_{d*} of the image $I_{n,n}$.

The decompression needs the codify of I_{d*} to obtain the y_h values beginning from the $n+\lambda n^2$ values; the optimum problem has to be again resolved in the variables y_h, the x_n values being constant. All the variable terms are randomly initialised and updated in subsequent MAIA iterations.

Fig.1a. The original image

Fig. 1b. The image produced by the MAIA NN

Figure 1 shows the image chosen for the application of the MAIA method; it is formed by 128x128 pixels, with 256 tonality of grey available for each pixel. Therefore, a completely connected network with 128 nodes and 16384 links with $\lambda=0.5$ has been used. The learning procedure on a computer with a Pentium 100 processor needs of about 10 seconds for obtaining the compression of the image, and about 1 second for the decompression, the level of the quality obtained being very satisfactory.

The decompression of the image can be improved by varying the level of precision in the computation of the optimum point of the objective function (Eq. 9). Figure 2 shows some decompressions of the image-object for some different values of ε, where ε represents the level of precision in the optimum calculation, that is when the value of the objective function is equal to ε the computation is stopped. As it can be seen, for $\varepsilon =0.05$ the decompression precision degree is very high; in fact, the image reproduced by the MAIA NN is very similar to the original one.

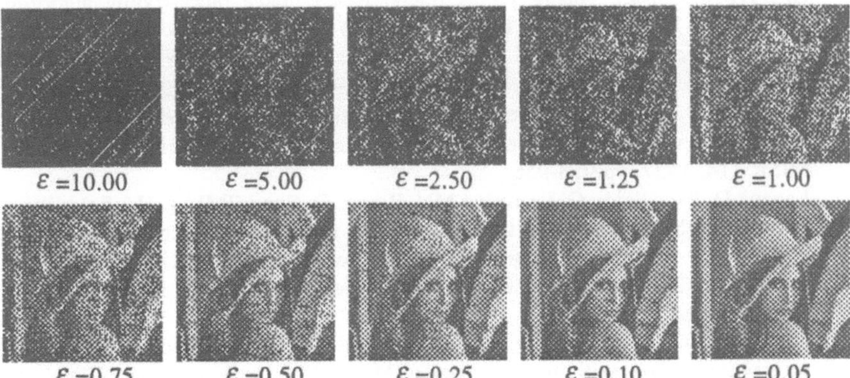

Fig. 2. Decompression (i.e. reconstruction of the image) during the optimisation.

Model	peah SNR
MAIA	39.67
Fractal compression	31.10
2LP	28.00

Tab.1. Comparison C/D models.

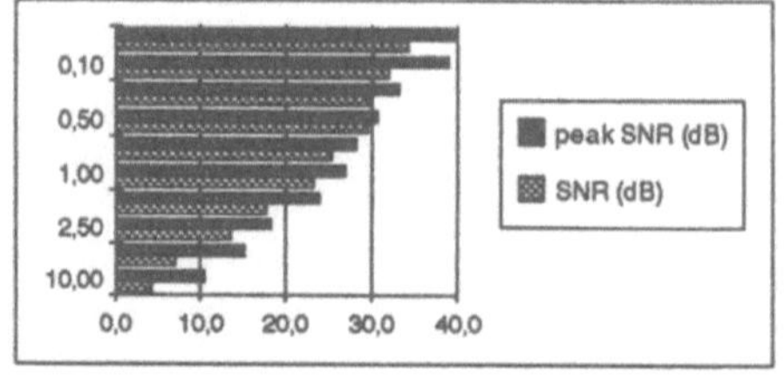

Fig. 4. SNR, Peak of SNR for different

The relationship between the signal and the noise when ε varies is showed in figure 3. For ε =0.05 the peak value of SNR assumes the value of 39.67 dB, the value of SNR being of 34.19 dB. Table 1 reports a comparison between the peak value of SNR obtained through the MAIA model and the fractal model of Jacquin and the neuronal model 2LP of Carrato et al. [2], for the same image.

The results obtained shows that the MAIA NN effects the image compression with a competitive effectiveness with regards to the SNR and the computation time. Furthermore, the image compressed through the MAIA NN actually has got a compression ratio 2:1. At present, a further development of the MAIA model is being studied in order to obtain better performances and possibly to improve the compression ratio.

References

1   Barnsley MF. Fractals Everywhere. Academic Press, New York, 1988

2   Carrato S, Marsi S, Sicuranza GL, Ramponi G. Two and three dimensional data coding. In: Proceedings 1° Italian Workshop on Digital Image Processing by NN. Fond. Bordoni, Roma, 1993.

3   Cottrell G, Munro P, Zipser D. Image Compression by BP: an example of extended programming. Advances in cognitive science 1988.

4   Jaquin AE. A novel fractal block-coding technique for digital images. In: Proceedings ICASSP. IEEE, 1990 pp. 2225.

5   Gelenbe E, Sunger M. Random nets and image compression. In: Proceedings ICANN 94, Springer-Verlag, London, 1994, pp. 1117-1120.

6   Gharavi-Alkhansari, Huang TS. A fractal based image block-coding algorithm. In: Proceedings ICASSP. IEEE, 1993, v.5, pp.345

7   Kangas J. Increasing the error tolerance in transmission of vector quantized images by self-organising map. In:Proceedings ICANN '95. EC2&Cie, 1995, 1:pp.287-291

8   Kolgomorov AN. On the representation of continuos functions od several variables by superposition of continuos functions of one variable and addition, Dokl. Akad. Nauk SSSR 114, 1957, pp.953-956

9   Laganà D, Pappalardo G, Postorino MN, Rosaci D, SarnèGML. A Hopfield like neural network in the simulation of traffic flows in a transportation network. In:Proceedings ICANN '95, Section Transportation. EC2&Cie, 1995,pp 36-41.

10   Pappalardo G, Rosaci D, Sarnè GML. Theoretical aspects of the MAIA Internal report 1995, University of RC.

11   Pappalardo G, Postorino MN, Rosaci D, Sarnè GML. Simulation of traffic flows in transportation nets with unsupervisioned MAIA NN. In: Proceedings WIRN 96, 1996.

Fast Spline Neural Networks for Image Compression

F. Piazza, S. Smerilli, A. Uncini
Dipartimento di Elettronica e Automatica,Università di Ancona
via Brecce Bianche - 60131 Ancona, Italy. E-mail: aurel@eealab.unian.it

M. Griffo, R. Zunino
Dipartimento di Ingegneria Elettronica e Biofisica, Università di Genova
via Opera Pia 11a - 16145 Genova, Italy. E-mail: zunino@dibe.unige.it

Abstract

The paper investigates the representation properties of a new neural network architecture based on adaptive activation function, called Generalized Sigmoidal Neural Network (GSNN) and analyzes the advantages of this model in the application domain of image compression. As a consequence, the basic network scheme is most similar to classical structures with the improved hidden-unit adaptive transfer function. In order to attain reliable results, the compression method is verified using standard testbed images, and the overall evaluation follows a comparative approach.

1 Introduction

A crucial aspect in constructing an image-compression system is identifying the correct data model. It is well known that neural network can be trained to form an internal representation of training data, and the various neural approaches to image compression can be classified according to the specific representation. The two reference paradigms relate to: 1) feedforward structures ultimately inspired to Multilayer Perceptrons [1], or 2) prototype-based schemata involving VQ methods [9]. In either case, the basic approach to neural image compression lies in splitting the processed picture into many subblock, which are treated as lower-dimensional patterns.

Feedforward structures follow a classical auto-associative schema, in which the number of hidden units is smaller than the data dimensionality. In training, the network learns a transformation rule between input blocks and their lower-dimensional coding, represented by the set of activation values of hidden nodes. Conversely, a VQ-compression system locates, for each block, one out of a fixed set of prototypes that best resembles transmitted information. The principles of operation also determine the different performances of the two approaches; MLP-based systems usually exhibit a satisfactory reconstruction quality at the cost of sometimes lower compression rates, as compared with VQ-based methods, whose limited vocabulary of prototypes attains remarkable compression performances but often conveys distortion and blockiness effects.

This paper focuses on qualitative aspects, and analyses the generalization and performance improvements that specific neuron modeling can yield in MLP-based compression schemata. It is well known that the behavior of a neural network as the Multilayer Perceptron (MLP) [1], depends on the chosen activation functions. Usually such functions are sigmoidal [1]; however, other functions, also depending on some free parameters, have been studied and applied [4-7]. The behavior of different classes of non-linear activation functions with respect to the computational capabilities of the MLP has also been widely studied (see for example [2]).

Recently, a new fast spline-based neural network with good generalization capabilities and suitable for real-time applications, has been introduced [7]. This spline-based NN is designed using a neuron, called generalized sigmoidal (GS) neuron, containing an adaptive parametric spline activation function [7]. This function has several interesting features: 1) is easy to adapt, 2) retains the squashing property of the sigmoid, 3) has the necessary smoothing characteristics, 4) is easy to implement both in hardware and as software simulation. The multilayer networks built with such neurons are still universal approximators and have usually a smaller structural complexity, maintaining good generalization capabilities.

The paper investigates the representation properties of spline-based neurons, and analyzes the advantages of this model in the application domain of image compression. As a consequence, the basic network schema is most similar to classical structures [10], with the improved hidden-unit adaptive transfer function. In order to attain reliable results, the compression method is verified using standard testbed images, and the overall evaluation follows a comparative approach. In particular, a spline-based network, a standard MLP, and a VQ-based system have been trained and tested on a common set of images; test pictures were not included in the training set. Experimental results demonstrate the improved reconstruction accuracy of the proposed model.

The paper first describes the specific spline-based network; then the compression schema is outlined, and finally experimental results on the visual domain are given.

2 The Generalized Sigmoidal Neural Net Architecture

A. The Generalized Sigmoid (GS)

The splines are, generally, smooth parametric curves, divided in multiple tracts; they have also the property of preserving the continuity of the derivatives at the joining points. In the planar case the graph $F_i(u)$ of the i-th curve span is represented by

$$F_i(u) = \left[F_{xi}(u), F_{yi}(u) \right]^T; \quad 0 \le u \le 1 \tag{1}$$

where u is the parameter (usually varying between 0 and 1), T is the transpose operator and the two polynomial functions $F_{xi}(.)$, $F_{yi}(.)$ describe the curve tract behavior in the two coordinates x and y. The i-th curve *spline basis functions* tract can be written in the form

$$F_i = \sum_{k=0}^{d} Q_{i+k} b_{i,d}(u) = \begin{bmatrix} \sum_{k=0}^{d} q_{x,i+k} b_{i,d}(u) \\ \sum_{k=0}^{d} q_{y,i+k} b_{i,d}(u) \end{bmatrix} \tag{2}$$

where $b_{i,d}(u)$ is the i-th element of the spline basis (a polynomial of degree d in the variable u) with $u \in [0, 1]$, and $Q_{i+k}=[q_{x,i+k} \; q_{y,i+k}]^T$ are the $(d+1)$ *control points* of the i-th curve tract: moving such points on the real plane will affect the shape of the curve. Using the Catmull-Rom spline base, the i-th curve span in (2) can be written as

$$F_i(u) = \begin{bmatrix} u^3 & u^2 & u^1 & 1 \end{bmatrix} \frac{1}{2} \begin{bmatrix} -1 & 3 & -3 & 1 \\ 2 & -5 & 4 & -1 \\ -1 & 0 & 1 & 0 \\ 0 & 2 & 0 & 0 \end{bmatrix} \begin{bmatrix} Q_i \\ Q_{i+1} \\ Q_{i+2} \\ Q_{i+3} \end{bmatrix} \tag{3}$$

with $u \in [0, 1]$. From (3) it is clear that every span is controlled by four control points: the second and the third will be interpolated, while the geometric tangent in these two points will be parallel to the segment passing through their adjacent control points as in Fig. 1

A Catmull-Rom-based spline curve has a continuous first derivative which allows to develop a backpropagation-style learning algorithm for neural networks that use such an activation function. After choosing the spline curve, we want to use it as an activation function in a neural network; this leads to the following two problems: *a*) how to choose the control points, and *b*) we need a *function*, not a plane curve.

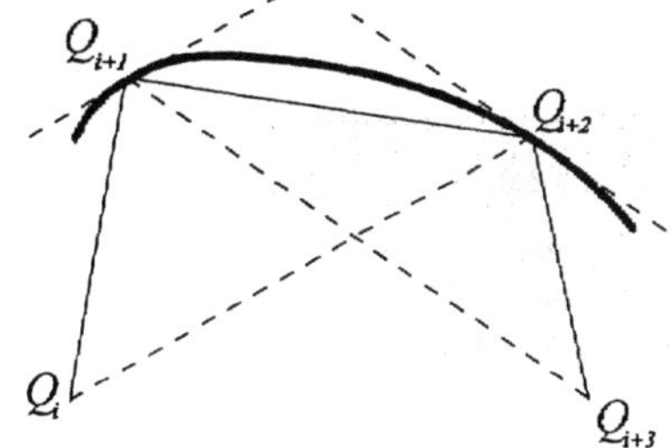

Fig. 1 - The i-th curve tract defined by its four control points.

The point *b*) is crucial: first we need to ensure that the curve is actually a function; then, given an abscissa value x_0, from (1) we calculate $u = F_{xi}^{-1}(x_0)$, which is substituted in $F_{yi}(.)$ to deliver the corresponding value y_0 on the curve. As $F_{xi}(u)$ is a cubic polynomial, we need to solve a third degree equation (or use a numerical algorithm) to calculate u.

The solution of this problem lies in the choice of the control points: if they are uniformly spaced along the x-axis (by a fixed step Δx), the variation diminishing property tells us that $F_{xi}(u)$ will not be a cubic polynomial anymore: it will be a *first degree* polynomial. Precisely, it is

$$F_{xi}(u) = u\Delta x + q_{x,i+1} \tag{4}$$

where $q_{x,i+1}$ is the abscissa value of the second control point of the i-th span (Q_i+1). For this reason we opted for the following choice of the spline control points: we take the sigmoidal function

$$F(x) = \frac{1-e^{-x}}{1+e^{-x}} \tag{5}$$

and, fixed a step Δx and an integer number n, we sample the function in $(n+1)$ uniformly spaced points along the x-axis ($Q_0,...,Q_n$), in an interval centered on the axis origin (i.e. the sampling interval middle point lies on $(0,0)$). Outside the sampling interval the spline curve will have a constant value: $q_{y,1}$ for the negative x coordinate, and $q_{y,n-1}$ for the positive x; also the first derivative will be constant outside the sampling interval: $(q_{y,1} - q_{y,0})/\Delta x$ and $(q_{y,n} - q_{y,n}-1)/\Delta x$ respectively for the negative and positive x coordinate.

This four points will never get modified, in order to maintain sigmoidal properties: we call this function Generalized Sigmoid (GS). This choice ensures also that a GS function will always be a *sufficiently kinky spline* (in the sense explained in [3]): such a neural network is then an universal approximator. As an implementation note, it is clear that the control points' abscissas are completely defined by the above procedure: this means that it is only necessary to store the $(q_{y,0},...,q_{y,n})$ values on a a lock-up table (LUT), to represent the GS function.

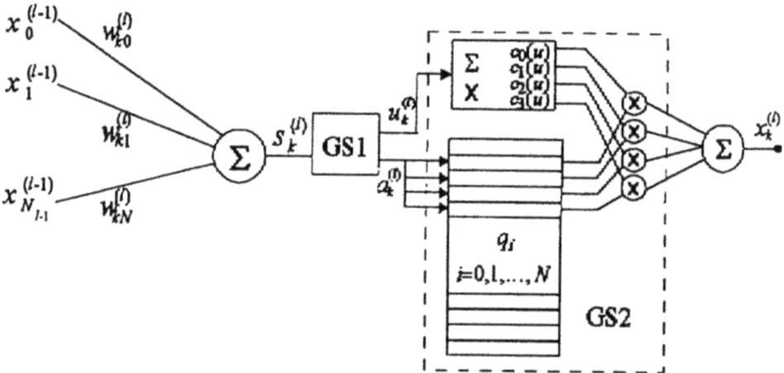

Fig. 2 - Generalized sigmoidal neuron (GS-neuron) based on the Catmull-Rom spline adaptive activation function with the internal structure of the GS2 block using the same notation proposed in [1].

B. The GS Neuron

The proposed neuron is composed of a classical linear combiner, which performs the weighted sum of the inputs, and, as Fig. 2 shows, of two blocks (GS1 and GS2) which realize the spline adaptive activation function.

The block GS1 performs the mapping of the linear combiner output to the parametric domain, while the block GS2 computes the neuron output by using the activation function's control points, stored in LUT, and the polynomial coefficients of Eq. (2).

C. Learning Algorithm for the GSNN

The learning algorithm for the GSNN is based on the classical backpropagation [1] where both the weights $w_{kj}^{(l)}$ and the local activation function free parameters

$Q_{ki}^{(l)}$ are adapted. Using the same notation proposed in [1], for the t-th step of the learning phase we have the follows recursive equation

FOR $l = M, \dots 1$

    FOR $k = 0, \dots N_l$

        FOR $j = 0, \dots, N_l\text{-}1$

$$e_k^{(l)} = \begin{cases} \left(d_k - x_k^{(l)}\right) & l = M \\ \sum_{p=1}^{N_{l+1}} \delta_k^{(l+1)} w_{pk}^{(l+1)} & l = M-1, \dots, 1 \end{cases} \tag{6}$$

$$\delta_k^{(l)} = e_k^{(l)} \left(\left. \frac{dF_{k,a_k^{(l)}}^{(l)}(u)}{du} \right|_{u=u_k^{(l)}} \right) \frac{1}{\Delta x} ; \tag{7}$$

$$\Delta w_{kj}^{(l)} = \mu \delta_k^{(l)} x_j^{(l-1)} ; \tag{8}$$

$$w_{kj}^{(l)}[t+1] = w_{kj}^{(l)}[t+1] + \Delta w_{kj}^{(l)}[t] \tag{9}$$

        FOR $m=0, \dots, 3$

$$\Delta Q_{k,(a_k^{(l)}+m)}^{(l)} = e_k^{(l)} \left(\frac{\partial F^{(l)}}{\partial Q_{k,(a_k^{(l)}+m)}^{(l)}} \right) = \mu_q e_k^{(l)} b_{k,m}^{(l)}\left(u_k^{(l)}\right); \tag{10}$$

$$Q_{k,(a_k^{(l)}+m)}^{(l)}[t+1] = Q_{k,(a_k^{(l)}+m)}^{(l)}[t+1] + \Delta Q_{k,(a_k^{(l)}+m)}^{(l)}[t] \tag{11}$$

(in Eq. (6), (7), (8), (9); the time index t is omitted) where Δ represents the local approximation of the function error gradient, d_k is the desired k-th output value, the terms μ and μ_q are the learning rate for the weights and the activation function parameters respectively, and the derivative of $F(u)$ is simply a second order polynomial.

The terms $\Delta w_{kj}^{(l)}[t]$ $\Delta w_{kj}^{(l)}[t]$ and $\Delta Q_{k,(a_k^{(l)}+m)}^{(l)}[t]$ are obtained by computing the output error derivative with respect to the weights and to the control points of the activation function respectively, as reported in [6]. For the control point adaptation (Eq. (11)), the parameter m ranging from 0 to 3 restricting the update to only 4 points. In this step we consider the parametric value u fixed (i.e. $u = u_k^{(l)}$), so the SG2 block is represented by a function of four variables (the control points).

3 Image-compression schema

MLP-based compression of visual information involves a classical auto-associative schema, in which inputs and outputs coincide, hence the network is forced to learn the best approximating rule by using a lower-dimensional hidden layer. In the described research, images are split into 8x8 pixel blocks, hence data dimension is

64. A block is reconstructed by using the (quantized) activation values of hidden units. Figure 3 illustrates the basic compression architecture.

This approach results in a *lossy* compression schema; MSE represents a typical (and suitable for training purposes) measure of the quality of reconstruction ability. It is worth stressing that the autoassociative schema is quite general, and applies independently of the specific activation function characterizing hidden units. This makes it possible a comparison between standard sigmoidal units and spline-based neurons.

The relevance of a feedforward compression schema lies in the simplicity and generality of the overall architecture. The basic nonlinearity is embedded into the neuron function; the layer organization allows one to exploit effective training procedures, and the network topology itself facilitates the structure's hardware implementation for real-time data processing.

The crucial problem underlying any example-driven approach to adaptive systems concerns their generalization ability, which ultimately relates to the generality and reliability of training data as compared with the actual (unknown) distribution of real data. In practice, generalization ability is estimated empirically by performing several runs on different raining sets and measuring performance on sample data not used for training. In the specific context of image compression, a few reference images are available and have been used as a benchmark for comparison with other approaches.

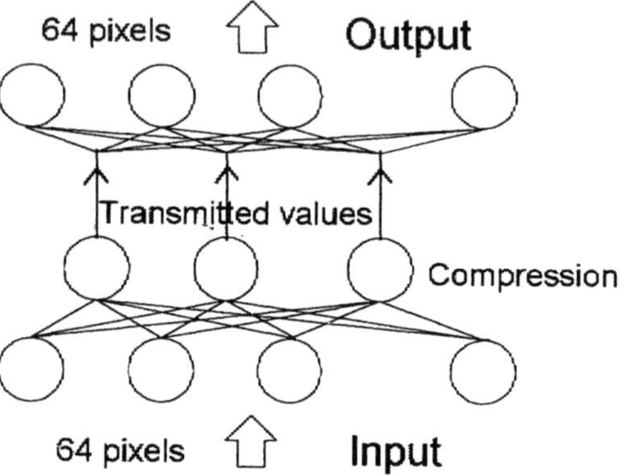

Fig. 3 - Feedforward networks for image compression

4 Experimental results

The experimental testbed used to validate the described technique involved several standard reference images; each picture included 512x512 pixels with 256 gray levels (8bpp), and was uniformly splitted into blocks of 8x8 pixels. As a consequence, each training image generated 4096 samples.

In the experiments, several networks composed by 64 inputs, 64 outputs (8x8 pixel) and two hidden layers are used. The first layer output, composed by 4, 7 or 8 neurons, represents the transmitted values (see Fig.3). In order to obtain a specific Bit-Per-Pixel (*bpp*) value, the first layer outputs (transmitted values) are quantized with fews bits (3, 4 as indicate in Table I)

In all simulations, test images were different from training ones to assess generalization performance. The evaluation of both training progress and test performance used MSE as a numerical measurement. It is well known, however, that this quantity is not sufficient to describe the actual adequacy of a compression schema, hence also a visual assessment was required.

The evaluation of the spline-based neuron model was performed first of all by a comparative analysis with similar techniques involving feedforward structures with classical sigmoidal units. The various simulation pointed out the remarkable improvement in performance brought about by the additional adaptivity of spline activations. The most relevant impact in this sense lies in the removal of blockiness effects due to the higher generalization ability of the network model.

In a further comparative step, the spline-based compression method was matched against a VQ-based schema: as expected, the latter approach provided a similar compression rate.

The images shown in Fig.4 provide a sample of test performance of the different methods used in the comparative analysis.

Table I Numerical results of the performed training runs (MSE is the Mean Square Error between the true images and the reconstructed ones as in [10], while the peak signal-to-noise ratio is defined as PSNR=10 $\log_{10}$ (256^2/MSE)).

Net	bpp	PSNR (dB)	MSE	N° Free Param.
Standard LVQ	0.699	31.999	41.034	--
GSNN-64-8-16-64 (4 Bit)	0.5625	30.955	52.193	2104
GSNN-64-8-16-64 (3 Bit)	0.4375	30.061	64.115	2104
GSNN-64-7-16-64 (4 Bit)	0.5	29.979	65.344	2019
GSNN-64-4-16-64 (4 Bit)	0.3125	28.780	86.111	1764
GSNN-64-4-16-64 (3 Bit)	0.25	28.133	99.946	1764
MLP-64-16-64 (4 Bit)	1.0625	26.533	144.461	2128

References

[1]   B. Widrow, M. A. Lehr, "30 years of adaptive neural networks: Perceptron, Madaline and BackPropagation", Proc. of the IEEE, Vol. 78, No. 9, Sept. 1990.

[2]   K. Hornik, M. Stinchcombe, H. White, "Multilayer Feedforward Networks are Universal Approximators", Neural Networks, Vol.2, pp. 359-366, 1989.

[3]   M. Stinchcombe, H. White, "Approximating and Learning Unknown Mappings using Multilayer Feedforward Networks with Bounded Weights", in Proceedings of IJCNN, San Diego, (CA) USA, 1990, pp. III 7-16.

[4]   F. Piazza, A. Uncini, M. Zenobi, "Artificial Neural Networks with Adaptive Polynomial Activation Function", Proc. of the IJCNN Int. Joint Conf. on Neural Networks, Beijing, China, November 1992, pp II-343-349.

[5]   M.S. Chen, M. T. Manry, "Conventional Modeling of the Multilayer Perceptron Using Polynomial Basis Functions", IEEE Trans. on Neural Net., Vol.4, No.1, pp.164-166, 1993.

[6]   F. Piazza, A. Uncini, M. Zenobi, "Neural Networks with Digital LUT Activation Function", Proc. of the IJCNN International Joint Conference on Neural Networks, Nagoya, Japan, October 1993, pp II-1401-1404.

[7]   S. Guarnieri, F. Piazza, A. Uncini, "Multilayer Neural Networks with Adaptive Spline-Based Activation Functions, World Congress on Neural Networks (WCNN'95), Washington, D.C., July 1995

[8]   E. Catmull, R. Rom, "A Class of Local Interpolating Splines", in R. E. Barnhill, R. F. Riesenfeld (ed.) CAGD, Academic Press, New York 1974, pp 317-326.

[9]   G.W.Cottrell, P.Munro, D.Zipser "Learning internal representation from gray-scale images: an example of extensional programming" Proc. 9 Annual Cogn.Sc. Soc. Congress, Wa. 1987

[10] Anguita D, Passaggio F, Zunino R "SOM-based interpolation to image compression" Proc. World Congress on Neural Networks WCNN'95, Washington, July 1995, vol. I, pp.739-742

Fig. 4. Reconstructed images of *Lena* (test) using 256 prototypes.

(a) Original image;

(b) standard *VQ (BR=0.699 bpp, MSE = 41.03, PSNR=31.99 dB)*;

(c) GSNN-64-8-16-64 *(BR= 0.5625 bpp, MSE = 52.193, PSNR=30.95 dB)*;

(d) standard MLP-64-16-64 *(BR= 1.0625 bpp, MSE = 144.461, PSNR=26.53 dB)*.

Age estimates of stellar systems by Artificial Neural Networks

L. Pulone

and

R. Scaramella

Osservatorio Astronomico di Roma, via dell'Osservatorio 2

I-00040 Monteporzio (Roma),Italy

Abstract

The interpretation of the colour–magnitude diagrams of galactic or extra-galactic clusters in terms of their age and chemical composition is a natural goal of stellar evolutionary theories. There are a variety of different methods which can be used to estimate the ages of stellar clusters. All of these techniques rely on comparing only some aspects of an observed CMD to theoretical isochrones, therefore not fully considering the available information. We report on preliminary results of a project whose aim is to evaluate the age of young stellar systems by the use of artificial neural networks. The neural structure we initially adopted is a simple backpropagation feed–forward network.

1. Introduction

Artificial neural networks (ANN), have been recently utilized in astronomy for a wide range of problems: adaptive optics (Angel *et al.* 1990), classification of young stellar objects (Adorf & Meurs 1988), morphological classification of galaxies (Storrie–Lombardi *et al.* 1992), star–galaxy classification (Odewan *et al.* 1992), faint object classification (Serra–Ricart *et al.* 1993), automated classification of stellar spectra (von Hippel *et al.* 1994). In the present study we explore a method for evaluating the age of stellar systems by an ANN algorithm. A simple back–propagation algorithm allows us to train a network on a subset of theoretical age–dependent patterns so as to estimate, using the observational counterparts, the age of stellar clusters.

In Sec. 2 we give a general description of the underlying stellar evolutionary framework. In Sec. 3 the adopted training pattern is shortly described. The ANN structure and preliminary results are summarized in Sec. 4.

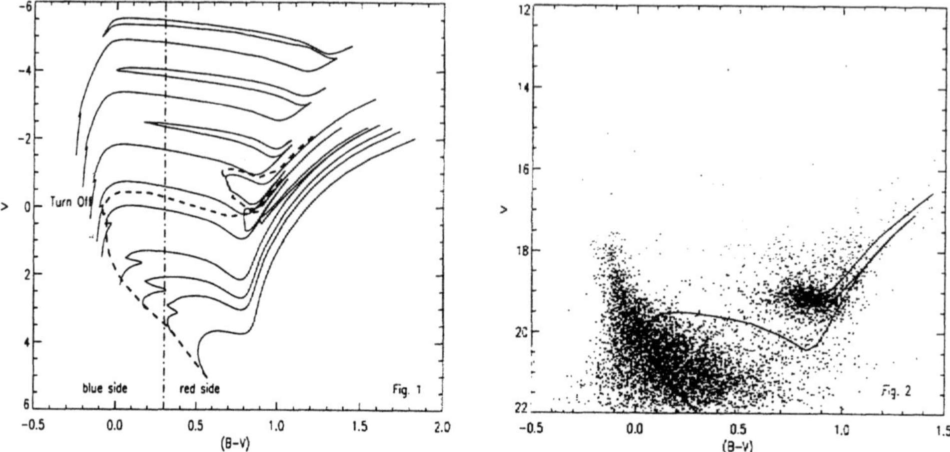

Figure 1: The CMD location of selected evolutionary tracks having metal content $Z = 0.006$, helium content $Y = 0.23$ and masses ranging from $M = 1M_\odot$ (bottom) to $M = 9M_\odot$. The isochrone of $380Myr$ is also indicated as an example (dashed line); the vertical dot–dashed line arbitrarily subdivides the so–called blue and the red side of the diagram.

Figure 2: The isochrones line of $780Myr$ is superimposed to the observed CMD of the stellar cluster $NGC419$

2. Dating Stellar Systems

We observe numerous physical aggregations of 10^5 to 10^7 stars called *clusters*, inside and outside our galaxy. The differences in the stellar clusters are expected to reflect differences in *quality* (chemical composition), in *quantity* (star masses) and in *evolution* (star ages).

A good hold on dating of stellar systems allows one to put constraints on the age of the universe and on the theories of formation of the galaxies (see e. g., Renzini, 1991).

The interpretation of the colour–magnitude diagram (CMD) of the stellar systems is therefore one of the main goals of stellar evolutionary theories. In particular the CMDs of young stellar clusters offer the unique opportunity to test the theoretical scenario emerging from the computations of the intermediate and high–mass stars, i. e. with mass $M > 2M_\odot$ (solar mass $M_\odot = 1.99 \cdot 10^{33}g$)

As recently reviewed in *The Globular Cluster–Galaxy Connection* (Smith and Brodie, eds. 1993), there is a variety of different methods which can be used to derive parameters of astrophysical interest, and in particular, the ages of the stellar clusters. All of these techniques rely on comparing some aspects of an observed CMD to theoretical *isochrones*.

The output of the stellar model calculations are evolutionary sequences, each referring to a star of given mass and chemical composition. By connecting points of the same age lying on the evolutionary lines of models of different mass we obtain the *isochrones*, i. e. the loci occupied, on the CMD, by stars of the same age and composition, but different masses (fig. 1).

Ultimately, cluster age–dating is based on the fact that the main sequence life-times decrease rapidly with increasing mass and that, during the hydrogen shell–burning phase, the stars evolve rapidly away from the main sequence to the right side in the CMD. This rightward evolution is so rapid for massive stars that there is no chance of finding stars in those phases; therefore the observed main sequence simply terminates at some maximum luminosity, i. e. the Turn–Off (TO) point. The luminosity of the TO is the clock which enables us to date the stellar clusters.

The typical result is obtained computing *isochrones* for different cluster ages and comparing them with the obervational data. Fig. 2 shows the CMD of the Small Magellanic Cloud (SMC) cluster $NGC419$ (Boehm *et al.* 1993) and its surrounding stellar field. The superimposed $790Myr$ isochrone well fits the red clump at magnitude $V = 19mag$. The blue sequence at $(B - V) = 0$ is the hydrogen–burning phase of the younger stellar field in which the cluster is contained. To superimpose the isochrone, we adopted the reddening $E(B - V) = 0.06$ and the true distance modulus $(m - M) = 18.85$.

Another dating technique is based on the comparison between the theoretical luminosity functions and the obervational ones.

From synthetic CMDs we directly obtain the theoretical luminosity function (LF), i. e. the function representing the number of stars $N(V)$ per magnitude–interval ΔV. Theoretical LFs are fairly insensitive to our current way of parametriz-ing the efficiency of the convective energy transfer, via the so–called mixing lenght parameter α. Therefore, the LF fit is an essential tool to date stellar clusters.

3. LFs as Training Sample

The purpose of the procedure is to teach a network to discriminate among a set of theoretical LFs, obtained through stellar evolutionary calculations, in order to provide a statistically meaningful answer when its input layer is fed with ob-servational data. All the stellar evolutionary tracks have been calculated with the updated release of the *FRANEC* software (FRascati Raphson Newton Evolutionary Code; Chieffi *et al.* 1996). The software environment we adopted for the simulations is the Stuttgart Neural Network Simulator (Zell & al. 1995).

The neural structure we initially adopted is a backpropagation feed–forward network.

This method will be an useful tool to evaluate spreads in ages of stellar clusters which have same distance modulus, i. e. Magellanic Clouds, M31, dwarf galaxies.

The adopted LF set has been obtained by evaluating stellar models assuming the chemical composition $Y = 0.23$. $Z = 0.006$ (Y is helium abundance in mass, Z is the abundance of the elements heavier than helium) and the stellar mass values $M = (1.0. 1.5, 2. 2.5, 3. 4. 5. 6. 7. 8. 9)\ M/M_\odot$. A set of isochrones have been computed

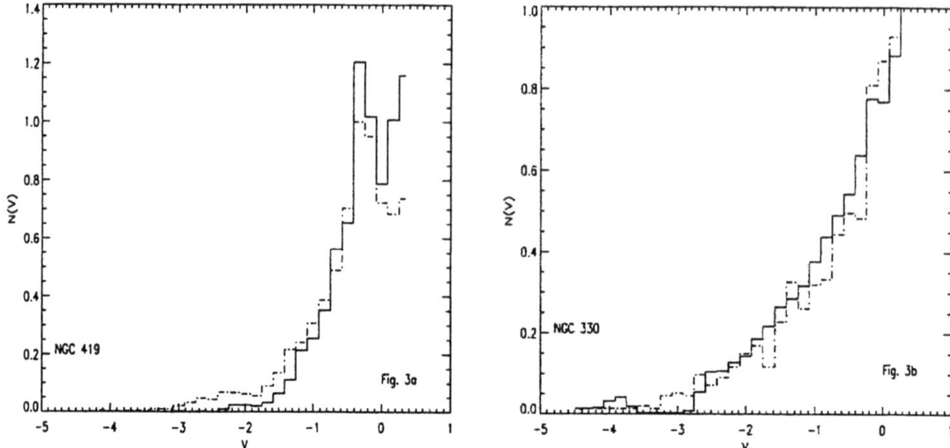

Figure 3: Panel *a* shows the comparison of the observed LF of $NGC419$ (dashed line) with the sum of the 540 and $790Myr$ theoretical counterpart as selected by the trained 60:15:26 neural network; panel *b* represents the theoretical $40Myr$ LF superposed to the observed one of $NGC330$ (dashed line)

and transformed into the observational plane. As a further step, starting from the isochrones, 100 synthetic CMDs have been evaluated for ages ranging from $50Myr$ to $1050Myr$, in steps of $10Myr$ ($1Myr = 10^6 years$). The adopted initial mass function $n(m)$ is the Salpeter's one with the number of stars $n(m) \propto m^\beta$ with index $\beta = -2.35$. The final product, for the adopted chemical composition, is a set of theoretical LFs whose morphology depends on the age. The LF set has therefore been subdivided in training and validation samples with the aim to stop the learning at the maximum generalization capability of the network.

4. Using Neural Networks

In successive refinements we implemented both single output node and multiple output nodes networks. As inputs we adopt $0.2mag$–wide bins of the LF, which feed 60 neurons. These are subdivided in two equal sets, respectively associated with the *red* side ($B - V > 0.3$) and with the *blue* side ($B - V < 0.3$) of the theoretical CMD (see fig. 1). This choice is motivated by the non–monotonic behaviour of the *red* LF of intermediate mass stars according to the age and to the lesser degree in which the age affects the *blue* side LF (Corsi *et al.* 1994). We utilized a single hidden layer containing 15 nodes. Checks on both the error of the training set and the error of the validation set have shown that, in the present problem, the number of the hidden nodes affects only the number of learning cycles to reach the limits of the network generalization capability. The single output node network acts as nonlinear least–squares optimization algorithm producing a mapping of the LF bins into a continuos age–classification function. The network has been tested with synthetic CMDs of star clusters. The learning function we used is the backpropagation with momentum term and flat spot elimination. The neural activations are updated in a topological order and the initialization of both weights and bias is accomplished with random values between -1 and 1.

The network 60:15:1 is able to evaluate the ages of simple stellar populations, that is, of chemically homogeneus clusters formed in a single star–formation episode.

As a further step we built up a second network 60:15:26 whose output layer is constituted by 26 neurons, one output node per age class. The age interval ranges from $40Myr$ to $1040Myr$ in $40Myr$ steps.

In the one output per class configuration, the output has to be interpreted as the probability for age class membership. After completion of the learning process on the same training set, the second architecture was able to discriminate between the observational LFs of two clusters of different age overimposed upon the stellar field of the parent galaxy SMC, namely NGC 419 and NGC 330.

In fig. 3 the final comparison is shown between the observed LFs (dashed lines) and those one restored from the network output (continuos lines): Work is in progress to explore the effects of pruning unnecessary links or hidden units. We also intend to make a detailed study of the effects of photometric errors on the neural network output, as well as the use of other architectures and input quantities.

References

[1] Adorf, H., M. & Meurs, E. J., À. 1988 in *Large-Scale Structure of the Universe: Observational and Analytical Methods*, eds. W. C., Seitter, H. W., Duerbeck & M. Tacke, Hidelberg, Springer–Verlag, 315

[2] Boehm, C., Bono, G., Fulle, M., Pulone, L. & Alcaino, G. 1994, Mem. Soc. A. It., 65, 3, 755

[3] Angel, J. R. P., Wizinovich, P., Lloyd–Hart, M. & Sandler, D. 1990, Nature, 348, 221

[4] Chieffi, A., Straniero, O., Limongi, M. 1996, *in preparation*

[5] Corsi, C. E., Buonanno, R., Fusi Pecci, F., Ferraro, R. F., Testa, V. & Greggio, L. 1994, Mon. Not. R. Astron. Soc., 271, 385

[6] Odewahn, S. C., Stockwell, E. B., Pennington, R. L., Humphreys. R. M. & Zumach, W. A. 1991, Astron. Journal, 89, 1451

[7] Renzini, A. 1991. in *Observational Test of Inflation*, eds. T. Banday & T. Shanks, Dordrecht, Kluwer

[8] Serra–Ricart, M., Gaitan, V., Delgado, S. D. & Perez–Fournon, I. 1991, in *Astronomical Data Analysis Software and Systems I*, eds. D. M. Worrall. C. Biemesderfer & J. Barnes, PASPC 25, 254

[9] Serra–Ricart, M., Gaitan, V., Garrido, Ll. & Perez–Fournon 1996, A&AS, 115, 195

[10] Storrie–Lombardi, M. C., Lahav, O., Sodre', L. Jr & Storrie–Lombardi, L. J. 1992, Mon. Not. R. Atron. Soc. SC, 259, 8

[11] von Hippel, T., Storrie–Lombardi, L. J., Storrie–Lombardi, M. C. & Irwin, M. J. 1994, Mon. Not. R. Atron. Soc., 269, 97

[12] Zell, A. & al. 1995, Stuttgart Neural Network Simulator, User Manual, version 4.0 (SNNS)

[13] *The Globular Cluster–Galaxy Connection*, 1993, eds. G. **H** .Smith & J. P. Brodie, ASP Conference Series, 48

SECTION 6

APPLICATIONS

Neural Nets for Hybrid On-Line Plant Control

M. Barabino, S. Bruzzo, F. Camastra, A.M. Colla

Elsag Bailey - Un'Azienda Finmeccanica S.p.A
Via G. Puccini, 2 - 16154 Genova (ITALY)

Abstract

This paper describes some experimental results obtained in the evaluation of neural net efficiency in continuous process control. To this purpose a neural controller was developed and integrated in the control system of an experimental laboratory (PRIAM Lab, developed within an Esprit RTD Project). A critical analysis of the performance of the controlled system showed very promising results.

1. Introduction

The use of neural techniques in process control looks appropriate when defining an accurate process model is difficult, for instance when different, possibly strongly coupled variables have to be regulated. Our case study was selected in the PRIAM laboratory, developed at ENEL CRA (Cologno Monzese, Italy) during the European project ESPRIT 6188 *"Prenormative Requirements for Intelligent Actuation and Measurement" (PRIAM)*. The control system of the experimental laboratory must reach and maintain operator assigned setpoints for four regulation loops: water level, pressure, flowrate and temperature.

Neural nets have been experimented as advanced control techniques in the pressure and flowrate loops. The problem was solved by a conservative approach, *leaving the previous control structure,* based on PID controllers, untouched. Neural nets were used to perform appropriate feedforward actions in order to improve the system performance without affecting its stability (by using setpoints instead of measured process variables). If necessary, the classical control scheme can be restored, as a degraded operating mode. This approach to neural net integration within a traditional control system is a strong point to make the adoption of neural technology both widely applicable and acceptable to users.

The paper compares the performances obtained by inserting the neural controller in pressure and flowrate control loops with those yielded by the traditional control system. The results with the neural controller are extremely promising.

2. The process

The physical process consists of a closed loop for water treatment. Four regulation loops are implemented: water level, pressure, flowrate and temperature. Each regulation loop is equipped with both conventional and "intelligent" instrumentation (the purpose of PRIAM was to define functional requirements of intelligent actuators and trasmitters connected via Field Bus).

Figure 1 shows the process scheme, the field devices and four regulation loops, together with the Maintainance System *(MS)* of the "intelligent" devices and the Process Control System *(PCS)*. A brief description of the four regulation loops follows.

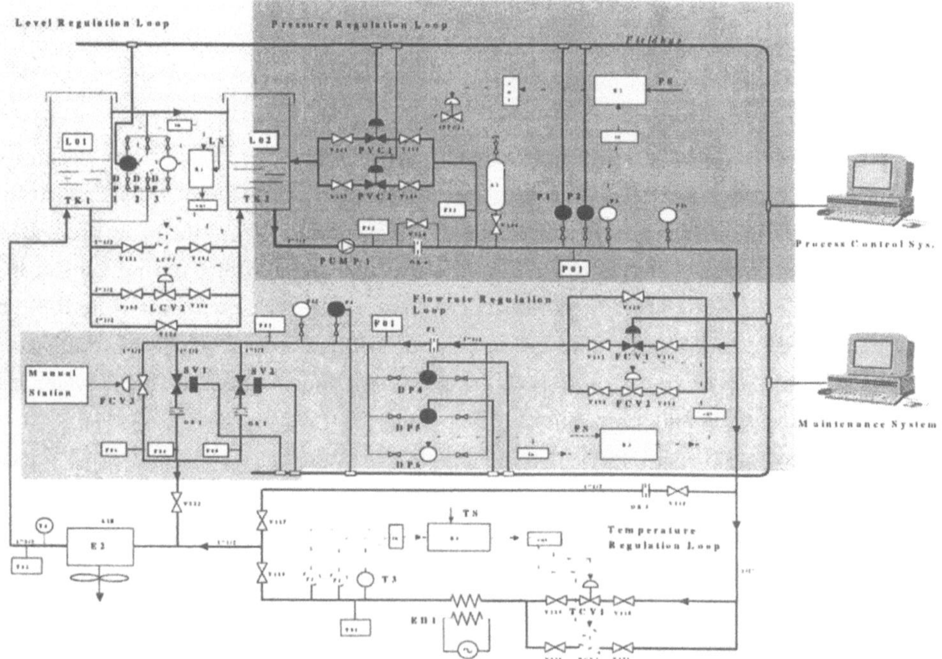

Figure 1 - *PRIAM Laboratory scheme*

Level regulation loop: level **L01** in tank **TK1** is measured by differential pressure sensors **DP1** and **DP3** and is controlled by pneumatic valve **LCV2**. The PCS evaluates the position of valve **LCV2** in order to maintain the level setpoint **LS** with water flowrate changes.

Pressure regulation loop: tank **TK2** feeds the pump, which is endowed with a recycling loop. The regulation loop keeps the pressure at the pump outlet constant with flowrate variations. The loop consists of twin coupled loops, with pressure sensors **P1** and **P2** and actuators **PVC1** and **PVC2**, whose position is controlled by the PCS in order to keep pressure **P01** at setpoint **PS**.

Flowrate regulation loop: flowrate **F01** is kept constant and equal to **FS** with load variations by the regulation loop, which consists of three differential pressure sensors (**DP4**, **DP5**, **DP6**), the PCS controller and two pneumatic valves **FCV1** and **FCV2**. This loop is based on pressure drops produced by two on-off valves (**SV1**, **SV2**) and one modulating valve (**FCV3**), which is manually regulated according to the functional characteristics of **FCV1** and **FCV2**.

Temperature regulation loop: the temperature loop operates on the water flow through the electrical heater **EH1**. The output temperature from **EH1** is measured by **T3** and reported to the PCS to control the position of valve **TCV1**. The cooling device **E2** ensures zero temperature balance in the main loop.

3. Neural controllers

Neural nets have been successfully applied for the solution of several problems in process control, such as *state estimation* [1,2].
The literature [4] describes different ways to integrate neural nets inside a conventional control architecture. The approach applied in the reported case study was very *conservative* towards the previous control structure: *neural networks are used to predict actuator positions* [5,6].

The solution scheme, shown in Figure 2, where the neural net predicts the actuator position in function of the setpoint and the disturbance, offers the following advantages:

- minimum impact on the conventional control structure;

- possibility to resume conventional control (degraded operation mode);

- off-line evaluation of neural network performance by comparison with the real actuator position;

- functional validation of actuators (namely, prediction that a given setpoint cannot be reached because of the actuator limits);

- robustness to error (prediction errors are compensated by the PID).

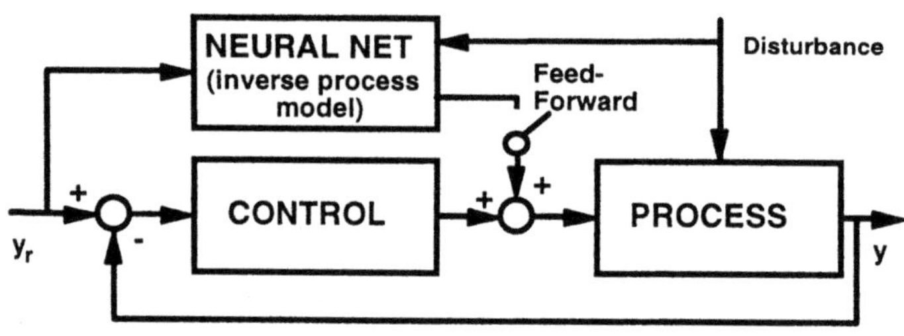

Figure 2 - *Advanced control strategy*

The neural controller was implemented based on a classical *Multi-Layer Perceptron (MLP)* with standard, feedforward topology, only one hidden layer and full connection between adjacent layers. The sigmoidal transfer function of both hidden and output nodes is the logistic one.

The cost function adopted for learning is the usual Mean Square Error. Learning is by pattern, that is, the weights and biases are updated after the presentation of each training sample [3]. The best neural net configurations for each loop were estimated by Cross-Validation. The neural nets used in the Pressure loop have only 5 hidden nodes, while those for the Flowrate loop have 11 hidden nodes.

The neural nets were implemented on the target control system, based on *Bailey's INFI90* ™. The basic control configuration and the neural modules were both implemented by means of library functions. The logistic transfer function was approximated in a piecewise linear way for efficiency sake.

The neural nets, including pre-processing and post-processing, run in 20 msec.

4. Experimental results

The experimental behaviour for a variation of the pressure setpoint was evaluated. Three control configurations were compared: conventional control (PID only), insertion of a neural net in the pressure loop (PID+1NN) and insertion of neural nets in both pressure and flowrate loops (PID+2NN).

The performances of the different control architectures, measured in terms of the *Mean Square Error* (*MSE*), are reported in Table 1. The figures emphasize the performance improvement of the controller (factor of *five*) due to the introduction of the neural modules.

Figure 3 reports the experimental results in terms of raise time.

Table 1 - *Experimental results with the three control schemes*

	PID	PID+1 NN	PID+2 NN
MSE	$1.1 \cdot 10^{-1}$	$2.2 \cdot 10^{-2}$	$2.1 \cdot 10^{-2}$

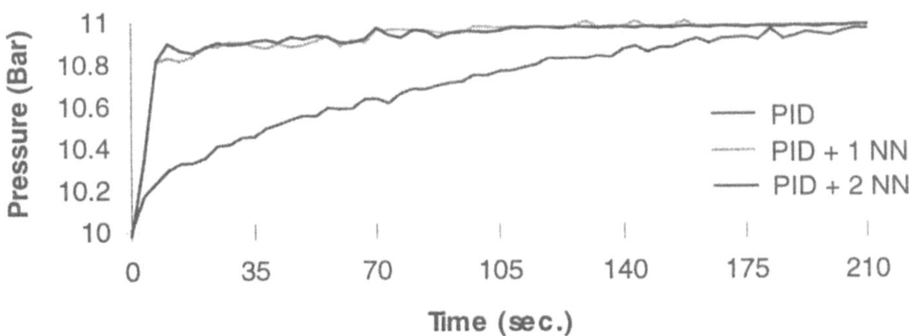

Figure 3 - *Experimental evaluation of raise time*

5. Conclusions

In this paper we presented an application of neural nets to continuous process control. The neural controller, inserted in the control system of the PRIAM laboratory, produced remarkable improvements towards the conventional control. The improvements have been obtained while preserving the traditional control architecture. The architecture has been enriched with new predictive functionalities. Moreover, an increased immunity towards disturbances can be observed.

Finally the substantial flexibility of neural methods with respect to process type ensures that neural controllers can be easily extended or modified to other similar applications.

Acknowledgements

We wish to acknowledge S. D'Imporzano, D. Laudato and A. Mella (Enel CRA) for their precious collaboration during the experimental data collection in the PRIAM laboratory.

References

1. M. Barabino, N. Bonavita, S. Bruzzo, F. Camastra, A.M. Colla, *Stato di una linea AP di preriscaldo: stima neurale ("Neural state estimation of a HP feedwater line")*, Automazione e Strumentazione (May 1995), pp. 97-103.

2. S. Bruzzo, F. Camastra, A.M. Colla, *Estimation of unmeasurable variable in a dynamical system by Resource Allocating Network*, Proc. of ISCAS 95 (Seattle, 29 April - 3 May, 1995), vol. III, pp. 1712-1715.

3. J. Hertz, A. Krogh, R.G. Palmer, **An Introduction to Neural Computation** (1991), Addison-Wesley.

4. K.J. Hunt, D. Sbarbaro, R. Zbikowski, D. Gawthrop, *Neural Networks for Control Systems - A Survey*, Automatica (1991), pp. 1083-1112.

5. J.G. Kuschewski, S. Hui, S.H. Zak, *Application of Feedforward Neural Networks to Dynamical System Identification and Control*, IEEE Transactions on Control Systems Technology, vol. 1, no. 1 (March 1993), pp. 37-49.

6. H. Wegmann, *Fuzzy Control and Neural Networks Industrial Applications in the World of PLCs*, Proc. 3rd IEEE Conference on Control Applications (Glasgow 1994), pp. 1245-1249.

Using Fuzzy Logic to Solve Optimization Problems by Hopfield Neural Model

Salvatore Cavalieri and Marco Russo

Institute of Informatic and Telecommmunications, University of Catania (ITALY)

Fax:+39 95 338280, Email:{mrusso,cavalieri}.iit.unict.it

Abstract

The solution of problems using a Hopfield model requires determination of the values of a certain number of coefficients linked to the surrounding conditions of the optimization problem itself. There have been no works in literature offering a general method for the search for coefficients which will guarantee optimal or close to optimal solutions. This paper proposes a method which allows automatic determination of Hopfield coefficients.

1   Introduction

The autoassociative memory model proposed by Hopfield [1, 2] has attracted considerable interest as a method for solving difficult optimization problems.

It has generally been found that although these networks do produce some useful results, a great drawback is that at times they provide non-valid solutions, or solutions referring to local minima.

The quality and percentage of valid solutions (henceforward referred to as robustness) of the solution provided by a Hopfield network essentially depends on the choice of the coefficients, here called Hopfield Coefficients (HCs), which link the energy function to be minimized to the surrounding conditions of the combinatorics problem to be solved.

No methods that allow unambiguous determination are currently available in literature.

Our aim was to devise a software tool that would autonomously search for HCs such as to ensure high percentages of valid, close to optimal solutions. This was achieved by using a fuzzy approach and transferring the know-how on the use of Hopfield networks we have acquired in various applications in several fields (see [3] for example) into a fuzzy knowledge base. The fuzzy system was obtained using the FuGeNeSys program [4] which adopts a supervised learning technique, the core of which is based on genetic algorithms.

As an application example, the results obtained with the well-known Travelling Salesman Problem (TSP) will be given. The tests show that it is actually possible to determine HCs such as to guarantee the desired robustness and/or quality of the neural solution. The results obtained by fuzzy tuning of the HCs will be compared with those existing in literature, more specifically those based on the set of HCs obtained by Hopfield and Tank in [2]. The comparison shows the clear quality and robustness increase we obtain with our approach.

2 The Fuzzy Tuning of Hopfield Coefficients

The aim of this section is to give a detailed description of the method we propose for automatic determination of HCs. It is based on close integration of the Hopfield Model and Fuzzy Inferences, as shown in Fig.1.(a).

The neural output (Hopfield Solution) is represented by the solution provided by the neural model, which can be valid or non-valid. The Solution_Check module has the task of assessing the validity of the neural solution and, if it is valid, its quality.

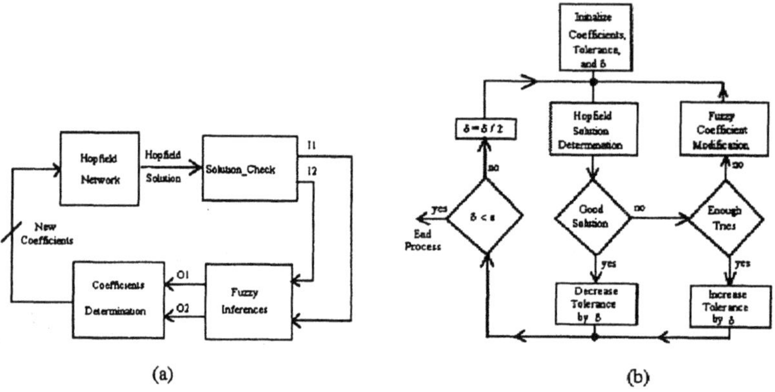

Figure 1: (a) Integration of Neural and Fuzzy Modules, (b) The Procedure for Determining a Suitable Set of HCs

In any optimization problem it is possible to identify the surrounding conditions (and therefore the associated HCs) which determine the validity of the neural solution. For example, in the case of the TSP, the neural solution is non-valid if at least one of the first three surrounding conditions, relating to the coefficients A, B, C (see [2]), is not met. Each valid solution has a certain degree of quality depending on the value of the solution determined as compared with the optimal solution. Even in such cases it is generally possible to identify the surrounding condition (and the HC) which will guarantee close to optimal solutions. Still referring to the TSP, the coefficient D (see [2]) is linked to the quality of the solution. Checking the validity of a neural solution to a generic optimization problem is quite straightforward as it consists of checking the value taken by each of the surrounding conditions. Verification of the quality of the solution, on the other hand, is much more difficult as the global minimum in any combinatorics problem is never known. In order to assess the quality of the solution, we consider a LowerBound of the optimal solution. This value can be determined for any optimization problem. In the TSP, for example, the LowerBound can be chosen as n times the minimum distance between two cities. Together with this value we consider a parameter called Tolerance. If the minimum of the energy function determined by the Hopfield network is lower than the Tolerance·LowerBound value, then the valid solution is also considered to be good. The Solution_Check module therefore has the task of checking the validity and quality of the solution provided by the neural network following the strategies outlined above.

The output of the Solution_Check module is made up of two parameters, indicated in the figure as I1 and I2. The first corresponds to the HC value determining non-validity of the solution or, if the solution is valid, the HC which mainly makes its quality poor. Parameter I2 is set as equal to the average of the remaining HCs. On the basis of these parameters the Fuzzy Inference module determines a new coefficient value for I1 (output O1) and varies the remaining coefficients (output O2). Choice of the inputs I1 and I2 was due to two reasons. The first depends on experience. In a heuristic, manual approach, the HCs are varied by determining the coefficient which has caused the solution to be non-valid or of poor quality. Its value is then increased according to the values assumed by the remaining coefficients. For this reason input I2 was considered to be significant. The second reason lies in the generalization of the approach presented. This choice allows the fuzzy approach to be applied to any optimization problem, as the inputs I1 and I2 and the outputs O1 and O2 are not directly linked to the number of coefficients in the problem to be solved.

Fig.1.(b) illustrates the general procedure for determining a suitable set of coefficients. The Tolerance parameter is initially set at a reasonable high value and is lowered by a pre-defined amount indicated with δ, if the network reaches a valid, good solution. If it does not succeed in providing a valid good solution with a certain Tolerance value, the latter is increased, still by an amount δ. Following each increase or decrease in the Tolerance parameter, δ is halved. Increases and decreases in the Tolerance parameter are therefore made on the basis of a binary search in order to speed the search process up.

The Enough Tries test refers to the attempt to find a valid, good solution with a certain Tolerance value. If no valid good solution is found after a pre-defined maximum number of iterations the Tolerance value is increased by δ.

Vice versa, if after this maximum number of tries at least one valid good solution is found, then the tolerance value is decreased by δ.

The process of increasing or decreasing the tolerance value ends when the parameter δ is lower than a previously defined threshold indicated with ϵ. In this way it is possible to identify the minimum Tolerance values below which no good solution can be found.

3   The Choice of the Parameters Used in HC Determination

The HC determination process described in the previous section is quite automatic, once certain parameters have been fixed. It is therefore necessary to make a few remarks about the way in which these parameters are to be established.

As mentioned previously, the LowerBound can be determined according to various criteria which are irrelevant to the algorithm itself. The only condition that has to be met is that the LowerBound be lower than the global minimum of the function to be minimized. It can be demonstrated that for any optimization problem it is always possible to determine a value that will satisfy this condition.

Choice of the parameters Tolerance, ϵ and δ is quite straightforward and not critical for the effects of the algorithm. More specifically, Tolerance has to be set to a reasonably high value to ensure that at least one good solution is reached. ϵ has to be as low as possible to make the search for an optimal set of HCs as exhaustive as possible. It should be noted, however, that too low an ϵ value could make the calculation time too long. Lastly, δ has to be chosen according to the Tolerance value, bearing in mind that it represents the first increase or decrease performed on Tolerance.

The only three parameters that really affect the results of the method proposed are IterDelta, IterMaxGood and PercIterMaxGood.

IterDelta is used to count how many consecutive non-valid solutions have been found before Tolerance is increased by δ.

IterMaxGood and PerIterMaxGood are used to count the number of valid good solutions. If this number is greater or equal than PerIterMaxGood and less or equal than IterMaxGood, the set of coefficients examined is considered to be both valid and good.

According to their meaning, these three parameters can affect the set of HCs determined as they are evidently correlated with the quality and robustness of the solutions.

More specifically, as IterDelta increases a greater number of attempts are made to find a good solution before increasing δ. This parameter thus directly affects the quality of the neural solution and an increase in it means trying to obtain better-quality solutions.

The other two parameters - IterMaxGood and PercIterMaxGood - are decisive in determining the robustness of the solution. As the ratio $\frac{PercIterMaxGood}{IterMaxGood}$ tends towards 1, in fact, HCs which make the neural network converge are always sought. It should be pointed out that an increase in PercIterMaxGood may cause a degradation in the quality of the solution. Asking for a large number of valid good solutions, in fact, may lead to a search for such a "flat" energy function minimum as to guarantee a large number of valid good solutions, But this minimum, which is difficult to get out of, may also turn out to be a local one.

The parameter IterMaxGood also affects the quality of the solution on its own. The greater the number of solutions found, the higher the probability of finding one very close to optimal.

Being able to set these parameters aribtrarily is, in our opinion, very important, as the user of an optimizer based on a Hopfield model may prefer a close to optimal solution obtained in a few attempts to the optimal one that can be obtained with a much larger number of tries. Vice versa, he may require the best solution whatever the price in terms of time.

4 Fuzzy Approach versus Heuristic Determination of Hopfield Coefficients

We want to show how the method we propose allows us to determine solutions which are much closer to optimal and much more robust than those provided by HC determination based on a classical heuristic search.

Secondly, we aim to demonstrate that appropriate use of the parameters IterDelta, IterMaxGood and PercIterMaxGood allows us to establish a priori the quality and robustness of the neural solution.

The results presented in this section, whose aim was to reach these two goals, refer to solution of the TSP.

The fuzzy logic-based algorithm was applied to a TSP featuring 10 cities with random distances between them. The minimum distance in the scenario was 0.05, so the LowerBound was set to $n \cdot 0.05 = 0.5$. The Tolerance was set to 5, δ to 4 and ϵ to 0.009. For the three parameters IterDelta, IterMaxGood and PercIterMaxGood the following three sets of values were chosen: {100,10,1}, {100,10,7} and {300,10,1}.

This choice was dictated by a desire to assess the influence of IterDelta on the quality of the solution and the influence of PercIterMaxGood on its robustness, as explained in Section 3. For each set of paremeters IterDelta, IterMaxGood and PercIterMaxGood the fuzzy-logic based coefficient tuning algorithm gave a set of HCs. The HCs $\{A = B, C, D\}$ determined were as follows: HCs#2={768,171,334}, HCs#3={893,134,444} and HCs#4={832,151,482}.

The solutions obtained on the basis of these three sets of HCs were then compared with those presented by Hopfield and Tank in [2], based on the following set of HCs: HCs#1={500,200,500}.

Below the comparisons between the four sets of HCs will be shown from two points of view: the robustness and the quality of the neural solution provided.

In Fig.2 we give the comparative results of robustness performance with varying numbers of cities. Two different TSPs were considered, one with 5 and one with 10 cities. 10 different random distance scenarios were considered for each problem and for each distance scenario 10 consecutive neural solutions were found. The tests were repeated for the four different sets of HCs seen previously, thus giving a total of 800 tests. In order to make the results as significant as possible, for each set of HCs the graphs show the percentage of non-valid solutions found in all the tests performed, both for the 10-city TSP and for the 5-city one.

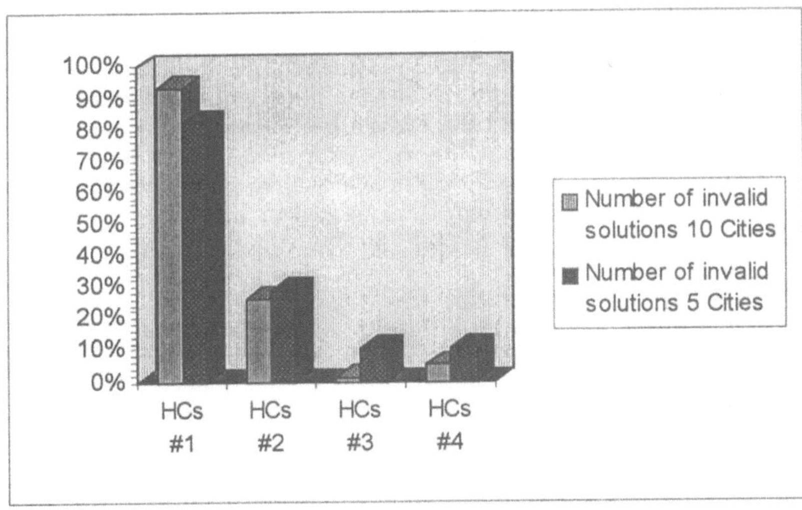

Figure 2: Robustness

An analysis of Fig. 2 shows that all three sets of coefficients determined using the fuzzy approach are more robust than those determined using the classical heuristic approach. It can also be seen that the most robust solutions are reached with HCs #3 and #4, which have the highest values for the $\frac{PerIterMaxGood}{IterMaxGood}$ ratio and the IterDelta parameter respectively. In particular set #3 has the greatest robustness. This was to be expected, as explained in Section V, since HC set #3 features the highest $\frac{PercIterMaxGood}{IterMaxGood}$ ratio.

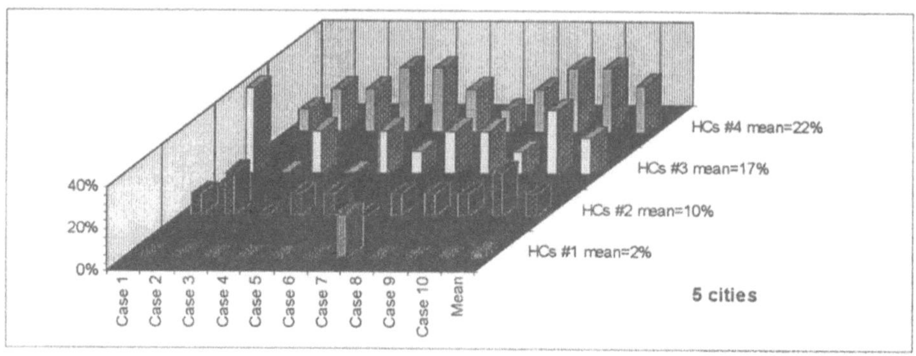

Figure 3: Quality

The aim of Fig.3 is to compare the quality of the solutions obtained with the four sets of coefficients. For each of the 10 distance scenarios considered the figure shows the average percentage of of optimal solutions found in 10 consecutive neural iterations. Alongside the 10 percentages the figure gives the mean of the 10 percentages for each set of HCs.

Here again the solutions obtained using the fuzzy approach are better than those reported in literature. More specifically, the percentage of optimal solutions reached with the parameters suggested by Hopfield and Tank is 2%, while with our coefficients we reached 10%, 17% and 22% respectively for HC #2, HC#3 and HC#4. It should be noted that the set of coefficients HC #4 has the highest IterDelta value and the highest percentage of optimal solutions. It should also be pointed out that the search for more robust solutions (set HC #3) causes on average a degradation in the quality of the solutions obtained t HC #4). This was to be expected, as explained in 3.

References

[1] J.J.Hopfield, "Neurons with Graded Response Have Collective Computational Properties Like Those of Two-State Neurons," in *National Accademy of Science*, pp. 3088–2092, May 1984.

[2] J.J.Hopfield and D.W.Tank, "Neural Computations of Decisions in Optimization Problems," *Biol. Cyber.*, vol. 52, pp. 141–152, 1986.

[3] S.Cavalieri, A.Di Stefano, and O.Mirabella, "Neural Strategies to Handle Routing in Computer Networks," *Int. Jour. of Neur. Sys.*, pp. 269–289, Sept. 1993.

[4] M.Russo, *Metodi Hardware e Software per Logiche di Tipo non Tradizionale*. PhD thesis, Univ. of Catania, Italy, Feb. 1996.

FIR NNs and Time Series Prediction: Applications to Stock Market Forecasting

Antonio d' Acierno,* Salvatore Palma and Walter Ripullone

I.R.S.I.P. - C.N.R.

via P. Castellino 111 - 80131 Napoli - Italy

Abstract

This paper deals with the Time Series Prediction problem, that is the prediction of a system future evolution given a certain knowledge about it and its past behavior; the neural networks based approach offers a new and a very promising concept of non linear prediction. The use of *FIR* neural networks and of the *Temporal Back–Propagation* learning algorithm are proposed for Stock Market forecasting; results on real-world series from "Milano Piazza Affari" Stock Exchange are shown and commented.

Keywords: Time Series Forecasting, FIR NNs, Temporal BP.

1 Introduction

The time series prediction problem can be simply stated as follows:

given a sequence $y(1), y(2), \ldots, y(N)$ up to time N, find the continuation $y(N + 1), y(N + 2), \ldots$

The series may arise either from discrete time systems or from the sampling of a continuous system, and be either pseudo–stochastic or deterministic in origin. Applications of prediction in signal processing range from adaptive line enhancers to differential pulse code modulation schemes for telecommunications. Prediction is also used in modelling turbolence and solar flux, forecasting the weather and managing stockmarket portfolios.

The standard prediction approach involves constructing an underlying model which gives rise to the observed sequence. In this sense (i.e. prediction as *system identification*) one is usually interested in estimating a single time step and, moreover, typically it exists an exogenous input which drives the system. When this solution cannot be applied since, for example, the underlying model is unknown, or it does not perform satisfactorialy, or one is interested in estimating several time steps, different approaches such as radial basis functions networks and neural networks can be used. Among neural networks based methods, we can use standard algorithms or neural paradigms specifically proposed for the problem at hand.

---

*Corresponding author: antonio@irsip.na.cnr.it

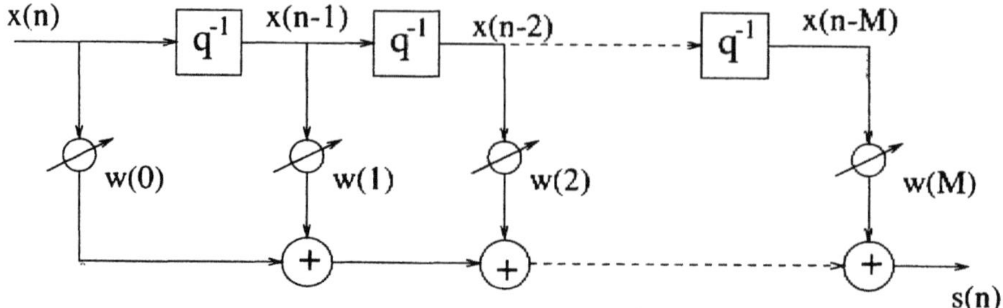

Figure 1: A simple connection in a FIR neural network.

In this paper we use a neural network paradigm specifically devised for time series prediction; we apply such a model to real world series from "Milano Piazza Affari" Stock Exchange and we show results for medium term forecasting.

2   FIR NNs and Temporal BackPropagation

FIR neural networks [1] have been proposed for time-series prediction; these networks are classical multilayer perceptrons whose connections are Finite-duration Impulse Response (FIR) filters (fig. 1).

Let $w_{ji}^l(k)$ denote the weight that connects neuron i in layer $l-1$ to neuron j in layer l (the index k ranges from 0 to M^l, where M^l is the total number of delay units of FIR filters between layers $l-1$ and l). The signal $s_{ji}^l(n)$ appearing at the output of the $i-th$ incoming synapse of neuron j is given by the linear combination of delayed values of the input signal $x_i^{l-1}(n)$ (convolution sum):

$$s_{ji}^l(n) = \sum_{k=0}^{M^l} w_{ji}^l(k) \cdot x_i^{l-1}(n-k) \tag{1}$$

where n denotes the discrete time step. Hence, summing the contributions of the complete set of p synapse, we may describe the output x_j^l of neuron j by using the following equation:

$$x_j^l(n) = \phi(\sum_{i=1}^{p} s_{ji}^l(n) - \theta_j) \tag{2}$$

where θ_j denotes the externally applied threshold, and $\phi(.)$ denotes the nonlinear activation function of the neuron.

The classical Back–Propagation cannot be applied as it is since, by *unfolding in time* the network in its static equivalent, we obtain a structure with duplicated weights so obtaining that *(i)* the necessary calculations expands and that *(ii)* the locally distributed process is lost since individual gradient terms must be recombined to obtain the total gradient for each unique filter coefficient.

To train the network, Wan proposed a generalised version of the standard Back–Propagation, for which he suggested the name *Temporal Back–Propagation*. The key concept is the following.

Consider the FIR filter shown in fig. 1; what happens in the backward step of the standard Back–Propagation algorithm, when error terms of, say, last hidden layer have to be evaluated using error terms of output neurons and connection backward? Wan suggested the concept of *reverse* FIR filter, where delay units are replaced with *advances* units.

The learning algorithm [1] can be now summarised as follows:

$$w_{ji}^l(n+1) \;=\; w_{ji}^l(n) - \eta \cdot \delta_j^l(n) \cdot x_i^{l-1}(n) \tag{3}$$

$$\delta_j^l(n) \;=\; -2 \cdot e_j(n) \cdot \phi'(x_j^L(n)) \quad for \quad l = L \tag{4}$$

$$\delta_j^l(n) \;=\; \phi'(x_j^l(n)) \cdot \sum_{m=1}^{N_{l+1}} \vec{\delta}_m^{l+1}(n) \cdot \vec{w}_{jm}^{l+1} \quad for \quad 1 <= l <= L-1 \tag{5}$$

where $\vec{\delta}_m^{l+1}(n) = [\delta_m^{l+1}(n), \delta_m^{l+1}(n+1), ..., \delta_m^{l+1}(n+M^l)]$ is the vector of propagated gradient terms, and η is the learning rate.

As it should be clear, this learning strategy that extends the classical Back–Propagation algorithm is non–causal, but it can be made causal simply by re–indexing variables; from the implementation point of view this implies that additional buffers have to be provided.

3   Open loop and Closed loop prediction

Figure 2 (left) illustrates the basic predictor training configuration for the FIR network. At each time step, the input to the network is the known value $y(k-1)$, and the output $\bar{y}(k)$ is the true series single step estimate, i.e.:

$$\bar{y}(k) = f_M(y(k-1)) \tag{6}$$

where f_M is an FIR network with total memory lenght M.

During the network training, the weights are adapted by using the Temporal Back–Propagation algorithm to minimize the squared error. This is the *open loop* adaption: the input and the desired response are both provided from the known training series and the actual output of the network is not fed back as input.

Once the network is trained, long-term iterated prediction is achieved by taking the estimate $\bar{y}(k)$ and feeding it back as input to the network (fig. 2, (right)), that is:

$$\bar{y}(k) = f_M(\bar{y}(k-1)) \tag{7}$$

This *closed loop* should allow to achieve predictions as far into the future as desired.

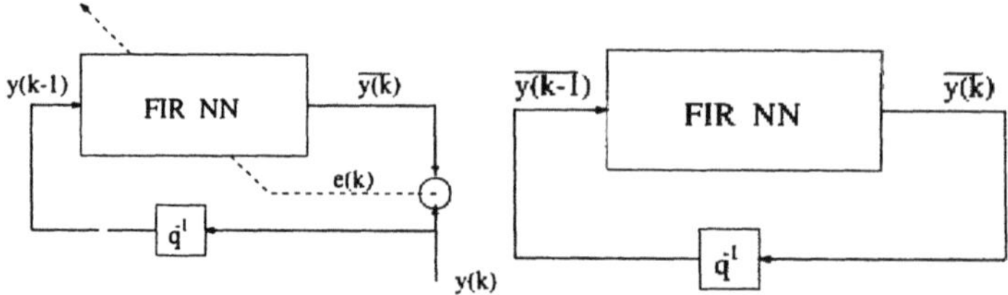

Figure 2: The open–loop and the closed–loop configurations.

For a linear system, iterating the prediction is of little value, being the response limited to a sum of dumped sinusoids; moreover, to ensure the closed loop stability the root of the regression polynomial must be monitored. The iterated dynamics of a neural system, on the other hand, is quite complex but will have output stability due to bounded activation functions which limit the dynamic range of the output.

The traning is based on single step prediction and so the accuracy of the long term iterated predictions cannot be guaranteed, because of accumulated errors. An alternative configuration which adapts the closed loop system directly might be useful; unfortunately gradient calculations within a feedback loop are significantly more complex and a modified Temporal Back–Propagation is under study.

4   Applications to Stock Market Forecasting

We applied *FIR* NNs to stock market forecasting; a parallel version of the code, developed having as target machine the Meiko CS-2 [2], has been used to speed up the computations. The predictive performance has been tested on 4 different stocks of "Milano Piazza Affari" Stock Exchange (*fiat, dalmine, alitalia, toro*); the used network has 2 hidden layers, each with 4 hidden units, and there are 10 delay units in the first level of connections, 5 delay units in the second and in the third level of connections, so that $M = 20$ (values were found simply by *trial and error*). The training phases have been performed using an on–line updating and assuming $\eta = 0.05$.

The size of the training set and of the test set used are shown in tab. 1. It could seem that training set and test set are not well balanced or, to be more precise, the test sets are too small; it is worth remembering that we are interested in *iterated* previson, whose quality decreases as we go far into the future because of propagated errors. If one is interested just in single-step prediction there is no reason to have small test sets.

The series have been rescaled in the range $[-1, 1]$ and then a simple low-pass filter (*mobile average*) was applied to avoid local spikes; the formula used for

Stock	Training set	Test set
fiat	1190	30
dalmine	1230	30
alitalia	1230	30
toro	1340	30

Table 1: Sizes of traning and test sets in the experiments.

the filter is the following:

$$NEW\_SERIE[i] = \frac{1}{K} \sum_{j=0}^{K} OLD\_SERIE[i-j] \tag{8}$$

with K being fixed at 10.

Figure 3 reports the results obtained. For the stock *fiat*, *dalmine* and *ansaldo* we have that there is a significant error in the predicted values but the global trend as well as the days at which there are maxima or minima are predicted almost exactly. For the stock *toro* we have that values have predicted almost exactly for about 25 days but it is loss the reversal of trend.

5   Conclusions

The sinergistic use of FIR NNs and Temporal Backpropagation seems a promising solution for forecasting stock market time series; in fact, despite to the simplicity (in terms of free parameters) of the used neural network, really interesting results were obtained using sligthly regularised real-world data. Since training is based on only single step predictions, the accuracy of the long term iterated predictions cannot be guaranteed after 30 days; this is mainly due to the propagation of the error related to the short term prevision. If we are interested just in a single–step prevision, we can use our predictor until the underlying conditions are constant. Figure 4, for example, shows how, with reference to the experiment on the *toro* stock, we can have a test set composed of 100 points without having any degradation in the performance.

References

[1] E. A. Wan, Finite Impulse Response Neural Networks with Applications in Time Series Prediction, PhD Thesis, Department of Electical Enginnering, Stanford University, 1993.

[2] A. d' Acierno, S. Palma, W. Ripullone, *FIR NNs and Temporal BP: Implementation on the Meiko CS-2*, WIRN96, This Volume.

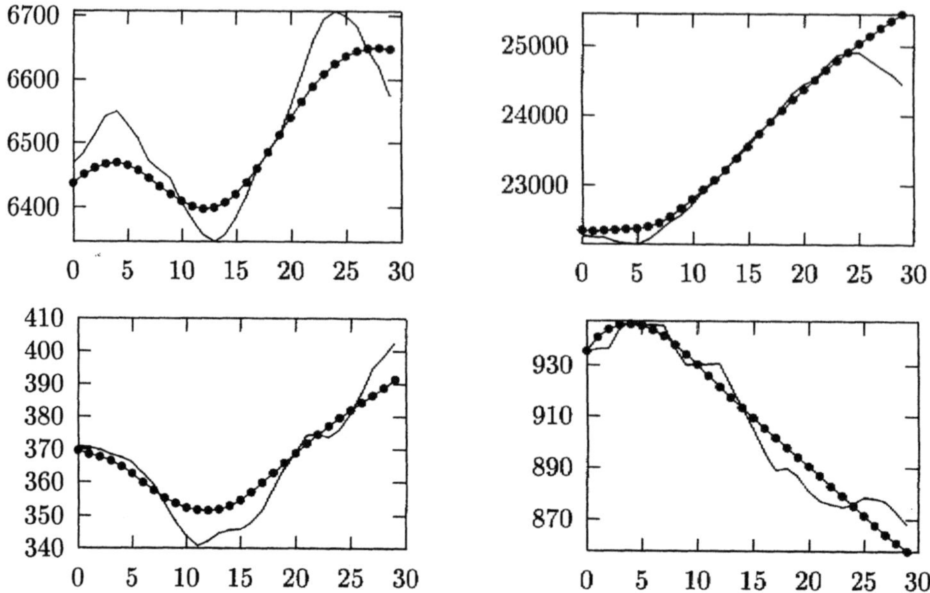

Figure 3: Obtained results with closed–loop prediction. From top left in clockwise: the *fiat* stock, the *toro* stock, the *alitalia* stock, the *dalmine* stock. Points represent predicted values.

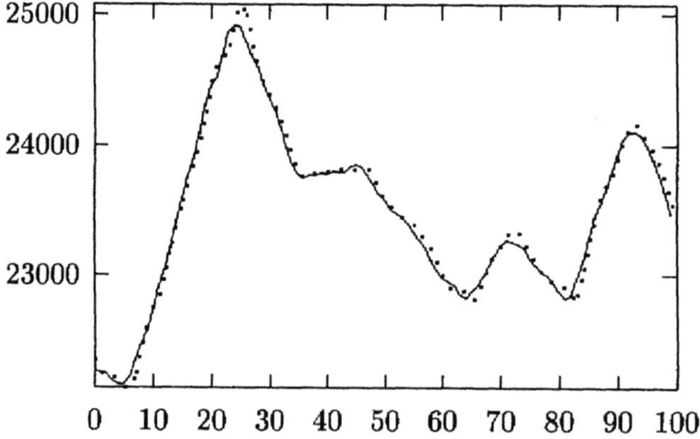

Figure 4: Obtained results. See text for details.

Simulation of Traffic Flows in Transportation Nets with Unsupervisioned MAIA Neural Net[1]

Giuseppe Pappalardo, Maria Nadia Postorino, Domenico Rosaci, Giuseppe M.L. Sarnè
DIEMA - Facoltà di Ingegneria - Università di Reggio Calabria
Via E. Cuzzocrea, 48 - Reggio Calabria, 89128 - Italia

Abstract

The planning of an urban transport system can take advantage of some new and innovative computational techniques. In this paper an unsupervisioned new neural network is employed to resolve the problem of traffic flow simulation. The results obtained are compared with those obtained through the use of conventional algorithms and techniques generally employed by the transport system analysts.

1 Introduction

In a more and more increasing number of cities the well known effects of the traffic congestion ask for the use of new tools able to control the transport system (TS) condition possibly in real time. For this reason the use of the Neural Networks (NNs) in the planning of transport systems is becoming more and more usual.

In this paper a unsupervisioned NN, based on a new theory called MAIA (Modulate Asynchronous Information Arrangement) [1], is used in order to obtain the traffic flow (TF) values on an urban transport network (TN), this problem being strongly characterised by the presence of both cyclic dependence among variables and random aspects. The use of the MAIA NN allows overcoming the difficulties generally found by using the supervised back-propagation (BP) NN [2] [3] [4] [5].

In the following, after an introduction on the MAIA theory and some information on the main traffic theory aspects, the results obtained through a unsupervisioned MAIA NN applied to a real case will be presented.

2 The MAIA Model and the MAIA Neural Network

The problem relating pairs of data can be defined as follows: "*given two sets X and Y of atomic elements (respectively of cardinality equal to K and H), the unknown relationship R between the elements of X and the elements of Y has to be reached*".

The relationship R is defined by its functional form (given in this case by the sequentially application of a finite number of elementary mathematical operations) and a suitable set of parameters. If in a first stage the functional form of R is

---

[1] In this work M.N. Postorino dealt with the transport aspects of the problem, while G. Pappalardo, D. Rosaci, G.M.L. Sarnè dealt with the Neural Network aspects.

supposed to be explicitly given, the previously problem can be expressed as follows: "a *set P of real numbers such that for each* $y \in Y$ *then* $y=R^P(x_1,x_2,...,x_K)$ *has to be reached,* R^P *being the functional form explicitly given and specified by the set of parameters P and the K elements* $\{x_1,x_2, ,x_K\} \in X$.". An average error ε (*a priori* fixed) can be made on the whole set Y in the calculation of each y defined as above. The proposed functional form R relates the elements of X to the elements of Y through a weighted graph G defined as follows: "*G:* $\forall x \in X$ *a node* $i \in G$ *exists on which the value of x is defined and* $\forall y \in Y$ *a link* $r \in G$ *exists on which the values of y is defined*". G is then formed at least by K nodes and H links and this condition is always possible. In fact, if a node is connected to all the others, $K(K-1)$ links are obtained, and if $K(K-1) \geq H$ the condition is verified, while in the other case K^* auxiliary nodes have to be introduced until $(K+K^*)(K+K^*-1) \geq H)$.

The Kolmogorov theorem [6] always assures the existence of at least an explicit functional form of R. This is undoubtedly of a great theoretical importance, but in the practice the solution can difficulty be obtained, and is strictly dependent on the relationship R that has to be reproduced.

In this note an heuristic solution of the problem is proposed. The R function is *ad hoc* constructed through the definition of some elementary mathematical functions that outline together the functional form of R. The problem is then simply converted in the estimation of the suitable parameters of these elementary relationships, i.e. those parameters that allow actually relating the sets X and Y through R. If H links $\in G$ are chosen, some $y \in Y$ element can be respectively associated to each of them, and an artificial transformation of the value of y is introduced; y is supposed to vary in the time through a cosinusoidal law defined by: $y_h(t) = y_h \cos(\omega_i + \phi_h)$, ω_i and ϕ_h being suitable parameters respectively of the node and the link. Furthermore, for each node $i \in G$ this functional form is defined:

$$O_i(t) = \mathrm{T}\left(x_i, y_{h_1^i}, y_{h_2^i}, \ldots y_{h_{M_i}^i}, t\right) = \tanh\left(a_i\left(x_i(t) + \sum_{j=1}^{M_i} y_{h_j^i}(t)\right) + b_i\right) \quad \textbf{(1)}$$

where T is a function of the values of x associated to the node i and of the M values y_{i_k} associated to the links connected to the node i, a_i e b_i being some suitable parameters of the hyperbolic tangent function. The hyperbolic tangent introduces a clipping in the sum of the periodical functions while the subsequent tuned circuits suitably mask the superior harmonic function. The transformation T has to be chosen such that: $O_i(t)=O_i\cos(\omega_i t)$, i.e. such that $O_i(t)$ is periodic cosinusoidal and ω_i is the *oscillation* frequency that is the same for all links starting from the ith node. Finally, for each link the following condition has to be verified:

$$y_h = O_{i_h} / w_h \quad \textbf{(2)}$$

where O_i is the *harmonic oscillation modulus* defined at the beginning of the link h and w_h the weight of the link h. Then Eq. 1 and Eq. 2 define the functional form R.

The fundamental hypothesis is that of establishing a relationship between $y_h(t)$ and $O_i(t)$, such that at each time all the $y_h(t)$ are the sum of a part proportional to $O_i(t)$ (a direct effect of the transformation at the node), a part proportional to the variation in the time of $y_h(t)$ (an effect of the short time memory of the link), and a part proportional to the sum of all values previously assumed by $y_h(t)$ till the time t

(an effect of the long time memory). These contributions, respectively called c_1,c_2 and c_3, can be expressed as function of a term that we call *link resistance*, α_h, for the hth link. c_1,c_2 and c_3 are inversely proportional to α_h, and they depend also on some other parameters as the following expression shows: $c_{1h} = 1/\alpha_h$; $c_{2h} = -\beta_h/\alpha_h$; $c_{3h} = -1/\gamma_h\alpha_h$. The MAIA law can then be expressed as:

$$y_h(t) = \left[O_i(t) - \beta_h\left(dy_h(t)/dt\right) - \int_0^t y_h(t) \middle/ \gamma_h \right] \middle/ \alpha_h \tag{3}$$

The Eq. 3 can be derived with respect to the time variable, and by organising the different terms the following differential equation is obtained:

$$\beta_h\left(d^2 y_h(t)/dt^2\right) + \alpha_h\left(dy_h(t)/dt\right) + y_h(t)/\gamma_h = dO_i(t)/dt \tag{4}$$

Through Eq. 4 the estimation of the parameters that define R are obtained. The Eq. 2 can also be expressed as:

$$\left(O_i(t)\middle/ w_h \cos(\omega_i t)\right) - y_h = 0 \tag{5}$$

The solution of the relationship R can then be obtained defining for each link $r \in G$ a system of the two equations (Eq. 4 and Eq. 5) in K variables.

The MAIA theory has been used to train a neural network; given the hypothesis made on $y_h(t)$ and $O_i(t)$ and the theory of the differential equations, the resolution of (Eq. 4) through a MAIA NN is true if the following conditions hold:

$$\phi_h = \tan^{-1}\left(\left(\beta_h\omega_i - \left(\gamma_h\omega_i\right)^{-1}\right)\middle/\alpha_h\right) \tag{6}$$

$$y_h = O_i \middle/ \sqrt{\alpha_k^2 + \left[\beta_h\omega_i - \frac{1}{\gamma_h\omega_i}\right]^2} \tag{7}$$

Using the Eq. 6 the weight can be expressed as:

$$w_h = \left[\sqrt{\alpha_k^2 + \left[\beta_h\omega_i - \left(\gamma_h\omega_i\right)\right]^2}\right]^{-1} \tag{8}$$

and from the comparison between Eq. 5, Eq. 8 and Eq. 6, it can be written:

$$\text{T}\left(x_i, y_{h_1^i}, \ldots, y_{h_{M_i}^i}, \alpha_{h_1^i}, \ldots, \alpha_{h_{M_i}^1}, \beta_{h_1^i}, \ldots, \beta_{h_{M_i}^i}, \gamma_{h_1^i}, \ldots, \gamma_{h_{M_i}^i}, \omega_{h_1^i}, \ldots, \omega_{h_{M_i}^i}, t\right)\left[y_h \cos(\omega_i t)\right]^{-1} = w_h \tag{9}$$

All these equations do not need any differential equation, and at each iteration the following objective function has to be minimised.

$$\sum_{h=1}^{H}\left|\text{T}\left(x_i, y_{h_1^i}, \ldots, y_{h_{M_i}^i}, \alpha_{h_1^i}, \ldots, \alpha_{h_{M_i}^1}, \beta_{h_1^i}, \ldots, \beta_{h_{M_i}^i}, \gamma_{h_1^i}, \ldots, \gamma_{h_{M_i}^i}, \omega_{h_1^i}, \ldots, \omega_{h_{M_i}^i}, t\right)\left[y_h \cos(\omega_i t)\right]^{-1} - w_h\right| \tag{10}$$

This function evaluates the global error made at the tth iteration in the resolution of the system formed by the Eq. 4 and Eq. 6. A further hypothesis on the relationship R is that of the preservation of the information at the node i, that can also be expressed as a Kirchhoff equilibrium between all the input quantities and all the output quantities. MAIA NN, in this condition, can work in unsupervisioned way through an energy function expressed by Eq. 10 and the Kirchhoff equilibrium.

In the Eq. 8 the term ω is always a variable whose initial value is randomly chosen, while the remaining terms are constant. If there is a correspondence between the elements of the NN and some physical elements present in the analysed problem (e.g. the NN topology and the TN graph are the same, the synapses identify the road links and the neurones the cross-roads) the terms α_h, β_h and γ_h can be conveniently identified with these physical elements. This substitution strategy allows considerably reducing the computing time and does not affect precision, provided the substitutions are really representative of the examined problem. All the variable terms are updated in subsequent MAIA iterations.

3 The MAIA NN Applied to the Mobility Problem

In this section some information about the modelling of a TS are given (for space problems, only the most important subjects are considered in a necessary concise way) and the results obtained through a unsupervisioned MAIA NN used for the simulation of the TF are presented.

An urban TS is generally represented by a connected, oriented and weighted graph (TN); it includes the set formed by nodes, links and cost functions. Cost functions are defined over network links; in general, they depend on the kind of problem, the examined system, and the physical and functional characteristics of the connections represented by the links [7] [8] [9]. A circular relationship exists among the TFs, the cost functions and the Origin/Destination (O/D) transport demand matrix [10] [11], where this matrix represents the traffic units moving from an origin to a destination point. In an equilibrium approach the relationships existing among demand, costs and flows are satisfied and under some general conditions the equilibrium solution exists and is unique.

The modelling of global mobility involves the knowledge of the transport demand among the O/D pair and the mathematical relationship relating link cost to flow on that link [12] [13]. The state of the TS is defined by the set of the link flows and the link costs at equilibrium. In order to determine the state of the system, the well know *assignment models* are used, whose solutions are often difficult to be obtained analytically [13]. Traditional algorithms have been devised for computing these solutions and are in widespread use in the transport research community; one of the best known, which has been selected for comparison with our NN approach, is the Stochastic User Equilibrium-Probit (SUE-P) [14], [13].

The MAIA NN is applied to the TN of Reggio Calabria (in the south of Italy) that is formed by 607 links and 304 nodes. The NN and the TN have the same topology, input and output to neurones are intended to model flows to and from TN nodes; Origin/Destination nodes are also represented as neurones and input/output to/from this nodes are the correspondent elements of O/D matrix.

Since the proposed MAIA NN is a unsupervisioned one it does not need a learning data set, but only one O/D demand matrix and a randomly initialised value of the variable parameter are sufficient to the MAIA NN for starting its operation.

The terms α_h, β_h and γ_h are assumed to be constant, and they represent the following physical characteristics of the examined problem: α_h depends on the cost function defined on the link, β_h describes the priority of the road h with respect to the other roads starting from node i ($\forall$ i: $\sum \beta_{h_i} = 1$); γ_h depends on the capacity of h (maximum admissible flow on it).

The convergence of the MAIA NN to stable configuration depends on the TN size; for the TN used in the described experiment the convergence is obtained in about 3300 iterations, while in the working stage the MAIA NN provides its results in one iteration. If the variation in the TN is considered (e.g. a new link is introduced when the learning is already ended), the existing configuration of the NN can be used in order to obtain the new NN, without accomplishing a complete learning (for example, if the TN variation refers to only one link, the learning will end in about ten iterations). On the contrary BP NNs are characterised by very long phases of NN parameter tuning (the complexity of the BP tuning phase strongly depends, in a more than linear way, on the TN size, and consequently the performances get worse) and each variation in the TN needs a complete learning. The performance of the MAIA NN has been compared with that of the SUE-P.

In fig. 1 the dispersion obtained by comparing the TFs produced by the MAIA NN in the working stage (after the learning step) with respect to the SUE-P algorithm TFs is represented. Several tests (about 2000) have been performed in order to assess the effect of variations in the link properties and the O/D demand matrix The MAIA NN always reaches the convergence, and the obtained results are always reliable; also, they often are the same as those provided by the SUE-P.

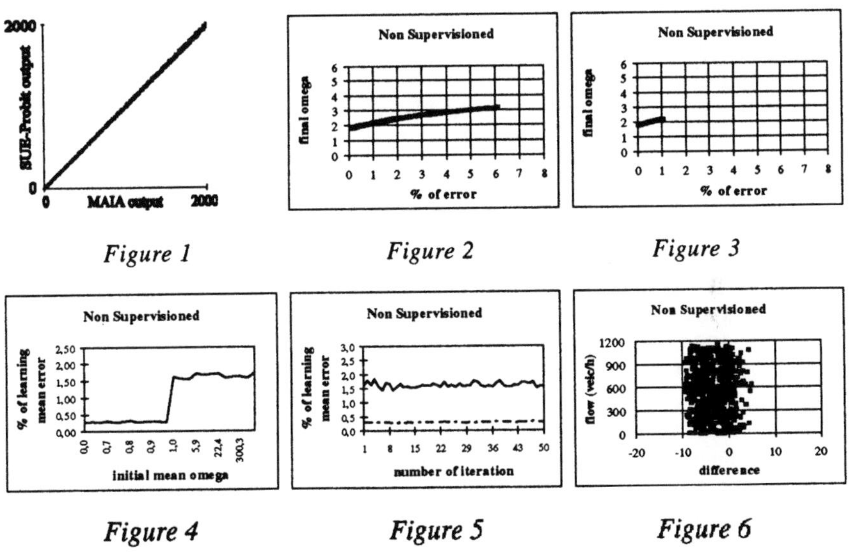

Figure 1	Figure 2	Figure 3
Figure 4	*Figure 5*	*Figure 6*

Fig. 2 shows the connection existing between the error associated to the simulation of TFs and the final ω; as it can be seen a suitable initialisation of the values of the parameters ω affects the size of the error made (see fig. 3); in fact, for this application, if the ω values are chosen in an interval defined around a suitable central value in the range $]0,1]$, the size of the error made decreases significantly. Furthermore, if in the simulation the central values of the initial ωs are modified the error curve present a step with a threshold point around the value 1, as it can be seen in the fig. 4. Fig. 5 represents the error for a set of simulations with central values both greater and less than the threshold value and random initialisation of the parameters ω; the fluctuation around a mean value depend on the ω random initialisation. The final errors made in the TF simulation are independent on the values of the TFs themselves, and this fact shows that the MAIA NN is better than the BP NN (in fact, with BP NNs the error made strictly depends on the value of the link flow). Finally, fig. 6 shows the difference between the MAIA NN TFs and the SUE-P algorithm TFs (± 10 units); in this case the difference shows the insensibility of the MAIA NN with regard to the variation in the TF values.

As the results obtained guarantee, the MAIA NN applied in the transport field has got very good computing properties, produces reliable results and in very short time and ,very important, works in a unsupervisioned way; if the ω parameters are suitably initialised they allow obtaining better performances than any type of traditional NNs can give. Furthermore, the MAIA NN is practically indifferent both to the size of the TN and the values of the TFs to be reproduced, and then it behaves with the same precision as regards all links.

At present further studies are being developed mainly for the understanding of the particular influence of the ω parameter in the learning stage and for improving the technique of initialisation of the ω values.

References

1. Pappalardo G, Rosaci D, Sarnè GML. Theoretical aspects of the MAIA Internal report 1995, University of RC.
2. Rumelhart D, McClealland JL and the PDP Research Group. Parallel distributed processing, vol 1, MIT Press, Cambridge 1986.
3. Postorino MN, Sarnè GML. Un'analisi delle capacità di apprendimento delle NN nella simulazione di mobilità. Reggio Calabria, Internal report n.6, University of RC, 1993.
4. Postorino MN, Sarnè GML. Dynamic simulation of traffic mobility through Neural Networks. To be published.
5. Sarnè GML, Postorino MN. Application of neural network for the simulation of traffic flows in a real transportation network. In: Proceedings ICANN 94. Springer-Verlag, London, 1994, pp831-833.
6. Kolgomorov AN. On the representation of continuos functions od several variables by superposition of continuos functions of one variable and addition, Dokl. Akad. Nauk SSSR 114, 1957, pp.953-956
7. Ford R, Fulkerson J. Flow in networks. Princeton University Press, Princeton, 1962.
8. Edmond J, Karp RM. Theoretical improvements in algorithm efficiency for network flow problem, Journal of the ACM, 1972; 19:248-264.
9. Zadeh N. Theoretical efficiency of the Eduard-Karp algorithm for computing maximal flows. Journal of the ACM, 1972;19:184-192.
10. Cascetta E. Metodi quantitativi per la pianificazione dei sistemi di trasporto. Cedam, Padova, 1990
11. Ortuzar J, Willumsen LD. Modelling Transport. John Wiley & Sons, New York, 1990
12. Ben-Akiva M, Lerman S. Discrete choice analysis. MIT Press, Cambridge, 1984
13. Sheffi J. Urban Traffic Assignment. John Wiley & Sons, NewYork, 1985.
14. Laganà D, Pappalardo G, Postorino MN, Rosaci D, Sarnè GML. A Hopfield like neural network in the simulation of traffic flows in a transportation network. In:Proceedings ICANN '95, Section Transportation. EC2&Cie, 1995,pp 36-41.

Towards Fog Forecasting in Meteorology by Means of Back-Propagation Neural Networks with Weighted Least-Squares Training

Antonello Pasini and Sergio Potestà

Servizio Meteorologico dell'Aeronautica, 2° CMR, Aeroporto "De Bernardi"
Via di Pratica di Mare, I-00040 Pratica di Mare (Roma). Italy

Abstract

A simple model for short-range visibility forecasting in meteorology. using back-propagation neural networks, was recently worked out by the authors. In this paper the attempt at adapting this model. in order to correctly forecast very low visibility values, is described. The theoretical errors associated with the different observed values of visibility are not constant, but depend on the values themselves. This fact leads to the modification of the cost function in order to achieve a weighted least-squares training. Improvements in the forecast are obtained and here described.

1 Introduction

In a previous paper [1] the application of a neural network model to the short-range forecast of meteorological visibility was discussed, referring to a case-study characterised by haze, but no fog. In this note we present the attempt at revisiting the model structure, in order to obtain a good forecast of intense fog episodes. and report preliminary results.

In the next section the neural model adopted in our previous works [1.2] is outlined. In section 3 some considerations on the noise associated with the observed visibility values lead to consider a weighted least-squares training through modification of the cost function. In section 4 a comparison between the performances of the two distinct training rules are presented in a case-study characterised by very low visibility values. Finally, in the last section brief conclusions are drawn.

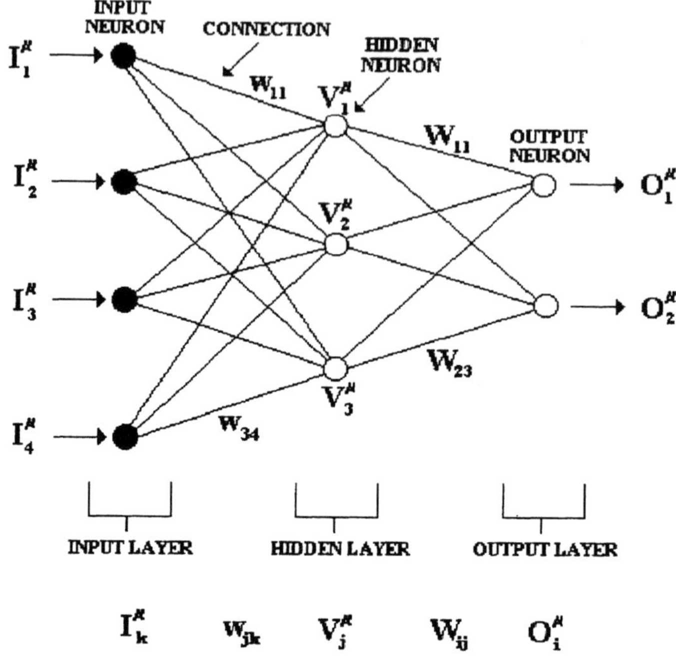

Fig.1. Structure of a feed-forward neural network with one hidden layer and notations used in this paper (see also ref.[3])

2 Outline of the model

The structure of our model is presented at full length in ref. [1], and to this paper the reader is invited to refer for details. Here we can say that the model is based on the software simulation of a feed-forward neural network with one hidden layer (fig.1) and on the weight updating through a back-propagation mechanism based on the minimisation of cost functions (for every pattern μ) of the following kind (O_i^μ = outputs, T_i^μ = targets):

$$E^\mu = \frac{1}{2}\sum_i \left(T_i^\mu - O_i^\mu\right)^2 . \tag{1}$$

The activation functions are sigmoids and the training goes on till the following condition is achieved:

$$\frac{\sum_{\mu,i=1}^{M_1,N} \left(O_i^\mu - T_i^\mu\right)^2}{M_1 N} < MSS\,, \tag{2}$$

where MSS (Mean Square Sum) is a previously fixed coefficient, M_1 is the number of training patterns and N is the number of output units.

In order to treat mutable situations like meteorological ones, training and test procedures must be adaptive as much as possible. Therefore we have chosen to supply the system with a "memory" of training cases which is fixed in width and is updatable at every new forecast, through the removal of the oldest training pattern and the insertion of the last available one. Thus we have obtained a "moving window training", which is very interesting in the field of meteorology because it perfectly simulates an operational forecast activity. The optimum width of this window was experimentally found to be about 2 months and a half of data; above this width the forecast performances of the network decrease, making us think of a typical influence of the seasonal change [1].

The network inputs are the normalised values of various meteorological variables, detected by one or more surface stations at time T_0, and the only output of the network is the value of visibility at one of these stations at a following time (for instance, at $T = T_0 + 3$ hours). Moreover, inputs are also the hour T_0 of the day (appropriately expressed through a sinusoidal function with maximum at noon and minimum at midnight) and the visibility time derivative with respect to the previous hour.

3 A weighted least-squares training procedure for fog forecasting

The model just outlined works well in forecasting visibility at stations which are not affected by fog episodes (visibility < 1000 m.), but shows systematical deficiencies when it must forecast the low visibility values typical of the Po Valley in winter: here a higher forecast accuracy is needed.

In our opinion, these deficiencies of the model are due to the peculiar detection of visibility, which can be instrumentally detected at low values, but which is subjectively observed at higher values. This situation does not allow us to work out a definite model of the theoretical errors to be associated with visibility observations. In any case, we can imagine that the theoretical absolute error is not constant along all the range of values, but there seems to exist a sort of direct proportionality between observed values and errors associated with them.

On the other hand, different cost functions represent different training criteria. Eq.1 leads to a least-squares training and it is well-known [4] that least-squares estimation is maximum likelihood for noise that is normally distributed with equal variance for all training cases and is not dependent on different cases. Thus, by using Eq.1, we can expect an optimum estimate for noise possessing these properties and non-efficient estimates for noise with other distributional features. If we make the heuristic hypothesis that the noise associated with visibility values is normally distributed but has greater variance for greater values, we can search for a cost function leading to a weighted least-squares training.

After some experimental tests, we have found that a cost function revealing good properties is the following:

$$E^{\mu} = \frac{1}{2}\sum_i \frac{\left(T_i^{\mu} - O_i^{\mu}\right)^2}{\sqrt{10 f(T_i^{\mu})}}, \tag{3}$$

where

$$f(T_i^{\mu}) = \begin{cases} T_i^{\mu} & \text{if } T_i^{\mu} \geq 0.1 \\ 0.1 & \text{if } T_i^{\mu} < 0.1 \end{cases}. \tag{4}$$

With reference to the notations of fig.1, the gradient descent terms in the formulae of weight updating become:

$$-\eta \frac{\partial E^{\mu}}{\partial W_{ij}(t)} = +\eta g_i'(h_i^{\mu})\frac{\left(T_i^{\mu} - O_i^{\mu}\right)}{\sqrt{10 f(T_i^{\mu})}} V_j^{\mu}, \tag{5}$$

$$-\eta \frac{\partial E^{\mu}}{\partial w_{jk}(t)} = +\eta \sum_i \frac{(T_i^{\mu} - O_i^{\mu})}{\sqrt{10 f(T_i^{\mu})}} g_i'(h_i^{\mu}) W_{ij} g_j'(h_j^{\mu}) I_k^{\mu}. \tag{6}$$

h_i^{μ} and h_j^{μ} are the weighted sums converging to the neurons of the output and hidden layers, respectively, and the g' are the sigmoid derivatives. This training allows a higher learning for low visibility values and a lower learning for high visibility values. This method is equivalent to maintaining Eq.1, making the learning rate η dependent on the target value. The weight updating rule includes also a term of "inertia" (momentum).

Together with the changes in the cost function, the condition of iteration stop (Eq.2) has been changed, in order to stop the convergency under a well-defined error medium threshold per cent. Therefore in the new model Eq.2 is substituted by

$$\frac{\displaystyle\sum_{\mu,i=1}^{M_1,N} \left(O_i^{\mu} - T_i^{\mu}\right)^2 / \left[f(T_i^{\mu})\right]^2}{M_1 N} < MSS. \tag{7}$$

Moreover, in order to measure the performances of the neural forecasts. a dispersion index WGC (Weighted Generalisation Coefficient) is calculated as follows:

$$WGC = 1 - \sqrt{\frac{\sum_{\mu,i=1}^{M_2,N}\left[(O_i^\mu - T_i^\mu) - \overline{(O_i^\mu - T_i^\mu)}\right]^2 / \left[f(T_i^\mu)\right]^2}{M_2 N}} \, , \qquad (8)$$

where M_2 is the total number of test patterns, N is the number of output units and $\overline{(O_i^\mu - T_i^\mu)}$ is the average error of the network. This coefficient resembles the Generalisation Coefficient (GC) used in our previous works [1,2], but it shows traces also of the rules indicated by the ICAO (International Civil Aviation Organisation) for the control of short-range visibility forecasts [5]. Because the data are normalised between 0 and 1, then $0 \leq WGC \leq 1$. The closer WGC is to 1. the better are the forecasts of the neural network.

4 Forecast results

A comparison test between the forecast performances of the two models is performed on a case-study concerning the very short-range visibility prediction at Milano Linate in winter 1992-'93. The moving window is fixed at 2 months. thus obtaining forecasts from January 1st to February 28th 1993. In these preliminary tests we consider in input only the data of Milano Linate at time T_0; the influence of the availability of a horizontal pattern of stations in this region was shown elsewhere [2].

After experimental tests on the 1-hour forecast in order to choose some free parameters (number of hidden neurons and MSSs), we obtain some interesting results. As an example, in fig.2 the new model's performance at $T = T_0 + 1$ hour in January is presented: you can see the very good forecasts for fog episodes. Actually, through all the period January - February '93, if we choose an error acceptability threshold of ± 250 m. for observed visibility < 1000 m. and of $\pm 25\%$ for higher values, the score of the model with least-squares training sets at 76%, while the score of the "weighted" model sets at 83%.

In terms of WGC, the situation is depicted in fig.3. Here we remark the consistent improvements in forecasting for $T + 1$ and $T + 2$ hours, while the $T + 3$ case shows inverted results. In connection with this latter case we must observe that here the model with weighted least-squares training does not converge at the fixed MSS (=0.0625, that is the $\pm 25\%$ of the target for visibility > 1000 m.), but stops the iterations at highest values because it cannot gain any more. This problem does not undermine our confidence in the good quality of a weighted least-squares scheme.

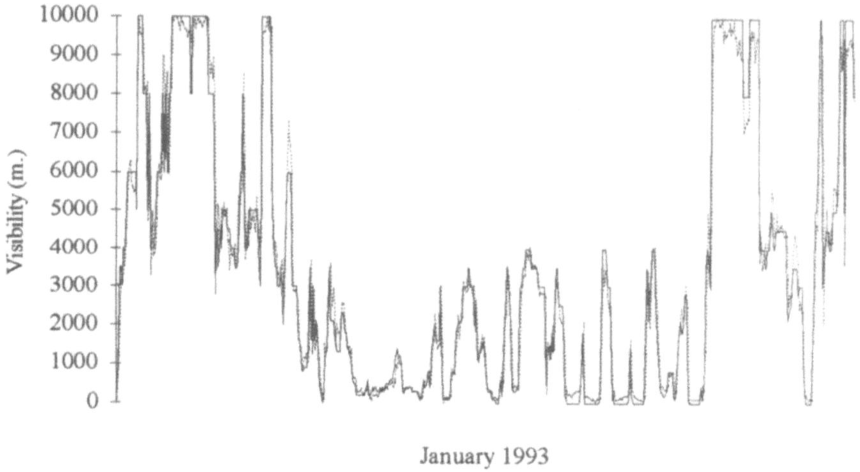

Fig.2. Forecast performances of the weighted least-squares training at $T = T_0 + 1$ hour (continuous line = observed visibility, dotted line = forecasted visibility)

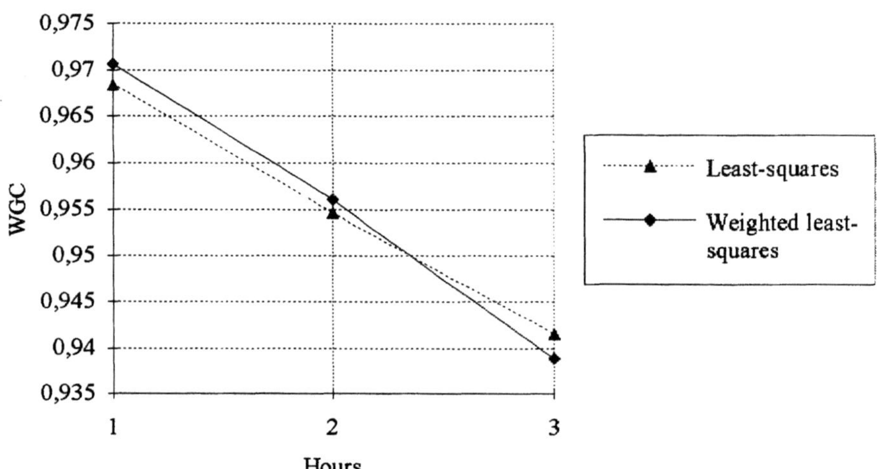

Fig.3. Comparison between the forecast performances of the two models

but leads us to think that, probably, a "softer" weighting is needed if we want to obtain good forecast results even at longer range.

5 Conclusions

In conclusion, the negative results obtained in applying the model outlined in section 2 to fog cases has led to change the cost function and the formula of iteration stop in a back-propagation scheme, with explicit reference to the peculiarity of visibility and its detection. The application of a new model with weighted least-squares training leads to encouraging results that induce us to go on with our theoretical and experimental investigation, in order to achieve a satisfying neural model for short-range fog forecasting.

Acknowledgements

We wish to thank P.P.M. Costa for his technical support.

References

1. Pasini A, Potestà S. Short-Range Visibility Forecast by Means of Neural-Network Modelling: a Case-Study. Il Nuovo Cimento 1995; 18C: 505-516
2. Pasini A, Potestà S. Neural Network Modelling: Perspectives of Application for Monitoring and Forecasting Physical-Chemical Variables in the Boundary Layer. In: Allegrini I and De Santis F (eds) Urban Air Pollution. Springer-Verlag, Berlin, 1996, pp 329-340
3. Hertz J, Krogh A, Palmer RG. Introduction to the Theory of Neural Computation. Addison-Wesley, New York, 1991
4. Bishop CM. Neural Networks for Pattern Recognition. Oxford University Press, Oxford, 1995
5. ICAO, Document on International Standards and Recommended Practice - Meteorological Service for International Air Navigation, 10th ed., Annex 3, 1986, p 59

Spectral Mapping: a Comparison of Connectionist Approaches

Edmondo Trentin, Diego Giuliani and Cesare Furlanello

Istituto per la Ricerca Scientifica e Tecnologica

Povo (Trento), Italy.

{trentin, giuliani, furlan}@irst.itc.it

Abstract

This paper presents two connectionist approaches to spectral mapping for speaker normalization. The first is based on a extended Radial Basis Functions network. The second approach is based on a slightly improved Multi-Layer Perceptron (MLP). The architectures of the models are briefly described, as well as their most computational features. Experimental results using 4 continuous speech, large vocabulary, speaker dependent recognition systems and 4 test speakers are reported. Only 5 utterances per speaker were used to train the normalization modules. The use of network-based normalization is shown to improve the performance of the speaker-dependent recognizers based on Hidden Markov Models. This also compares favorably with the results obtained adopting a standard linear-regression model. In particular, the generalized MLP gave a 16.9% average *word error rate* (WER), that represents a considerable 52% WER reduction with respect to the baseline system alone, resulting in a viable solution for the non-linear, multivariate regression problem under consideration.

Keywords: speech recognition, speaker normalization, word error rate, Multi-layer Perceptron, Radial Basis Functions, multivariate regression.

1 Introduction

Speaker normalization [2, 8, 6] can be formulated as a regression problem between the multidimensional acoustic spaces corresponding to different speakers. The acoustic space of a speaker is represented by means of a set of feature vectors, obtained by acoustic analysis of the speech signals acquired from the speaker himself. The aim is to build a vectorial function, namely a *spectral mapping*, capable to map the acoustic patterns belonging to a novel speaker, called *new speaker*, onto the corresponding acoustic patterns of another speaker, called *reference speaker*, whose speech data were used to train a speaker-dependent speech recognition system. Application of the estimated spectral transformation on the acoustic data of the new speaker, before sending them to the recognizer, should allow the speaker-dependent (SD) recognition system to reach acceptable performance, from an applicative point of view.

Different techniques are available to deal with the present regression problem. The most popular and easy to implement relies on the linear regression model. In this case, a linear relationship is assumed between the acoustic spaces. Given two sets of feature vectors, representing acoustic observations

Spk. Id.	Gender
anco0	*female*
bian0	*female*
gisv0	*male*
saor0	*male*

Table 1: *Reference speakers list.*

Spk. Id.	Gender
ilco0	*female*
dafa0	*male*
gian0	*male*
rofi0	*male*

Table 2: *Test speakers list.*

from the new and the reference speaker respectively, a linear transformation is estimated in order to minimize the average squared error between these acoustic vectors.

Here the spectral transformation is realized with neural networks. The networks were trained on the time-aligned pairs of feature vectors of the speakers involved. These were subsequently used, at the recognition step, as an acoustic front-end for the speech recognizer. The results, as reported in Section 4, are compared either with the baseline system performance, i.e., word error rate obtained with the speaker-dependent recognition system without normalization at all, or adopting the linear regression model for normalization. Two major connectionist classes were used and compared, namely a generalized *Radial Basis Function* (RBF) architecture [11], and an improved *Multi-Layer Perceptron* (MLP) [7]. Both of the models were trained using a limited amount of acoustic material from each speaker (5 utterances). Although both of the networks introduced a significant improvement in terms of word recognition rate, the MLP behaved better in this task.

This paper is structured as follows. Section 2 provides a general background on the speaker normalization problem and briefly describes the experimental environment. In Section 3 the neural network models are presented. Section 4 illustrates the results obtained on an intensive experimental task involving 8 speakers. The results obtained with the connectionist approaches are compared with those of the baseline system for a continuous speech, large vocabulary task. Section 5 draws conclusive remarks.

2   Experimental environment

This work involved 8 speakers, 4 of whom played the role of reference speakers (see Table 1) for which SD recognition systems were trained, while the other 4 speakers were used as test speakers (see Table 2).

Speaker dependent versions of the recognition system were obtained using the speaker dependent part of APASCI database [1]. For each of the reference speakers, the available collection of 520 utterances was used to train the corresponding speaker specific recognition system. The latter was based on Hidden Markov Models (HMMs). A set of 34 context independent acoustic-phonetic speech units were modeled with left-to-right HMMs. Output distribution probabilities were modeled with mixtures of Gaussian probability densities. In this case, each mixture consisted of 16 Gaussian components for a total of about 1,500 Gaussians in the system.

For test purposes, speech material from 4 test speakers was collected. The recognition task consisted of dictation of fragments of newspaper articles taken from the financial Italian journal *Il Sole 24 Ore*. It is a continuous speech task with word dictionary size of 10,000. For each test speaker, two sets of utterances were collected: the adaptation (training) set and the test set. On average, each utterance presented a duration of 12 sec corresponding to 19 words. The test set and the adaptation set consisted of 30 utterances and 5 utterances, respectively. Adaptation and test sets of each speaker were acquired on different days in an office environment. Texts were randomly chosen from a large selection of texts from the *Il Sole 24 Ore* journal. As a consequence of this, the acoustic-phonetic content of the adaptation sets resulted unbalanced, and different from speaker to speaker. Furthermore, the overall acoustic quality of the speech material in the adaptation sets can be considered quite low. For details about the recognition system see [5].

All the speech signals employed for both training and test phase were processed as follows. Each signal was preemphasized by a digital filter having transfer function $H(z) = 1 - 0.95z^{-1}$ and then blocked into time frames by appling a 20 ms Hamming window every 10 ms. For each frame, spectral analysis was performed and 8 Mel Scaled Cepstral Coefficients (MSCC) [3] together with the log-energy were extracted. MSCC and log-energy were normalized on an utterance by utterance basis. MSCC and log-energy together with their first and second order derivatives were arranged in a feature vector with 27 components. At recognition stage, the normalization modules were applied just on MSCC and the log-energy, while derivatives were computed from transformed parameters.

In this work, given a reference and a test speaker, the training set for the normalization modules was obtained according to the following procedure. Using the HMMs of the reference speaker, each utterance in the adaptation set of the test speaker was aligned against the model concatenation corresponding to the uttered text. Alignment was performed using the Viterbi algorithm [12] in such a way that each input feature vector was assigned to a single mixture component. The means of the Gaussians can be seen as synthetic patterns in the acoustic space of the reference speakers and the alignment algorithm puts in correspondence each input vector with the Gaussian mean which becomes the corresponding target vector.

3   The Network Models

Two network architectures were considered for the normalization task. The first was a *Generalized Resource Allocating Network* (GRAN) [6], a generalization of the RBF network [11], able to recursively allocate new kernels, i.e., hidden units, whenever needed, according to Platt's procedure [10]. The second network model was an improved version of the popular MLP [9, 7] trained with the *backpropagation* algorithm. This Section describes the main features of these models as they were implemented here.

The RBF realizes a multivariate regression by combining the contribution from two different terms. The first term is obtained by evaluating the local basis functions (or *kernels*), encapsulated in the hidden units of the network, on the input pattern. The activation values of the kernels are then linearly combined using the weights of the second layer of the network as linear coefficients. The second term is a direct linear transformation of the input. Whenever an offset b is also considered, the resulting regression equation takes the following form:

$$\mathbf{f}(\mathbf{x}) = HG(\mathbf{x}) + L\mathbf{x} + \mathbf{b} \tag{1}$$

where H is the matrix of the weights of the second layer of the network and $G(\mathbf{x})$ denotes the vector of the values of the local basis functions, evaluated on the current input pattern $\mathbf{x}$. The matrix L defines a linear transformation and is estimated via Singular Value Decomposition (SVD), along with the offset $\mathbf{b}$. Also the weights H are initialized with linear algebra techniques over a set of pre-allocated kernels. Kernel locations can be found using an iterative, partitioning clustering algorithm, such as the k-means or *isodata* [4]. In the present task, the normalized Elliptical Gaussian kernels resulted effective, in conjunction with precomputed linear term L and offset $\mathbf{b}$.

It is worth noting that all the patterns fed through the RBF were previously normalized by projecting them into the space of their Principal Components (PC), and dividing each coordinate by the corresponding eigenvalue. The PC were computed on the whole training set, on a speaker-by-speaker basis. Note that no feature reduction was carried out, since all the PC were retained: this procedure allowed for an homogeneous data treatment, constraining individual values into normalized ranges.

The second network model was an improved MLP. A two layer architecture was chosen after a preliminary experimental phase on another database, used to tune up all of the network's free parameters, before starting the actual training on the database presented herein. Sigmoidal activation functions were used in the hidden layer, while the output units were linear, thus computing the function

$$y_i(\mathbf{x}) = \sum_{j \in J} w_{ij} \sigma_j \big(\sum_{k \in K} w_{jk} x_k \big) \tag{2}$$

where y_i is the i-th output, J denotes the set of all hidden units, w_{ij} is the weight of the connection between units j and i, K is the set of all input units, and x_k is the k-th component of input vector $\mathbf{x}$. $\sigma_j(.)$ is the sigmoidal activation function associated to hidden unit j. The sigmoids had a variable smoothness, that was estimated using cross-validation techniques during the preliminary experiments. No linear term, e.g. skip-layer connections between input and output layer, was used. A diagonalization unit was introduced both in the input and in the hidden layer, fully connected with the units of the next layer. The optimization was based on the usual *delta rule* [9] with backpropagation. A *momentum rate* [7] was introduced in the rule in order to speed up the learning. This was crucial to reduce the number of epochs on the dataset. A gradient descent technique was also used to iteratively adjust the values of the bias summed to each input of the sigmoidal functions, in an on-line manner. Finally, an *ad hoc* algorithm was used to provide an adaptive learning rate, resulting in improved performance.

4   Experimental Results

A set of recognition experiments was carried out in order to compute the normalization modules. First, each SD recognition system was tested using speech material from the test speakers without performing feature normalization. Table 3 provides WERs for each pair of reference and test speakers. An average WER of 35.6% was obtained. Performance is definitely unsatisfactory, particularly when test and reference speakers are of different gender.

Ref. spkr./Test. spkr.	ilco0	dafa0	gian0	rofi0
anco0	16.9	35.4	33.4	52.8
bian0	21.1	53.0	70.8	75.0
gisv0	16.0	28.2	20.4	42.1
saor0	28.2	20.9	27.3	27.8

Table 3: *WERs for the test speakers using the SD recognition systems. Average WER 35.6%.*

For each pair of reference and test speakers, 5 utterances were used to train the normalization modules. Table 4 shows the results obtained when the linear regression model was used for normalization. In the present case the WER is improved to a significant extent. Singular Value Decomposition technique was used to estimate the linear feature map used to normalize the acoustic spaces.

Ref. Spkr./Test Spkr.	ilco0	dafa0	gian0	rofi0
anco0	13.2	22.6	22.5	19.9
bian0	14.6	20.2	23.2	24.2
gisv0	13.1	24.3	19.7	24.8
saor0	27.4	28.2	20.1	17.6

Table 4: *WERs obtained adopting a normalization module based on linear transformation. Average WER 21.0%.*

Performance using the first connectionist approach are reported in Table 5. The GRAN model was used to build the regression. 48 elliptical Gaussian kernels were allocated using a k-means clustering algorithm, and a linear term (see eq. 1) was adopted. SVD on both of the terms was computed in order to initialize the weights of the network and the linear coefficients. A gradient-based optimization technique was used to train the model. This approach results effective, lowering, on average, the WER to 18.7%.

Finally, table 6 shows the results adopting a normalization module based on MLP networks. The average WER is 16.9%, which corresponds to a 52.5% WER reduction with respect to the baseline. Such a performance is definitely better than that obtained adopting the previous normalization modules.

5 Conclusions

Speaker normalization is an effective way toward more robust speech recognition systems. Two connectionist approaches to speaker normalization have

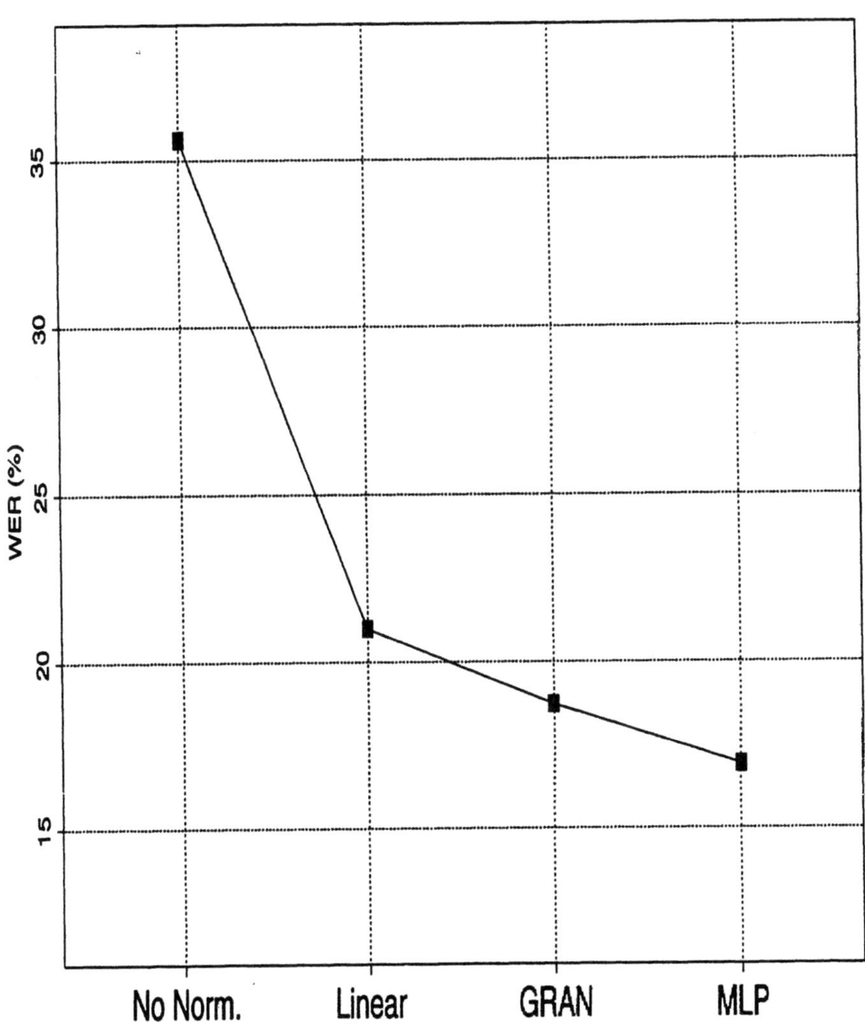

Figure 1: Average WERs for the different approaches.

Ref. Spkr./Test Spkr.	ilco0	dafa0	gian0	rofi0
anco0	15.0	20.0	17.1	20.3
bian0	14.5	21.4	19.7	20.6
gisv0	11.5	19.8	18.0	14.2
saor0	20.7	23.8	23.0	18.8

Table 5: *WERs obtained adopting a normalization module based on RBFs. Average WER 18.7%.*

Ref. Spkr./Test Spkr.	ilco0	dafa0	gian0	rofi0
anco0	10.3	20.0	21.5	15.9
bian0	12.0	18.1	18.3	13.3
gisv0	13.6	18.3	19.4	13.3
saor0	18.8	22.2	19.6	15.9

Table 6: *WERs obtained adopting a normalization module based on MLP. Average WER 16.9%.*

been investigated, namely a generalized RBF model and an improved Multi-Layer Perceptron architecture. A summary of features of these networks has been outlined. Intensive experiments on a continuous speech, large vocabulary task have shown remarkable improvements in recognition performance when the networks were used as acoustic front ends for a recognizer based on HMMs. In particular, the speaker normalization module based on MLP shows a 52% WER reduction with respect to the baseline system performance. A comparison of the results obtained with the different approaches is illustrated in Figure 1.

Acknowledgments

The contribution of Stefano Merler is gratefully acknowledged.

References

[1] B. Angelini, F. Brugnara, D. Falavigna, D. Giuliani, R. Gretter, M. Omologo, "Speaker Independent Continuous Speech Recognition using an Acoustic-Phonetic Italian Corpus", *Proc. ICSLP*, Yokohama, September 1994, Vol. 3 , pp. 1391–1394.

[2] F. Class, A. Kaltenmeier, P. Regel, K. Troller, "Fast Speaker Adaptation for Speech Recognition System", *Proc. of ICASSP 90*, vol. 1, pp. 133-136, April 1990.

[3] S. B. Davis, P. Mermelstein, "Comparison of Parametric Representations for Monosyllabic Word Recognition in Continuously Spoken Sentences",

IEEE Trans. on Acoustic, Speech and Signal Processing, 28 (4), 357 – 366, August 1980.

[4] R. O. Duda, P. E. Hart, "Pattern Classification and Scene Analysis", Wiley, New York, 1973.

[5] M. Federico, M. Cettolo, F. Brugnara and G. Antoniol, "Language modelling for efficient beam-search", *Computer Speech and Language*, 9, pp. 353-379, 1995.

[6] C. Furlanello, D. Giuliani, and E. Trentin, "Connectionist speaker normalization with Generalized Resource Allocating Networks," in *Advances in Neural Information Processing Systems 7* (G. Tesauro, D. S. Touretzky and T. K. Leen, eds.), MIT Press, pp. 867-874, 1995.

[7] J. Hertz, A. Krogh, R. G. Palmer, "Introduction to the Theory of Neural Computation", Addison-Wesley, Redwood City, 1991.

[8] X. D. Huang, "Speaker Normalization for Speech Recognition", *Proc. of ICASSP 92*, vol. 1, pp. 465-468, March 1992.

[9] Y. H. Pao, "Adaptive Pattern Recognition and Neural Networks", Addison-Wesley, Redwood City, 1989.

[10] J. Platt, "A Resource-Allocating Network for Function Interpolation", Neural Computation, vol. 3, n. 2, 213-225, 1991

[11] T. Poggio, F. Girosi, "A theory of networks for approximation and learning", A.I. Memo No. 1140, 1989

[12] L. R. Rabiner, "A Tutorial on Hidden Markov Models and Selected Applications in Speech Recognition", *Proc. of IEEE*, vol. 77, no. 2, pp. 267-295, October 1989.

SECTION 7

ARCHITECTURES AND ALGORITHMS

A Reconfigurable Analog VLSI Neural Network Architecture with Non Linear Synapses

G. M. Bo, D. D. Caviglia, M. Valle, R. Stratta, and E. Trucco

Department of Biophysical and Electronic Engineering
University of Genoa, Via all'Opera Pia 11/A, 16145 Genova, Italy
ph: +39 10 3532287, fax: +39 3532795, e_mail: gian_bo@dibe.unige.it

Abstract

A reconfigurable analog VLSI neural network architecture is presented: it is composed of non linear synapses and linear neurons and it implements a Multi Layer Perceptron network which can be trained by using the standard Back Propagation algorithm. The reconfigurability allows to change the interconnections between synapses, thus programming the topology of the network without any modification of the off-chip wiring. The architecture features a greater robustness with respect to noise and to errors introduced during the computations if compared to others presented in the literature.
Indexing term: Multi Layer Perceptron network, reconfigurable systems, analog VLSI neural networks.

1. Introduction

Reconfigurable neural systems are desiderable for many different reasons:

- general problem-solving enviroments: different applications need networks with different topology, thus a single architecture with programmable topology can be suited to solve many applicative problems;
- correction of offsets (analog VLSI implementation): the network can be programmed by using extra synapses to compensate offsets introduced by analog circuits;

- test and isolation of defects: a reconfigurable system allows to test any individual block of the architecture to locate defective circuits that can by-passed.

Many different approaches to the implementation of neural networks in analog VLSI technology have been presented in the literature [2, 8-11]; usually these architectures feature a fixed topology; as a consequence it is not possible to change the number of weights (for example, by increasing the number of hidden neurons) in the network, for a better tailoring to the applicative problem.

The aim of our work is to develop a general purpose analog VLSI neural architecture, capable to be reconfigured (by using switches in the interconnections between synapses and neurons) without any modification of the off-chip / on board wiring.

To obtain an efficient hardware implementation, non linear synapses and linear neurons have been used. By properly connecting the computational cells, it is possible to build a Multi Layer Perceptron network which can be trained by using the Back Propagation algorithm.

In Section 2 a description of the hardware architecture is given; in Section 3 is briefly discussed the analog VLSI implementation of the architecture.

2. The Hardware Architecture

Some examples of reconfigurable neural networks have been presented in [1, 2, 3]. In [1] and [2] a multichip architecture is presented: a switch chip, a neuron chip, and a synapse chip are used. The main disavantage of this approach is the complexity of the overall architecture.

In [3] a single reconfigurable neural network chip has been presented; this architecture uses a distribuited approach: the main computational cell is a distribuited neuron-synapse which consists of an OTA-based multiplier and two non linear resistive loads. A drawback of the architecture is that the voltage drops at the reconfiguration swicthes could affect the neural computation.

In this Section, we present a reconfigurable hardware architectute that uses a single chip (containing neurons, synapses and configuration switches) as basis module but without the drawback evidenced in [3]. The architecture is based on two main cells: non linear synapses and neuron-switchboxes.

In Fig. 1 a schematic representation of the non linear synapse is shown. This cell can be implemented by using an OTA-based multiplier [4]; it is worth noting that the input voltage V_{in} and the output current I_{out} are accessible on ortogonal directions. The weight and the control signals are not shown for the sake of simplicity.

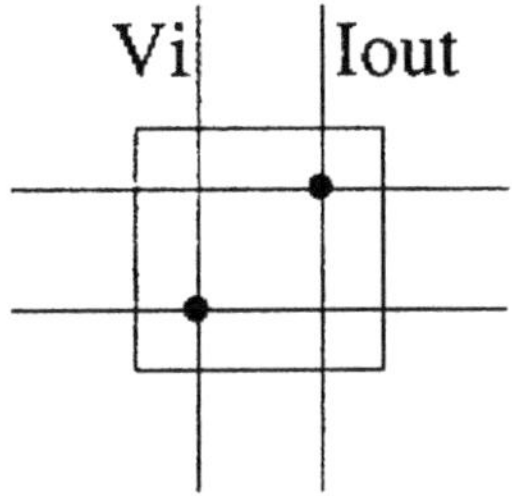

Fig. 1: Schematic representation of a non linear synapse.

In Fig. 2 the neuron-switchbox (and its schematic representation) is shown; it consists of a linear resistor, which implements a linear neuron transfer function, and of a matrix of switches (signals S_j, j=1,...,4) which control the flow of data.

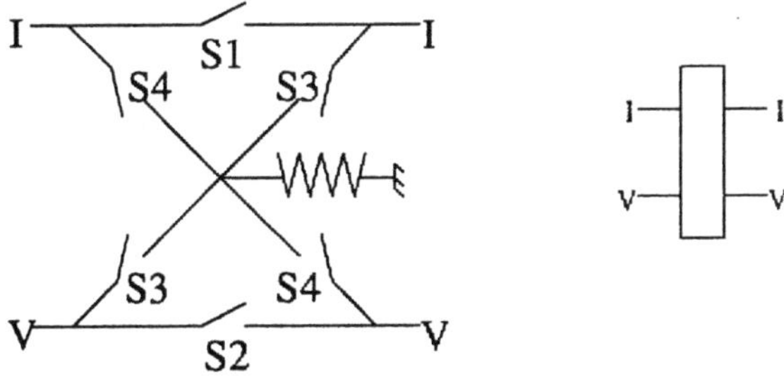

Fig. 2: Circuit and schematic representation of the neuron-switchbox.

Let us consider for example two non linear synapses connected through a neuron-switchbox (see Fig. 3). To compute the sum of the two output synaptic currents (I_1 + I_2), the S_1 switch is closed, while the other switches are open; to distribuite an external input voltage to the two synapse inputs (V_1 and V_2), the S_2 switch is closed, while the other switchws are open. The S_3 and S_4 switches are used when the output current of one synapse has to be converted into a voltage through the neuron and then feed the other synapse (i.e. th two synapses are connected through the neuron); one should close S_3 or S_4 depending on the signal versus flow.

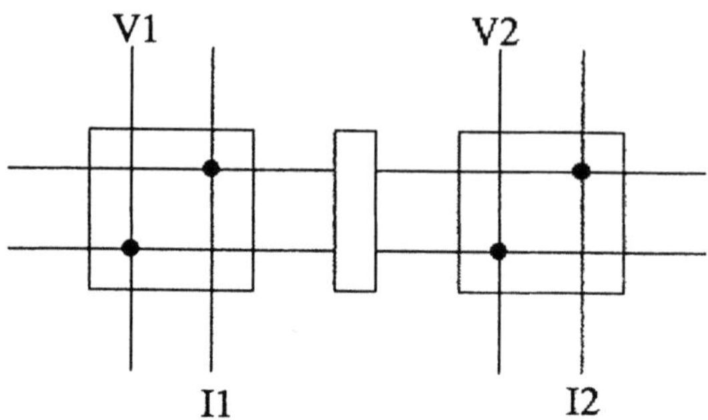

Fig. 3: Two synapses connected through a neuron-switchbox.

It is worth noting that the switches used to program the interconnections between cells, do not introduce errors (due to the voltage drop) when they are closed as is [3]. In fact, along the synaptic outputs, the neural information are mapped on currents, and then the low voltage drops introduced by the reconfiguration switches do not affect the current values. On the other hand, along the synaptic inputs, the neural informations are mapped on voltages: along these interconnections no current is flowing after the transient time, and no voltage drops are present.
An example of the organization of synapses and neuron-switchboxes is illustrated in Fig. 4: a reconfigurable neural network with 4 x 4 non linear synapses is shown. Extra sets of neuron-switchboxes can be provided on the periphery of the array of units of reconfiguration, to realize a reconfigurable input / output data flow. In this way, due to the modularity of the network, more neural chips can be connected together to implement larger neural networks. This network can be programmed to obtain for example a 3 inputs, 4 hidden neurons and one output neuron.

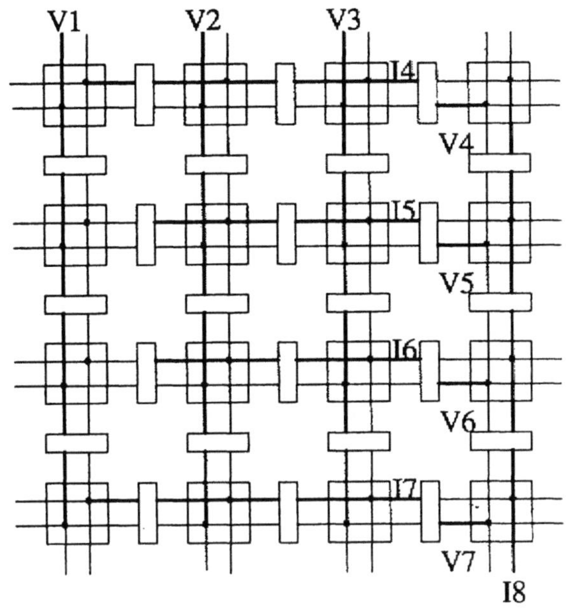

Fig. 4: A reconfigurable neural network with 4 x4 synapses.

In Fig. 4 the signal flow when the network is configured in this way is evidenced. V_1, V_2, and V_3 are the three input signals; I_4, I_5, I_6, and I_7 are the four sum of the synaptic output currents of the first layer. These currents are converted into voltages by the four hidden linear neurons, obtaining V_4, V_5, V_6, and V_7, which fed the secon layer of synapses. The current I_8 is the sum of the synaptic output currents of the second layer of the network; this current represents the output of the network and it can be converted into a voltage by using a peripherical linear neuron.

One can notice that in the architecture shown in Fig. 4, it is possible change the interconnections between single synapses; we can say that this architecture is based on a *one synapse unit of reconfiguration*. A better organization of the architecture could be obtained by using for example a *4 x 4 synapses unit of reconfiguration*, in which it is possible change the interconnections between tiles of 4 x 4 synapses [3].

3. The Analog VLSI implementation

In the previous Section, two basic neural computationl cells have been introduced and their use in a reconfigurable MLP network has been discussed. With respect to the multichip solution presented in [1, 2], and the single chip solution presented in [3], the proposed reconfigurable neural network feature better robustness with respect to noise on the synaptic input signals; moreover the neural information mapped on currents and voltages is non affected by the voltage drops across the reconfiguration switches.

For the analog implementation of the computational cells one can refer to the many analog implementations of neural networks presented in the literature.

The linear resistor neuron can be built by using "high resistive polysilicon" [5] or by using circuits with MOS transistors biased in the linear region [6] or in the saturation region [7].

The synapse is implemented by using an OTA-based multiplier, in order to compute the hyperbolic tangent of the normalized input voltage signal (see Fig. 5).

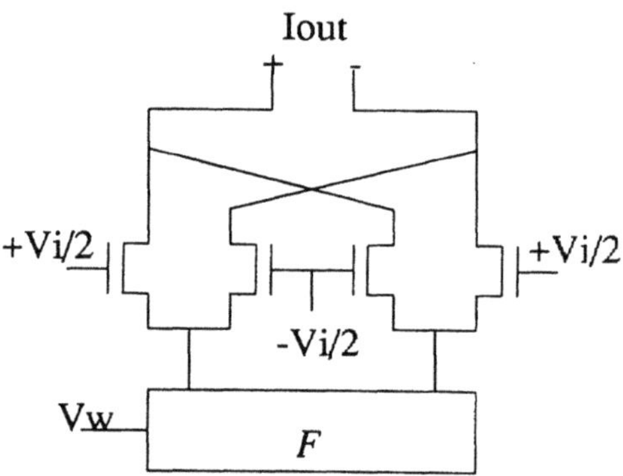

Fig. 5: Circuit representation of a non linear synapse.

In Section 2 we suppose to map the weight control variable onto a voltage (V_w), thus a bias circuit is necessary to convert this voltage into a differential current (the block F of Fig. 5). In [4] a simple bias circuit is presented. Another solution consists in using a transconductor as in [3]; the weight voltage is dinamically stored by using a capacitor close to the synapse: an external weight refresh architecture is necessary.

4. Conclusions

In this paper, we present an analog VLSI architecture for reconfigurable Neural Networks. The need of reconfigurability is desirable mainly for the following reasons: i) to have general problem-solving environments; ii) to compensate the offsets in analog circuits and to effectively test single circuits inside the network.
Even if may analog VLSI implementations have been presented in the literature, few of them address the topic of on-chip reconfigurability. We started from the solution proposed in [3], but we present an original architecture that aims to solve some of the circuit design issues (i.e. precision of computation) evidenced in the cited paper. It is worth noting that the configurable overall network architecture is a MLP-like network where the basic building blocks are linear neuron and non-linear synapse circuits. To obtain a correct functionality, a proper I/O interface can be used: we can use dedicated on-chip analog circuits or an I/O software interface, running on a host computer. The learning can be achieved by using a chip-in-the-loop technique: the errors and the weight updates are computed by the host computer, while the neural chip performs the feed-forward computation.
Future work will concern the design and validation of a test chip.

Acknoeledgements

The authors wish to thank Prof. Patrick Garda of Université Pierre at Marie Curie (Paris) and Dr. Eric Belhaire of Institute de Electronique Fondamentale of CNRS (Orsay, France) for useful discussion and suggestions.
This project has been partially founded by the "Galileo" programme.

References

[1]   S. Eberhardt, T. Duong, and A. Thakoor. Design of Parallel Hardware Neural Network Systems from Custom Analog VLSI 'Building Blocks' Chip. In: Proc. Joint Conf. Neural Networks, 1989, vol. 2, pp. 183-190.

[2]   P. Muller et al. A Programmable Analog Neural Computer and Simulator. Advances in Neural Information Processing System I, vol. 1, pp 712-719, D. S. Touretsky, Ed. San Mateo, Ca: Morgan Kaufmann, 1989.

[3]   S. Satyanarayana, Y. P. Tsividis and H. P. Graf. A Reconfigurable VLSI Neural Network. IEEE Journal of Solid State Circuits, 1992; 27:67-81.

[4]   M. Valle, D. D. Caviglia, and G. M. Bisio. An Analog VLSI Neural Network With On-Chip Back Propagation Learning. Accepted for publication on Analog Integrated Circuits and Signal Processing Journal, Kluwer Academic Publishers.

[5]   R. S. Soin, F. Maloberti, and J. Franca (Editors). Analogue- Digital Asics. Peter Peregrinus Ltd., 1991.

[6]   H. Kobayashi, J. L. White, and A. A. Abidi. An Active Resistor Network for Gaussian Filtering of Images. IEEE Journal of Solid State Circuits, 1991; 26:738-748.

[7]   K. Bult, and H. Wallinga. A Class of Analog CMOS Circuits Based on the Square-Law Characteristic of an MOS Transistor in Saturation. IEEE Journal of Solid State Circuits, 1987; 22:357-365.

[8]   P. Masa, K. Hoen, and H. Wallinga. High Speed Neural Network for High-Energy Physics. In: Proceedings of the Fourth International Conference on Microelectronics for Neural Networks and Fuzzy Systems,1994, pp.422-428.

[9]   M. Verleysen, P. Thissen, J. Vos, and J. Madrenas. An Analog Processor Architecture for a Neural Network Classifier. IEEE Micro, 1994; 14:16-28.

[10] E. A. Vittoz. Analog VLSI Implementation of Neural Networks. In: Proceedings of ISCAS 19090, pp. 2524-2527.

[11] G. M. Bo, D. D. Caviglia, and M. Valle. A Current Mode CMOS Multi-Layer Perceptron Chip. In: Proceedings of the Fifth International Conference on Microelectronics for Neural Networks and Fuzzy Systems, 1996, pp.103-106.

Off-Chip Training of Analog Hardware Feed-Forward Neural Networks through Hyper-Floating Resilient Propagation

G. M. Bollano, M. Costa, D. Palmisano, E. Pasero
Dept. of Electronics, Politecnico di Torino
Torino, Italy

Abstract

Any attempt of implementing abstract functional relationships on silicon is likely to considerably increase the complexity of analog circuits for Feed-Forward Networks (FFNs). We developed an alternative model that can make direct use of the native computational properties owned by elementary electronic devices. A practical framework is described to train such analog FFNs "off-chip". This is especially useful whenever the weight storage elements cannot be re-programmed "on the fly" at a high rate. We finally give some examples of very simple architectures based on floating gate transistors that substantially diverge from the usual "ideal" behaviour but fully comply with the proposed model.

1   Introduction

Most Artificial Neural Network (ANN) models have been developed to mimic the collective behaviour emerging from biological or physical systems characterized by a high number of simple but strongly interconnected computing elements. Since these systems typically carry out some form of analog information processing, analog hardware implementation of ANNs not only bears a remarkable practical relevance, but it is also considered a key theoretical topic.

Several ANNs have been expressly conceived from the beginning in terms of electronic devices: for instance, Hopfield nets equipped with continuous-valued units [2] can be directly represented in terms of resistors, capacitors, non-linear voltage amplifiers with symmetrical outputs and external bias current sources. This, however, does not apply at all to Feed-Forward Networks (FFNs): mainly because of the chief role played by the learning phase, here the modelling process just aimed at embodying the underlying computational features into a coherent and simple *mathematical* framework.

Now, such a coherence can hardly be preserved if we attempt to translate only part of that framework into a physical context. To illustrate this point, let us suppose we are interested in designing a system based on an analog FFN for applications that do not specifically demand for on-chip learning capabilities: we might then plan to compute the weights off-chip according to the usual abstract model. But since we are not dealing with a digital architecture, we cannot just "put their values" somehow into the network: rather, we need a faithful physical interpretation of their formal meaning. Although subsequent fine weight tuning with chip-in-the-loop alleviates some problems, in any case we are forced to implement predefined, exogenous functions on silicon with sufficient precision. Otherwise stated, in so doing we conveyed towards the circuit itself part of that complexity we had put aside during the learning phase.

Several solutions can be found in literature which do not suffer from these shortcomings. For instance, in [7] a VLSI architecture is proposed with full on-chip learning capabilities, while in [5] training with chip-in-the-loop is performed from scratch. However, in neither arrangement analog EEPROMs could be efficiently used as weight storage elements, mainly because of the low programming speed inherent in present technology. This might be a limitation in view of the development of highly compact circuits, since both memory and computing capabilities involved in synaptic processing can be implemented on a single floating gate device [3].

We therefore adopted a different approach that revaluates off-chip training from a slightly different perspective. That is, we developed an alternative model with the purpose of simplifying analog design while keeping the basic computational features of FFNs. Although the resulting mathematical framework is more complex than the original one, in practice such complexity entirely weighs on the off-chip training procedure.

In section 2) we outline the properties of the proposed model. Here both neurons and synapses are directly viewed in terms of generic non-linear electronic devices, although they in fact represent more abstract building blocks where for instance dc voltages and currents could in principle be replaced by other physical quantities. Once specific realizations of those elements are given, their behaviour is summarized by means of appropriate I/O tables obtained from circuit simulations or direct measurements.

In section 3) the question is addressed of how such analog FFNs can be trained in practice. We explain why ordinary back-propagation is unsuited for this purpose and suggest an alternative approach based on the extension of a very simple adaptive algorithm known as Resilient PROPagation (RPROP) [4].

In section 4) we give some examples of analog architectures based on floating gate devices. In particular, we start from a simple circuit we recently developed [1] and show how the proposed guidelines allow to further reduce its complexity.

2　Analog Building Blocks for FFNs

In the operation of abstract FFNs two formally independent assumptions are made that constitute a real challenge for analog designers. That is:
- Synapses behave as perfectly linear multipliers;

- There is no mutual interaction (i.e. coupling) between any pair of synapses.

While the former assumption just meets a parsimony requisite about the model, the latter has several deep implications, namely:

1) The inner state of each neuron can be computed through the linear superposition of all the incoming contributions (provided that, of course, such contributions have to be summed: notice that summation and superposition have different meanings in presence of couplings);

2) Neurons must behave as if they were ideal generators of their outputs (or else synapses fed by the same neuron would mutually interact to set the value of their common input).

The final consequence is that both neuron and synaptic processings are *local*, each one at its own level. Locality is in fact the very basic property shared by most ANN models.

Now, let us see what happens if both assumptions are relaxed, but feature 2) is retained: then neurons clearly keep on performing local computations, whereas their *afferent synapses* in general do not. However, mutual interactions do not extend over the entire network, but are restricted within every single neuron. All things considered, we came to the definition of a broader, more biologically plausible FFN model.

In the following we describe the generic analog building blocks that can be treated by the software simulation package we are currently developing. They can account for a wide class of architectures, and some others (like fully differential circuits) can be added with minor efforts.

Figure 1 shows the analog counterpart of the modified formal neuron: this is viewed as a three terminal device whose dc voltages are set in response to the dc currents I_1 and I_2 flowing into its input summing lines. We chose that configuration because summing lines can be directly viewed as carriers of excitatory vs. inhibitory contributions. Notice that, according to the above guidelines, V_o has to be interpreted as the open-circuit output voltage. For a real implementation to be consistent with the model, synapses should dissipate negligible dc power at their input port, or else the neuron must include an output buffer stage that delivers the required current to the load (this is always the case for output neurons).

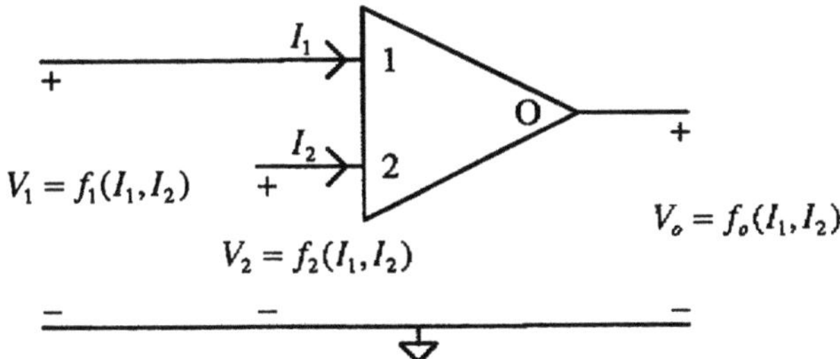

Figure 1: The Analog Neuron

Figure 2 shows the analog counterpart of the modified formal synapse: this delivers a current to one of the summing lines that depends on the input voltage V_{in}, on an adjustable parameter W and on the voltage V of the line itself to account for finite output resistance. Synapses are organized in pairs whose elements share the same input but are connected to different summing lines. So throughout Figure 2 index i identifies the pair, while subscript x identifies the line. Whenever the neuron does not behave as a differential amplifier, then transconductances $\partial g_x/\partial V_{in}$ of paired synapses typically have opposite signs.

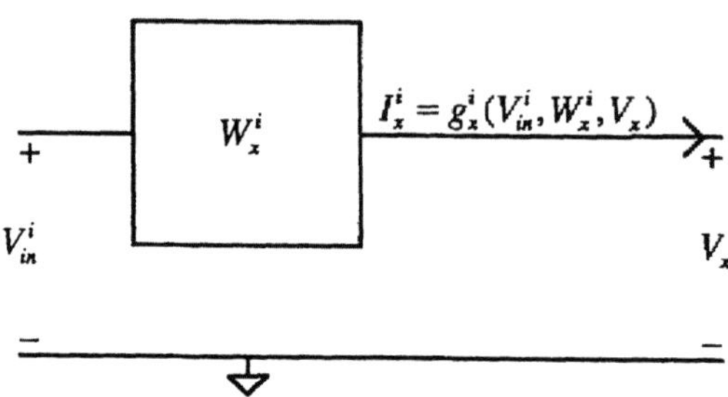

Figure 2: The Analog Synapse

Once specific realizations of those building blocks are given, their static behaviour is thus described in terms of few functional relationships. In practice, such functions along with their derivatives are approximated from appropriate I/O tables obtained through circuit simulations or direct measurements. A linear interpolation scheme on rectangular meshes has been adopted to simplify calculations and to allow direct identification of mesh indexes with the values taken by input variables on sample points.

3 The Off-Chip Training Procedure

As a result of the mutual interaction among synapses afferent to the same neuron, the feed-forward phase actually carries most of the computational burden involved in the training procedure. In fact for every neuron the following system of equations:

$$V_x = f_x\left(\sum_j g_1^j(V_{in}^j, W_1^j, V_1), \sum_j g_2^j(V_{in}^j, W_2^j, V_2)\right) \quad (x \in \{1,2\})$$

has to be solved with respect to the unknowns V_1 and V_2 to compute the operating point (its existence, uniqueness and stability need of course to be verified in advance). Among the many methods that can be used for the purpose, we chose to turn the problem into the minimization of the "loss" function $(V_1-f_1)^2+(V_2-f_2)^2$ and then apply the same learning algorithm we will describe later to seek for "optimal" values of the summing lines voltages.

Upon completion of the feedforward phase, error signals are propagated backwards throughout the network according to the following formulas:

$$\frac{\delta V_o}{\delta W_1^i} = \frac{\partial g_1^i}{\partial W_1^i}\frac{\Gamma_{12}}{\Delta}; \quad \frac{\delta V_o}{\delta W_2^i} = \frac{\partial g_2^i}{\partial W_2^i}\frac{\Gamma_{21}}{\Delta}; \quad \frac{\delta V_o}{\delta V_{in}^i} = \frac{\partial g_1^i}{\partial V_{in}^i}\frac{\Gamma_{12}}{\Delta} + \frac{\partial g_2^i}{\partial V_{in}^i}\frac{\Gamma_{21}}{\Delta} \tag{1}$$

$$\Gamma_{xy} = \frac{\partial f_o}{\partial I_x}\left(1 - \frac{\partial f_y}{\partial I_y}\sum_j \frac{\partial g_y^j}{\partial V_y}\right) + \frac{\partial f_o}{\partial I_y}\frac{\partial f_y}{\partial I_x}\sum_j \frac{\partial g_y^j}{\partial V_y} \quad ((x,y) \in \{(1,2),(2,1)\}) \tag{2}$$

$$\Delta = \left(1 - \frac{\partial f_1}{\partial I_1}\sum_j \frac{\partial g_1^j}{\partial V_1}\right)\left(1 - \frac{\partial f_2}{\partial I_2}\sum_j \frac{\partial g_2^j}{\partial V_2}\right) - \frac{\partial f_1}{\partial I_2}\frac{\partial f_2}{\partial I_1}\sum_j \frac{\partial g_1^j}{\partial V_1}\sum_j \frac{\partial g_2^j}{\partial V_2} \tag{3}$$

being the rest identical to the ordinary delta rule.

Up to this point it seems that additional non-linearities and mutual interactions did not give rise to dramatic changes in the original framework. The key problem in fact resides in the core of the learning phase, that is in the weight updating policy. To illustrate this topic, let us consider an abstract FFN: given a loss function L, the generic weight W does not directly appear in the usual expression of the corresponding gradient component $\partial L/\partial W$. In fact the value of W may affect *any* gradient component just because it contributes to the outcomes of the neurons belonging to the succeeding layers. As a result, on the average the relative magnitude of $\partial L/\partial W$ mainly depends on the "distance" of W from the output layer.

The present situation is quite different: since analog synapses are allowed to behave as highly non-linear devices, now the magnitude of $\partial L/\partial W$ can be greatly influenced or even completely dominated by the value of W itself. In this case a single learning rate cannot account for the very wide dynamic range spanned by internal error signals.

There exists, however, a learning algorithm known as Resilient PROPagation (RPROP) [4] that is inherently insensitive to such scaling problems. It shares many features with SuperSAB [6] in that each weight has its own local learning rate, whose value is adjusted according to a very similar adaptation scheme. But in RPROP weight updating does not depend on the magnitude of the gradient: in fact the amount of change is determined only by the value of the local learning rate, while its direction opposes – as usual – the sign of the corresponding gradient component (where $sign(0)=0$ is assumed for consistency).

Although RPROP is well suited for our purposes, nonetheless some additional precautions have to be taken to prevent the network from getting stuck far away from any proper local minimum of the loss function. In fact, static I/O relationships describing electronic devices are often flat over broad regions, and that might severely slow down the learning process or even "freeze" some weights. However, not always absolute flatness reflects physical reality: it can also be due to round-off errors in numerical simulations, or to the necessity of filtering out fluctuations within instrumental uncertainty, or simply because we are allowed to make any useful approximation that has negligible effect on the system operation. In any case, for all practical purposes no harm is done if we assume that even in "flat" regions output quantities actually vary by an unappreciable amount, provided that we are able to guess the correct trend: the only information RPROP needs to explore the weight space.

We therefore introduced an "hyper-floating" representation of reals where the occurrence of infinitesimal quantities is handled by means of an additional exponent. Derivatives are accordingly evaluated as ordered pairs (x,m), where x and m are respectively a regular floating point number and an unsigned integer representing such exponent. The following formula describes how the usual finite difference scheme is modified to compute the derivative f' of a generic function $f:R\rightarrow R$ on a set of sample points in hyper-floating notation:

$$f'(x_i) \cong \begin{cases} \left(\dfrac{f(x_{i+1})-f(x_{i-1})}{2\Delta x},0\right) \Leftrightarrow f(x_{i+1}) \neq f(x_i) \vee f(x_{i-1}) \neq f(x_i) \\[2em] \left(sign(f(x_{i+j})-f(x_{i-j})),j-1\right) \Leftrightarrow \\ \Leftrightarrow \left(\forall k:|k-i|<j, f(x_{i+k})=f(x_i)\right) \wedge \left(f(x_{i+j}) \neq f(x_i) \vee f(x_{i-j}) \neq f(x_i)\right) \\[2em] (0,0) \Leftrightarrow \forall k \neq i, f(x_{i+k})=f(x_i) \end{cases}$$

As it can be seen, the second element in the pair substantially provides a measure of the extension of "flat" intervals. The above recipe can be easily generalized to multivariate functions whenever sample points are arranged into a rectangular mesh.

Once elementary derivatives have been evaluated, they must be added or multiplied to compute the gradient of the loss function. This can be accomplished according to the following rules:

$$(x,m)\cdot(y,n) = (xy,m+n)$$

$$(0,m)+(0,n) = (0,\min\{m,n\}); \quad \forall x \neq 0, \quad (x,m)+(0,n)=(x,m)$$

$$\forall x \neq 0, y \neq 0, \quad (x,m)+(y,n) = (x\theta(m-n)+y\theta(n-m),\min\{m,n\})$$

$$\left(\theta(a)=\begin{cases}1 \Leftrightarrow a \geq 0 \\ 0 \Leftrightarrow a < 0\end{cases}\right)$$

In particular, hyper-floating addition has been defined to keep track exclusively of contributions coming from lower order terms. Although as stated in (1) also division is involved in the back-propagation of the error signal, yet it has not been included among the basic operations because for consistency we are forced to assume $\Delta=(x,0)$ for some $x\neq0$ (singularities cannot be cured by simply changing notations), so that division can be directly derived from multiplication.

Upon availability of the generic gradient component in the form (x,m), $sign(x)$ is passed to RPROP and the rest of the learning phase is carried out as usual.

4   Analog EEPROM-Based Implementations of FFNs

To highlight the advantages arising from the proposed framework, we will first briefly describe an analog architecture for FFNs we recently developed [1] and then show how it can be progressively simplified according to the above guidelines.

The original basic circuit is outlined in Figure 3: both weight storage and synaptic computation rely on floating gate devices (the programming circuitry is not shown for clarity). Contributions of either sign are conveyed on a single summing line, so that with reference to Figure 1 input voltages can be expressed as $V_1=V_2=f(I_1+I_2)$. The feedback resistor of the inverting neuron not only is used to smooth the output squashing function, but also to keep the summing line voltage as close as possible to $V_{dd}/2$. In fact, at that time we were not able to efficiently account for mutual interactions: therefore synapses have been treated as (non-linear) transconductance amplifiers with zero output conductance, and in the corresponding I/O tables V_x was tied at $V_{dd}/2$. Such an assumption is clearly unrealistic, but it can be justified at the system level in that, as long as the neuron gain is high, actually $V_x \cong V_{dd}/2$; and if the neuron saturates, then its output voltage is weakly affected by even large errors in the estimation of the total current flowing into the summing line.

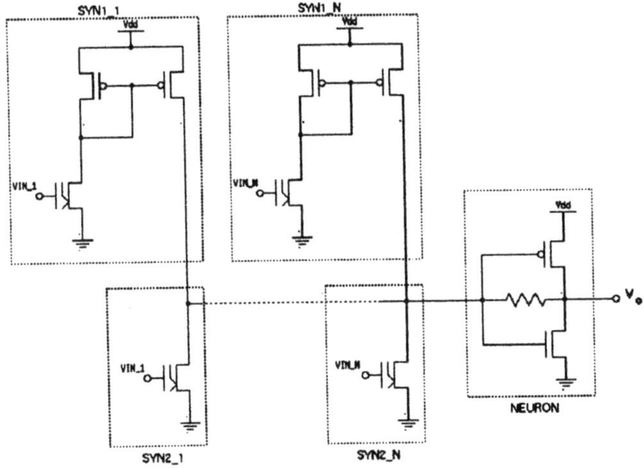

Figure 3: The Original Basic Circuit

Explicit insertion of synaptic couplings into the model eliminates the need of "ad hoc" assumptions together with eventual supporting circuitry. For instance, in Figure 4 we lumped into a single device all the current mirrors previously used to supply inhibitory contributions. In this way both kinds of synapses are in fact identical, and these can be arranged in more compact and regular structures. However, we can go further on. In fact, it is now apparent that the feedback inverter can be removed without giving up the essential features of neural computation.

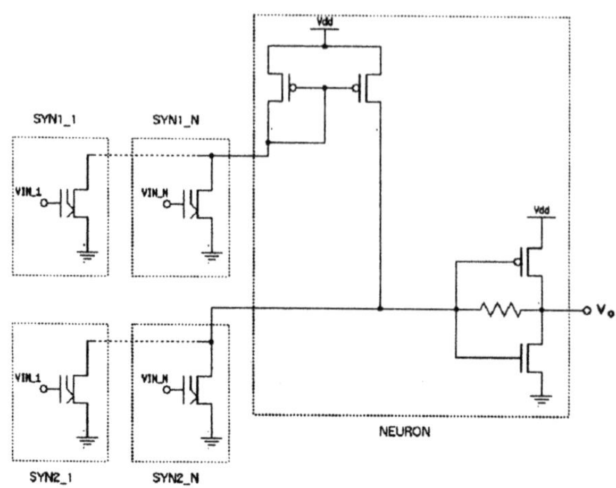

Figure 4: Solution with Single Current Mirror

The resulting "neuronless" architecture is outlined in Figure 5. It bears its name by the fact that the output is directly connected with one of the summing lines, although a residual neuron still formally exists.

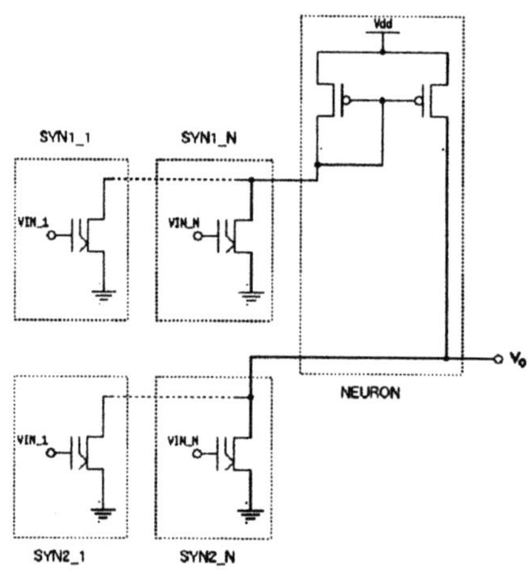

Figure 5: "Neuronless" Analog FFN

5 Conclusions

Direct analog hardware implementation of detailed abstract ANN models is sometimes a slightly misleading issue. Any correct design should be based on a clear distinction between fundamental and accessory features. In compliance with the suggestions coming from both biological systems and neuromorphic architectures, we proposed a comprehensive analog model for FFNs where the native computational properties owned by elementary electronic devices play a central role. In this way, departures from "ideal" behaviours can be turned into valuable resources. However, these ideas would remain unpracticable without an effective training procedure. Resilient Propagation constitutes a real breakthrough towards the solution of the many problems arising from the adoption of non-linear, coupled synapses. Of course, subsequent fine tuning with chip-in-the-loop is mandatory to take process variations and drifts into account.

References

1. Costa M, Palmisano D, Pasero E. NESP: an analog NEural Signal Processor. In: Proc IEEE International Symposium on Circuits and Systems. Seattle (WA), 1995, pp 2189-2192
2. Hopfield JJ. Neurons with Graded Response Have Collective Computational Properties like those of Two-State Neurons. Proc National Academy of Sciences 1984; 81: 3088-3092
3. Kramer A, Sin CK, Chu R, Ko PK. Compact EEPROM-Based Weight Functions. In: Lippmann RP, Moody JE, Touretzky DS (eds), Advances in Neural Information Processing Systems 3. Morgan Kaufmann Publishers, San Mateo (CA), 1991, pp 1001-1007
4. Riedmiller M, Braun H. A Direct Adaptive Method for Faster Backpropagation Learning: the RPROP Algorithm. In: Proc IEEE International Conference on Neural Networks. San Francisco (CA), 1993
5. Satyanarayana S, Tsividis YP, Graf HP. A Reconfigurable VLSI Neural Network. IEEE Journal of Solid-State Circuits 1992; 27(1): 67-81
6. Tollenaere T. SuperSAB: Fast Adaptive Back Propagation with Good Scaling Properties. Neural Networks 1990; 3(5): 561-573
7. Valle M, Caviglia DD, Bisio GM. An Experimental Analog VLSI Neural Network with On-Chip Back-Propagation Learning. Journal of Analog Integrated Circuits and Signal Processing 1996 (in press)

A Learning Strategy which Increases Partial Fault Tolerance of Neural Nets

Salvatore Cavalieri and Orazio Mirabella

Institute of Informatic and Telecommunications, University of Catania (ITALY)
Fax:+39 95 338280, Email:{cavalieri,omirabel}.iit.unict.it

Abstract

The paper deals with the problem of fault tolerance in multilayer networks. Although they already possess a reasonable capacity for fault recovery, it may be insufficient in particularly critical applications. Studies carried out by the authors have shown that traditional back-propagation algorithm may entail the presence of one or more weights with a much higher value than the others. A fault in that weight/s may therefore lead to a substantial decrease in the network's fault tolerance. The authors propose a learning algorithm which updates the weights, distributing their values as uniformly as possible in each layer. Tests performed on benchmark test sets will show the considerable increase in fault tolerance obtainable with the proposed approach as compared with traditional algorithms, and with another approach to be found in literature.

1 Introduction.

The use of artificial neural networks (ANNs) in applications requiring high reliability makes it necessary to improve their inherent fault tolerance. Investigations carried out in [1] [2] have shown that ANNs are not always fault tolerant. Literature provides examples of different directions of research in the attempt to make ANNs more robust to faults. They can essentially be classified as on-line fault tolerance [2] and recovery through relearning [3]. In this paper the authors focus on the former recovery strategy, as it guarantees the lowest fault recovery times.

Literature presents various on-line fault recovery strategies. The most exhaustive analysis of these strategies is to be found in [4] where all possible causes of faults are considered. In [4] the authors propose a partial fault tolerance strategy based on the replication of hidden units. Although this solution guarantees high fault tolerance rates it has the great drawback of require the bias currents of the output units to be scaled by g, where g is the number of replication. Although this does not affect a software implementation of the feedforward network, the consequences in a hardware implementation are far from negligible. The approach, in fact, requires higher values for the current flowing to the output nodes with a consequently greater dissipation of power, proportional to the number, g, of replications introduced. Moreover the replication strategy implies only discrete number of neurons in the hidden slabs to be considered. This may produce an excessive waste of memory in a software implementation or an excessive power dissipation in a hardware implementation.

In this paper we propose an innovative partial fault tolerance strategy for multilayer perceptron networks which overcomes the limits outlined above. It consists of a learning algorithm which distributes the weight values uniformly over each layer of the network. The idea is based on an accurate analysis of the distribution of weights in feedforward networks trained with backpropagation algorithms, made by the authors. This analysis showed the non-uniform distribution of weight values in each layer during the learning stage. It may therefore happen that a limited number of weights have much higher values than others. A fault in these weights would thus be critical for the fault tolerance of the network as a whole.

The fault tolerance strategy will be described and its performance will be assessed and compared with the fault-tolerance performance of the traditional learning algorithm and the replication- based solution.

2 Neural Network Topology

The network considered in the paper is multilayer perceptron networks of sigmoidal units with supervised learning algorithms. The number of layers is indicated with L and includes the hidden and output slabs. The inputs and weights (including biases) are real values ranging in the interval $(-\infty, +\infty)$. The outputs of the i-th neuron belonging to slab k ($k=1,..,L$ where $K=0$ is the input slab, $K=1$ is the first hidden slab and $K=L$ is the output slab) are:

$$O_i(k) = f(sum_i(k)), f(x) = \frac{1}{1 + e^{-x}}$$

and:

$$sum_i(k) = \sum_{j=1}^{config(k-1)} w_{i,j}(k) \cdot O_j(k-1) - bias_i(k) \quad (1)$$

Here config(k-1) is the total number of neurons in slab k-1, $w_j,i(k)$ is the weight from unit i (belonging to slab k-1) to unit j (belonging to slab k), and $bias_i(k)$ is the bias current of the i-th neuron in the slab k.

3 A Learning Algorithm to Improve Fault-Tolerance

The aim of this section is to illustrate the learning algorithm we used to improve fault tolerance in the feedforward network. It is essentially based on the traditional backpropagation algorithm, as described in [5]. The modifications introduced aim to produce a levelling in the weights for each slab. During the learning phase a weight is updated according to the gradient descending rule described in [5], only if the new value it assumed does not exceed a certain threshold. Two different thresholds were considered in the paper: the first refers to the maximum values for the weights in each slab, while the second refers to the average weight value for each slab. From the tests carried out it was seen that a good trade-off between convergence times and fault tolerance enhancement is achieved by adopting the second threshold only for the output slab and the first one for the hidden slabs. It was seen, in fact, that indiscriminate adoption of the second threshold in all the slabs led to very high convergence times.

Fig.1 shows the procedure WeightModification (k,i,j) for modification of the weight $w_{i,j}(k)$, in Pasca-like notation.

```
Procedure WeightModification(k,i,j);
begin
 temp_wi,j(k):=wi,j(k);
 wi,j(k):=wi,j(k)+alfa*deltaj(k)*outi(k-1)+beta*old_wi,j(k);
 alfa_m:=alfa;
 beta_m:=beta;
 if ( ((k=L) and (abs(wi,j(L) > average_weights(L)))
       or ( (k<L) and ((abs(wi,j(k)) > abs(max_weight(k)) ) ) ) then
 begin
  alfa_m:=0;
  beta_m:=0;
 end;
 wi,j(k):=temp_wi,j(k)+alfa_m*deltaj(k)*outi(k-1)+beta_m*old_wi,j(k);
 old_wi,j(k):=wi,j(k)-temp_wi,j(k);
end;
```

Figure 1: WeightModification Procedure

The WeightModification procedure modifies each weight according to the learning rate and momentum values, alfa and beta respectively. As can be seen, the updating of the weight $w_{i,j}(L)$ is inhibited if its new absolute value is higher than the absolute value of the average for the weights (average_weights(L)) belonging to the output slab. As concern the updating of the weight $w_{i,j}(k)$ where k=1,..,L-1, it is only inhibited if the new absolute value for the weight $w_{i,j}(k)$ is higher than the maximum value for all the weights in layer k (max_weight(k)). After each Forward Phase the average_weights(L) and max_weight(k) (k=1,..,L-1) are updated.

4 Fault-Tolerance Evaluation

Assessment of the fault tolerance of the learning algorithm we propose will be compared with that stated in [4]. The performance of both the replication-based solution and the solution proposed in the paper will then be compared with the inherent fault tolerance of feedforward networks trained with the classical algorithm described in [5]. In order to obtain comparable results, the hypotheses adopted in [4] will be considered during the tests performed, as outlined below.

Fault Model. The fault typology considered refers to a single fault in the weights (including biases) in each layer. It is assumed that each weight or bias can, on the occurrence of a fault, assume the following values: $-\infty, +\infty$ and 0. As was done in [4], we only analyze a single fault in a weight or bias. Multiple faults are not considered, although we are already working on analysis of such a scenario.

Training Test Sets. The tests performed in [4] refer to Sonar Classification [6], Two-Spiral [6] and Vowel [6] benchmark tests. We tested the algorithm proposed using the same test set. For reasons of space only the Sonar Classification results will be shown.

Network Architecture. With reference to the Sonar Benchmark test, the network is characterized by 2 output units. Experiments were run with a single hidden layer with 42, 60, 84, 120, 126 and 180 units. Choice of these hidden units depended on the need to obtain a scenario as close as possible to those considered in [4], which measures the inherent fault tolerance of a network, trained with a classical backpropagation algorithm, with 42 and 60 hidden units. By replicating the hidden units, fault tolerance was assessed for networks with 84 and 120 units in the hidden layer (obtained by single replication of the hidden units of networks with 42 and 60 hidden units respectively) and with 126 and 180 units (replicating the hidden units of networks with 42 and 60 units twice).

Fault Tolerance Performance. For each pattern belonging to the Sonar Benchmark test set all the possible types of fault were assessed for each weight. The neural network output on the occurrence of a fault was assessed as incorrect if it differed from the expected value by +/- 0.5. The fault tolerance values were assessed using the following formula:

$$\frac{1 - NumerOfError}{NumberOfOutputs \cdot NumberOfPatterns \cdot NumberOfFaults}$$

where :

$$NumberOfFaults = NumberOfWeights \cdot TypeOfFaults$$

Fig.2.a shows the fault tolerance values obtained with the traditional algorithm (in the following it will be called Traditional Learning), the one proposed in [4] (Replication) and the one proposed here (Modified Learning). The comparison refers to networks with 42, 84 and 126 hidden neurons. It should be pointed out that, in the case of the replication-based algorithm, the networks with 84 and 126 neurons were obtained by replicating the 42 neurons in the hidden layer once or twice. The fault tolerance values obtained with the replication-based algorithm, referring to a network with 42 hidden nodes, therefore coincide with those for Traditional Learning. In the case of Traditional or Modified Learning, on the other hand, the networks with 84 and 126 neurons were trained independently of the network with 42 hidden units. Comparison shows that the learning algorithm we propose increases fault tolerance as compared with that of both the traditional algorithm and the replication-based one. In particular, even with only 42 hidden units our algorithm guarantees fault tolerance values of 99.8, which are almost comparable with those obtained by replicating the 42 units once.

Fig.2.b shows the same comparisons, this time considering networks with 60, 120 and 180 hidden neurons. As before, in the case of the replication-based algorithm the networks with 120 and 180 hidden neurons were obtained by replicating 60 hidden units once or twice. Once again, comparison shows that our learning algorithm enhances fault tolerance as compared with that of both traditional learning and the replication-based algorithm. Even more than in Fig.2.b, it can be seen that with only 60 hidden units our algorithm guarantees fault tolerance values of 99.89, which are almost comparable with those obtained by replicating the 60 units only once.

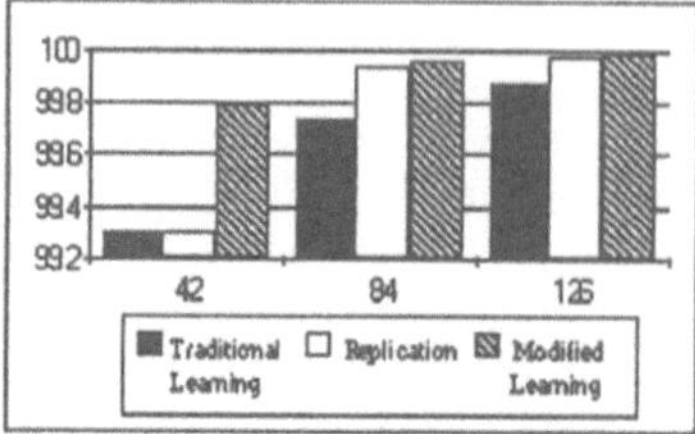 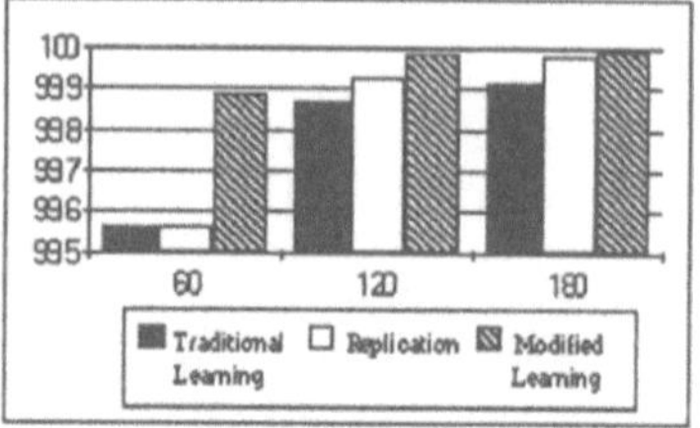

Figure 2: a,b - Fault Tolerance for All Kind of Faults

The comparisons made so far do not, however, seem to be complete, as they do not show up the real advantage of the solution proposed here. During training with any traditional algorithm for feedforward networks, the values of the weights in each layer are not uniformly distributed: there may be some weights with much higher values than others. A fault in one of these weights may seriously affect the fault tolerance of the network as a whole. This problem exists both in networks trained with the traditional algorithm and in those obtained by replication. The tests conducted previously do not show up this drawback as the fault tolerance values obtained were calculated by averaging the percentage of errors for each fault occurrence. It therefore seems of some interest to assess the decrease in fault tolerance in the event of a fault in an extremely critical weight. To do so, we assessed the fault tolerance values following a fault in the most critical weights (i.e.the weights featuring the highest absolute values). For each network trained with the traditional algorithm, with the modified algorithm or obtained by replicating the hidden units, we simulated a fault in all the most critical weights (i.e. with the hightest absolute value) belonging to the output layer, by setting their values to zero. Then for each single fault we evaluated the fault-tolerance performed by the network. Among these faults, the lowest fault-tolerance value was taken into account. Fig.3.a and Fig.3.b show the fault tolerance values obtained with networks of 42, 84 and 126 neurons and 60, 120 and 180 hidden neurons respectively. As can be seen, a fault in the highest weight causes a considerable decrease in fault tolerance with both the Traditional Learning algorithm and the replication-based one. Although the fault tolerance values obtained with our algorithm are slightly lower than those shown in Fig.2.a and Fig.2.b, they are still much higher than those obtained with the other two algorithms, and are quite acceptable.

5 Final Remarks

The learning algorithm proposed is based on the simple idea of balancing the weights in each layer of the feedforward network. The solution was tested using benchmark test sets available in literature. The tests carried out show that the algorithm proposed presents no particular convergence problems. In addition, they allowed us to choose the levelling strategy which would best guarantee

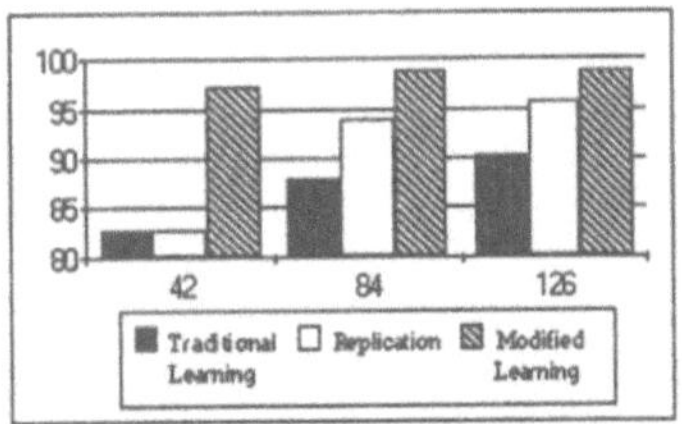

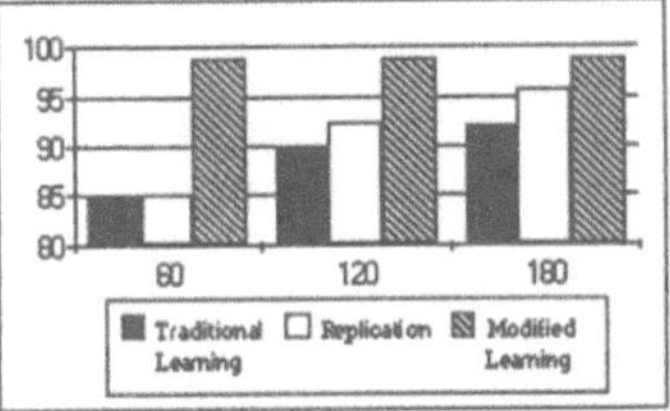

Figure 3: a,b - Fault Tolerance for the Most Critical Neuron Fault

low convergence times. The fault tolerance values obtained showed a great increase as compared with the traditional learning algorithm. This increase was also observed as compared to the best-known on-line fault tolerance solution in literature, based on redundancy of the hidden layer neurons. Particular attention was paid during the tests to the effect on the feedforward network's recovery capacity of a fault in one of the most critical weights (i.e. one with a much higher value than the rest). The tests performed showed that the traditional algorithm and the replication-based one may undergo a considerable decrease in fault tolerance following a fault in a single particularly critical neuron. The solution proposed in the paper does not seem to suffer from this drawback, however, thanks to the uniform distribution of weights in each layer. Further studies will concentrate on fault tolerance following multiple faults, which has rarely been dealt with in literature.

References

[1] M.J.Carter, The Illusion of Fault-Tolerance in Neural Nets for Pattern Recognition and Signal Processing, Proc.Technical Session on Fault Tolerance Integrated Systems, University of New Hampshire, Durham, 1988.

[2] M.J.Carter, F.J.Rudolph and A.J.Nucci, Operational Fault-Tolerance of CMAC Networks, Neural Information Processing Systems 2, D.S.Touretzsky, Ed.Morgan Kaufman, 1990

[3] T.Petsche, B.W.Dickinson, Trellis Codes Receptive Fields and Fault Tolerant Self-Repairing Neural Networks, IEEE Transaction on Neural Networks, 1, 151–166, 1990.

[4] D.S.Phatak. I.Koren, Complete and Partial Fault Tolerance of Feedforward Neural Nets, IEEE Transaction on Neural Networks, vol.6, no.2, March 1995.

[5] D.E.Rumelhart, J.L.McClelland, Parallel Distributed Processing, vol.1:Foundations, MIT Press,1986.

[6] S.E.Fahlman et.al., Neural Nets Learning Algorithms and Benchmarks DataBase, mainteined at the Computer Science Dept., Carnegie Mellon University.

An Integrated Neural and Algorithmic System for Optical Flow Computation

A. Criminisi, G.A.M. Gioiello*, D. Molinelli, F. Sorbello

DIE - Dipartimento di Ingegneria Elettrica, Università di Palermo

Viale delle Scienze, 90128 Palermo, Italy

*DIE and CRES - Centro per la Ricerca Elettronica in Sicilia

Via Regione Siciliana 49, 90046 Monreale, Italy

Abstract

Motion detection plays a central role in several visual environments: knowledge of object velocities and trajectories is fundamental in scene interpretation and segmentation. This task appears a simple problem, but detecting moving objects is very difficult, in fact this is a problem that cannot be considered completely solved today [1] [2] [3].

In this paper we present a novel method that uses two different approaches: a "neural" one and an algorithmic one. In fact, a Multilayer Perceptron is used in the first stage, in order to detect some motion areas in the scene [5] [6]; a matching algorithm is then used to obtain a sparse optical flow and to compute the epipolar geometry of the moving camera [7] [8]; and, finally, a refinement algorithm is used to produce a denser optical flow field. Thus this method can extract features automatically from moving objects in a scene discarding stationary ones. This approach seems to be very useful for tracking and motion segmentation.

This work was developed in the context of JACOB project, to achieve the automatic retrieval of images based on motion [9].

1 Introduction

In the last recent years several approaches based on the automatic extraction of salient low level features from images and videos have been attempted, borrowing methodologies and tools from classic areas of computer vision and image analysis. In particular, motion information seems to be very useful to achieve a scene segmentation, and especially when we are dealing with robot navigation, object tracking, image compression or visual databases. The task of segmenting and recognizing moving objects in a scene is very difficult and today remains unresolved notwithstanding past attempts. In this paper a new approach is introduced that merge a neural net preprocessing to a traditional algorithm for matching points of interest extracted from the scene. In the first stage a three layers perceptron locates some interesting moving areas in the scene and then a traditional correlation technique, together with epipolar geometry information, tries to match extracted moving features to achieve a robust optical flow. Our purpose, in fact, is to accomplish an automatic moving object tracking system, and we have chosen the output of a motion detection network as input for a motion estimation algorithm.

2 The neural preprocessing

Even if a simple frame difference might be used as a measure of pixel motion, such an approach is too noise sensitive. Our approach consists of a more robust solution, involving a very simple computation on each point in the image plane. A motion measure is computed for each pixel, in accordance with (1):

$$m(x, y) = E\{(f(x, y, t))^2\} - (E\{f(x, y, t)\})^2 \qquad (1)$$

In this formula x and y are the image plane coordinates and E indicates the mean over time on four subsequent frames. Values are normalized in the range $[0..1]$. The normalization function acts as stated in (2):

$$\tilde{m}(x, y) = \sqrt{\frac{m(x, y) - min}{max - min}} \qquad (2)$$

where $m(x, y)$ is the normalised motion measure of the pixel (x, y) and min and max are respectively the minimum and maximum value of $m(x, y)$ over the entire image plane. A better representation was obtained using the square root to raise the importance of the lowest values of the original normalised motion measure map. This motion measure map $m(x, y)$ represents the input for the neural network we used.

The net processes each image 16 bits at once, arranged as a 4*4 matrix, thus obtaining a final $(dimx/4) * (dimy/4)$ motion map. The proposed architecture is a Multilayer Perceptron, consisting of an input layer, one hidden layer, and the output layer. A hardware version has been developed, nevertheless we used a software version of this net [5] [6].

The net was trained to discriminate if motion occurred within the generic 4*4 input matrix. A specific training set was developed to train the net. This set consists of 4*4 matrices of numbers whose values refer to the entity of the motion in the same portion of the image. The net learned very well to discriminate true motion and noise effects. The input layer consists of 16 neurons; instead the first layer consists of 32: each neuron has 16 weights, one for each pixel of the input pattern. The post-synaptic activation function is the logistic one. The second layer consists of 2 neurons, as many as the output classes are. Each second layer neuron has 32 weights, one for each first layer unit. The activation function of the second layer is the logistic one. The output activations are numbers in the range $[0..1]$. Powell's minimization algorithm is used [4].

Neural preprocessing task shows interesting properties as: high elaboration speed, high recognition rate, easy VLSI hardware implementation, small amount of memory resources needed.

3 Feature matching

We apply neural preprocessing twice, in fact we use the net to detect motion areas in two frames that are four or more frames far. So we obtain two files storing motion areas and their activations. We choose only the best areas, those with an activation greater than 0,8. But in those 4 by 4 patches only one pixel is selected showing the greatest gradient response, a point that is likely on an object edge. Now we obtained a collection of moving edge pixels in the first frame and a similar collection in another frame of the same sequence. These are the input of the matching algorithm.

To achieve this aim we use a software developed by INRIA istitute researchers, named *image-matching* [8]. This algorithm is composed by several steps: firstly it searches possible matches, inside a search window, using a correlation score evaluated using a 7 by 7 mask. It achieves a first matching step, therefore many candidate matches are found for every point in the first frame. The search window is 40 by 40 pixels wide. If $\mathbf{m_1}$ and $\mathbf{m_2}$ are two matches respectively on first and second frame, their correlation score is computed according to (3).

$$Score(\mathbf{m_1}, \mathbf{m_2}) = \frac{sum}{(2n+1)(2m+1)\sqrt{\sigma(I_1)\sigma(I_2)}} \tag{3}$$

where

$$sum = \sum_{i=-1}^{n} \sum_{j=-m}^{m} [I_1(u_1 + i, v_1 + j) - \overline{I_1(u_1, v_1)}] \cdot$$

$$\cdot [I_2(u_2 + i, v_2 + j) - \overline{I_2(u_2, v_2)}]$$

and

$$\overline{I_k(u, v)} = \frac{\sum_{i=-1}^{n} \sum_{j=-m}^{m} I_k(u + i, v + j)}{(2n+1)(2m+1)}$$

is the average at point $\mathbf{m} = (u, v)$ of I_k and $\sigma(I_k)$ is the standard deviation of the image I_k in the neighborhood of (u, v).

$$\sigma(I_k) = \sqrt{\frac{\sum_{i=-1}^{n} \sum_{j=-m}^{m} I_k^2(u, v)}{(2n+1)(2m+1)} - \overline{I_k(u, v)}} \tag{4}$$

Now, using a relaxation technique, the algorithm resolves matching ambiguities. For this purpose we let some constraints, such as uniqueness and continuity, propagate themselves during calculation through the neighbourhood of candidate matches. When the process reaches a minimum of a total energy function it terminates and the third step begins. The third step is one of the most crucial; it calculates *epipolar geometry* of the moving camera. This is an unstable task especially if we have not many matches or if they are uncertain, but we use a "least median of squares" criterion to discard false matches (*outliers*) and to recover robustly the *fundamental matrix* that describes epipolar geometry of the scene. If $\mathbf{m_1}$ and $\mathbf{m_2}$ are two matches and $\mathbf{F}$ is the fundamental matrix the following equation is valid:

$$\mathbf{m_2}^{\mathbf{T}} \mathbf{F} \mathbf{m_1} = 0 \tag{5}$$

This matching algorithm results to be very insensitive to noise for several kinds of image sequences and thus very useful to our aims.

4   Optical flow thickening

Now we have obtained a very sparse, but correct, optical flow. We have the matched pairs and, above all, the fundamental matrix. So, for each point in the first frame, we can calculate the related epipolar line in the second frame and vice-versa. This method uses several geometrical and intensity constraints such as: epipolar geometry, computed optic flow, correlation score, body rigidity, uniqueness.

Given a point its corrispondent one is certainly very close to its epipolar line. Each pair of candidate matches are caracterized by a gaussian weight related to distance between the point and the epipolar line. Besides, we use a correlation score to measure the similarity of the pair. From interesting points in the first frame we select those that satisfy the constraint of having coherent neighborood motion vectors.

For a point with more than 3 close coherent vectors a very little search area is computed in the second frame; x and y dimensions of this area are functions of the incoherence of close motion vectors. Therefore new candidate matches are found, inside the search window, for each fundamental point. These candidates are stored in a long list with their correlation score and the epipolar weight function. Finally, for each pair a total weight is computed multipling epipolar weight and correlation score. The list obtained is given as input to a relaxation algorithm that in a few iterations finds true matches disambiguating candidates. During this step some total weights are increased by moving points, pixel by pixel, inside the search area and recomputing the correlation score and the distance from the epipolar line; instead, those that present many instances in the list are reduced. So, after a maximum of 10 iterations the algorithm terminates and matches are updated.

In this final step we obtain a 30 to 90 percent improvement in number of matches. This refinement task is very insensitive to noise, thanks to the coherence constraints adopted.

5   Experimental results

We report one frame and the computed optical flow for some test sequences. In the "rubic" and "hammer" sequences a stationary camera is used, instead a moving one in the "garden" sequence.

In the "hammer" sequence a traslational motion of the hammer and the tennis ball is perfectly detected (fig. 1). In the "garden" sequence the tree is faster than background because of perspective effect; in its optical flow vectors related to the tree are the longest (fig. 2). We can notice that in the "rubic" sequence our system found only moving points discarding stationary background (fig. 3). The basement is faster than the cube because of rotation.

The whole task takes short CPU time, it results sufficiently fast, in fact it requires only a few minutes to run in a SUN20 SPARC Station.

6 Conclusions and future works

In this paper a new mixed neural and algorithmic approach has been proposed to extract motion information in a sequence of monocular images. This approach consists of a simple neural net based preprocessing and a complex matching algorithm. Our method is very robust to noise, fast and independent from analysed sequences. A future step is clustering motion vectors in the output velocity field according to their affine motion parameters, in order to achieve a first segmentation task, fundamental in developing an object tracking system.

References

[1] S. Negahdaripour, S. Lee, Motion recovery from image sequences using only first order optical flow information,International Journal of computer vision, 9:3, 163-184 (1992).

[2] A. Singh, P. Allen, Image-Flow Computation: An estimation-theoretic framework and an unified perspective, CVGIP-IU, vol. 56, no. 2, September 1992.

[3] J. Weber, J. Malik, Robust computation of optical flow in a multi-scale differential framework, IEEE Fourth International Conference on Computer Vision 2/93.

[4] M.J.D. Powel, Restart Procedures for the Conjugate Gradient Method, Mathematical Programming, Vol. 12, pp. 241-254.

[5] A.Abruzzo, G.A.M.Gioiello, M. La Cascia, F.Sorbello, Motion Detection From Image Sequences Using A New Fully Digital VLSI Neural Architecture, EUFIT95 - August 26-31, 1995, Aachen, Germany.

[6] A.Abruzzo, G.A.M.Gioiello, M. La Cascia, F.Sorbello, A New Fully Digital Neural Architecture For Motion Detection From Image Sequences, EANN95 - International Conference on Engineering Applications of Neural Networks, August 21-23, 1995, Helsinki, Finland.

[7] Q.T.Luong,R.Deriche,O.D.Faugeras, T.Papadopoulo, On Determining the Fundamental Matrix, Technical Report 1894, INRIA, Sophia-Antipolis, France, 1993.

[8] Z. Zhang, R. Deriche, O. Faugeras, Q.-T. Luong, A Robust Technique for Matching Two Uncalibrated Images Through the Recovery of the Unknown Epipolar Geometry, Artificial Intelligence Journal, Vol.78, pages 87-119, October 1995. Also Research Report No.2273, INRIA Sophia-Antipolis.

[9] M. La Cascia, E. Ardizzone, JACOB: Just a content-based query system for video databases, IEEE International Conference On ACOUSTICS, SPEECH AND SIGNAL PROCESSING, May 7-10, 1996, Atlanta, Georgia (USA).

 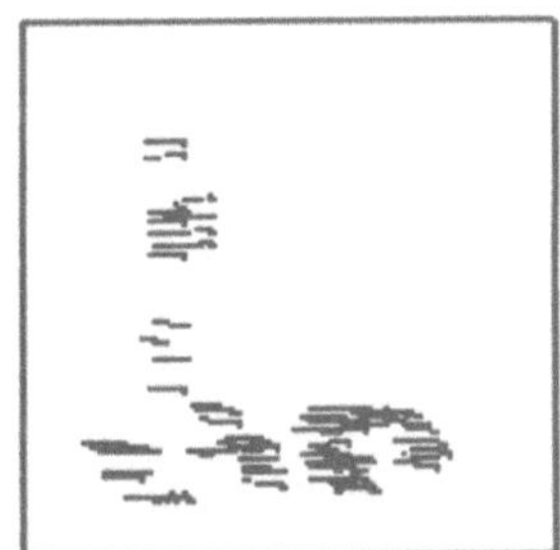

Figure 1: One frame and the optical flow for *hammer* sequence

 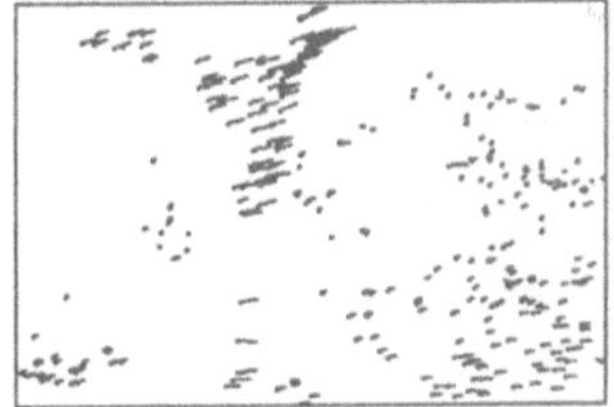

Figure 2: One frame and the optical flow for *garden* sequence

 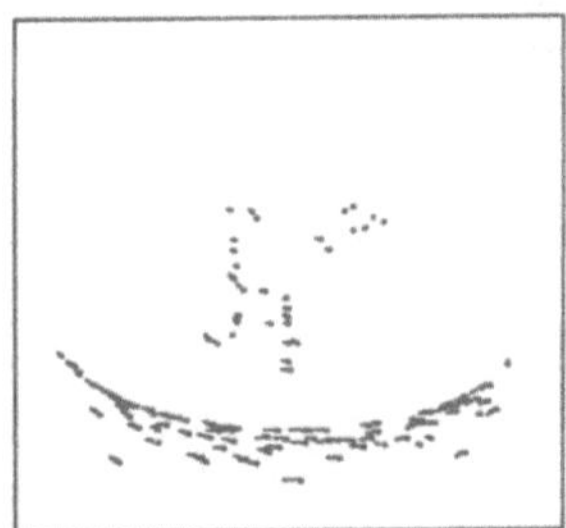

Figure 3: One frame and the optical flow for *rubic* sequence

FIR NNs and Temporal BP: Implementation on the Meiko CS-2

Antonio d' Acierno,* Salvatore Palma and Walter Ripullone

I.R.S.I.P. - C.N.R.

via P. Castellino 111 - 80131 Napoli - Italy

Abstract

This paper describes a parallel implementation of the Temporal Back-Propagation, a learning algorithm proposed for *FIR* neural networks, on the Meiko CS-2 MIMD parallel computer. The mapping scheme here used (proposed elsewhere) is based on a collective operator (*global sum*), and allows to obtain an efficient parallel code which is easy to port from the sequential one. Implementation results are shown and commented.

Keywords: FIR NNs, Temporal BP, MIMD Parallel Computer.

1 Introduction

The behaviour of an Artificial Neural Network (ANN) is determined by parameters denoted *weights* to be evaluated using learning procedure. This procedure tend to be very time–consuming and it is therefore obvious to try to develop faster learning algorithms and/or to capitalise on the intrinsic parallelism of these systems in order to speed up the computations since massively parallel computers today available have the potential to provide the computational power for high–speed simulations.

Although learning algorithms typically involve only local computations, however, the output of an unit usually depends on the output of many other units. Thus, without careful design, an implementation of such algorithms on a massively parallel (distributed memory) computer can easily spend the majority of its running time, say 75% or more, in communication rather then in actual computation. Hence, the mapping problem is not trivial and deserves attention, since the communication problem is obviously crucial.

This paper describes a parallel implementation of the *Temporal Back–Propagation* (TBP) [1], a learning algorithm proposed for *FIR* ANNs, on the Meiko CS-2 MIMD parallel computer. For a brief and informal description both of FIR networks and of TBP, see [2].

2 The Parallel Implementation

The parallel implementation here studied is based on that proposed in [3] for classical feed-forward networks, that has been extended [4] in order to deal with

---

*Corresponding author: antonio@irsip.na .cnr.it

feed-forward networks with two or more hidden layers.

Our mapping scheme is based on the use of a *vectorial global sum* operator and the porting of a sequential standard Back-Propagation on the Meiko CS-2 just requires than one or two lines of code have to be added [4], depending on the parallelism used (synaptic parallelism or neuron parallelism [5]). By considering, for the sake of simplicity, networks with just 1 hidden layer the mapping can be simply described as follows (see fig. 2).

Let Ni be the number of input neurons, Nh the number of hidden neurons, No the number of output neurons, $W1[Nh][Ni]$ the connection matrix between input and output layers and $W2[No][Nh]$ the connection matrix between hidden and output layers; morover let us suppose that *(i)* the memory of each processor is unlimited and that *(ii)* the number Nh of hidden neurons equals the number P of processors to be used. Starting from these hypotheses, we can suppose that each process knows all the training patterns and handles an hidden neuron. As a first step, each process evaluates the activation value of the handled hidden neuron; supposing that the $i-th$ process knows the $i-th$ row of $W1$, there is not communication and the load is perfectly balanced.

To evaluate the output value of output neurons, each process in our implementation (say the process s) calculates the vector $J(s)$, representing the product of the output of hidden neuron s and the $s-th$ column of $W2$; suppose that such a column is handled by process s. Once the vectors J have been evaluated (and this evaluation is performed in a fully parallel and perfectly balanced way), they are summed through a *read, sum and send* algorithm; the processes are thus logically organised as a tree rooted by a *master* process. The master completes the evaluation of the net inputs to output neurons (by subtracting biases, if any), applies the activation function and evaluates the error terms of output neurons. Such error terms are then broadcast to *slave* processes.

Once the error terms of output neurons have been received, the error of the handled hidden neuron must be evaluated. This step only requires that the $s-th$ process knows the $s-th$ column of $W2$ to be executed; since this hypothesis is verified, there are not data to be communicated and the load is perfectly balanced. Finally, each weight must be updated; since each process knows all it needs to evaluate the weight changes for the connections it knows (and handles), we have again that there is not communication and the load is perfectly balanced.

The parallel implementation is realised using, as parallel statement, only the *gdsum* function, that realises a combination of vectors. This function works as follows: suppose that each of P processes in a parallel application holds a vector X of dimension n and let X_{old}^{j} the vector in process j; after the *gdsum(X,n)* call, the $i-th$ component of vector X_{new} in each process equals $\sum_{j=1}^{P} X_{old}^{j}(i)$.

Because of the structure of the *gdsum*, we have that the work assigned to the *master* process has to be replicated and is executed by each process; this clearly requires that each process has to know the input part as well as the output part of the whole training set. It is worth noting that, starting from a sequential code able to deal with a number of hidden neurons fixed at *run-*

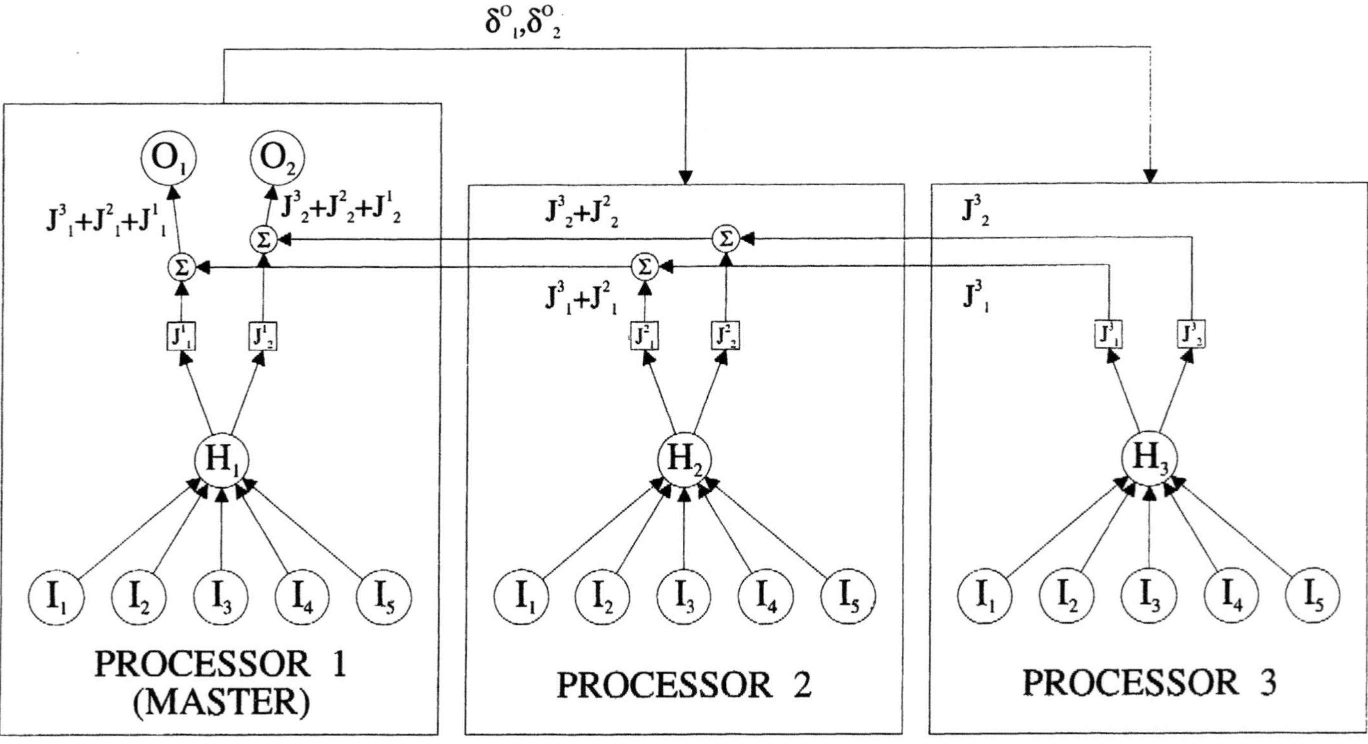

Figure 1: The mapping scheme with 3 hidden layers and 3 processors.

time, what we need to obtain the parallel code is to add a *gdsum* call before evaluating the error terms of output neurons. A collection of P processes, each simulating a network with Ni input neurons, Nh/P hidden neurons and No output neurons, will then simulate the neural network we are dealing with.

The main drawbacks on the proposed mapping scheme are represented by the following facts: *(i)* it requires that each process knows all the training set and *(ii)* it cannot exploit all the parallelism for networks with input and output layer significantly larger than the hidden layer (e.g. feed-forward ANNs designed for solving compression problems). Both problems can be solved by inverting the scheme and using:

- synapse parallelism to evaluate the activation values of hidden neurons;

- neuron parallelism to evaluate the activation value of output neurons;

- synapse parallelism to evaluate the error terms of hidden neurons.

Using such a second scheme, the porting of the sequential code on the parallel platform is obtained by adding a *gdsum* statement in the forward phase (before evaluating the output of hidden neurons) and a *gdsum* statement in the backward phase, before completing the evaluation of error terms of hidden neurons. Of course, both schemes can be easily generalised to networks with two or more hidden layer.

The problem we have to deal with in the case of FIR networks is the following. When the number of hidden layers is odd (say 1) the first mapping scheme can be applied as it is, since its structural problems does not apply. In fact the whole training set is simply composed by a mono-dimensional array (so that its storage does not represent a problem) and the hidden layer is the biggest one (since both the input layer and the output layer have just a neuron). When, as in the application we are dealing with [2], the number of hidden layer is even (say 2), we have that, since our mapping scheme needs that synapse parallelism and neuron parallelism have to be alternatively used for different layers, the maximum number of usable processors equals the dimension of the smallest layer where neuron parallelism is used. Both starting with neuron parallelism and with synapse parallelism, we obtain $P^{max} = 1$ that simply means no parallelism at all.

To avoid this problem we have to *unfold in time* the first level of connections, so obtaining an equivalent network with M^1 input neurons. It is worth noting that the unfolding in time of just a layer of connections does not introduce duplicated weights. In this case, we can start with neuron parallelism for input layer, obtaining the desired network partitioning. Once the code for the *unfolded in time* network is available, we can apply the second mapping scheme so that, to obtain the parallel code, just 3 lines of code have to be added [4].

P	net_1	net_2	net_3	net_4
2	0.79	0.93	0.8	0.94
3	0.71	0.89	0.67	0.90
4	0.55	0.81	0.54	0.85
5	0.50	0.75	0.45	0.79
6	0.42	0.68	0.41	0.77
7	0.40	0.58	0.34	0.73
8	0.33	0.54	0.31	0.70

Table 1: The efficiency as P vary for some networks.

3   Experimental Results

We tested the performance of our parallel algorithm on two networks with 1 hidden layer and on two networks with 2 hidden layers.

The first network (net_1) has 1 hidden layer with 15 units with 8 delay units in the first level of connections, and 2 delay units in the second level of connections. This network has been used for the prediction of the Markey–Glass chaotic serie [1]. The second network (net_2) is a bigger one, with 24 neurons in the hidden layer, and has, respectively, 25 and 15 delay units in the first and in the second level of connections.

The other networks (net_3 and net_4) have two hidden layers. The network net_3 has 12 neurons in each of two hidden layers and has 20 delay units in the first level of connections, and 5 delay units in the second and in the third level of connections. This network has been used for the *Santa Fe Institute Time Series Predictrion Competition* to predict the chaotic pulsations of an NH_3 laser, outperforming all other suggested methods [1]. The network net_4, with 24 neurons in each of two hidden layers, has, respectively, 25, 15 and 5 delay units in the first, in the second and in the third level of connections.

We used a cluster of 8 *SuperSpark* processors and tab. 1 shows the performance obtained; it is worth noting that we are using a very coarse–grained approach, so the small number of used processor should not be surprising.

By considering networks with just an hidden layer also for a simple net (net_1, simple in terms of free parameters) we have an efficiency greater than 0.5 even using 4 processors. By increasing the complexity of the network, since the communication overhead remains unchanged, we have an increasing efficiency.

The analisys of results is slightly more complex when networks with two hidden layer are considered. By assuming an efficiency of 0.5 as accepted worst case, we have that, for net_3, at most 4 processors can be used, because of the simplicity of the network in terms of free parameters. As the complexity of the neworks increasess, so does the maximum number of usable processors (having fixed the worst case efficiency); for net_4 we can use all the power we have ($P = 8$), obtaining an efficiency ≈ 0.7. The efficiency, in other words, increases as the complexity of the network increases. This is a well known phenomenon but, by using our scheme, it is amplified since the communication overhead simply does not depend on some parameters.

4 Conclusions

We have implemented a parallel version of the Temporal Back-Propagation, used for FIR neural networks, that have been proposed for time series prediction. The mapping scheme we used, allows to port the sequential code on the parallel platform in a trivial way. The obtained performance seems very promising, at least when the number of free parameter is sufficiently high.

An interesting question under investigation is the following. When networks with two hidden layers are considered, the overhead due to the communication for the used mapping scheme just depends on the dimension of the first hidden layer. Given a network that solves a problem, to improve performance the question is :

"is it possible to obtain another network with the same number of free parameters, that solves the problem at hand at least with the same accuracy and with a number of neuron in the first hidden layer as low as possible?"

References

[1] E. A. Wah, Finite Impulse Response Neural Networks with Applications in Time Series Prediction, PhD Thesis, Department of Electical Enginnering, Stanford University, 1993.

[2] A. d' Acierno, S. Palma and W. Ripullone , FIR NNs and Time Series Prediction: Applications to Stock Market Forecasting, WIRN '96, This Volume.

[3] A. d' Acierno and R. Vaccaro, The Back-Propagation Learning Algorithm on Parallel Computers: A Mapping Scheme, Neral Nets WIRN '93, E.R. Caianiello Ed., World SDcientific Publishing Co., Singapore,1994.

[4] A. d' Acierno and S. Palma, The Back-Propagation Learning Algorithm on the Meiko CS-2: Two Mapping Schemes, LLNCS 1067, HPCB'96, PP. 890-897.

[5] T. Nordstrom and B. Svensson, Using and Designing Massively Parallel Computers for Artificial Neural Networks, Journal of Parallel and Distributed Computing, 14, 3, 1992, pp. 260-185.

Some Comments and Experimental Results on Bayesian Regularization

Michel de Bollivier

Neural Networks Laboratory (NNL), E.C. Joint Research Centre
Via E. Fermi, 21020 Ispra, Italy

Domenico Perrotta

NNL and, also, Lab. de l'Informatique du Parallélisme
Ecole Normale Supérieure de Lyon
69364 Lyon Cedex 07, France

Abstract

This work is an experimental attempt to determine whether the Bayesian paradigm can improve Multi-Layer Perceptrons (MLPs) learning methods. In particular, we experiment here the paradigm developed by D. MacKay (1992). Bayesian MLPs are used on three public classification databases and compared to other methods. The paper also comments on some practical points of MacKay's work, having in mind future applications.

1   Introduction

This paper presents an experimental attempt to determine whether the Bayesian paradigm can improve Multi-Layer Perceptrons (MLPs) learning methods, applied to classification tasks.

Two drawbacks of the MLPs learning procedure are common to most of the classifiers. The first drawback is related to the computation of the given classifier parameters (weights in the case of MLPs). Usually, the parameters are estimated using a data set (the training set) and the phenomena of overfitting can sometimes be observed. This means that it is possible to get 100% accuracy on the training set while getting very poor performances on some data set apart by the experimenter (the test set). It can thus be very probable that performances would also be very poor on future data. The methods used for estimating parameters are usually iterative without limit or stop criteria. In the case of MLPs, this can lead to computing parameters which fit too closely the training set. When using the classifier at different stage of the learning iterative process on the test set, it is possible to see that its performances on the test set sometimes reach a maximum before decreasing. Then finding a stopping criterion or a criterion for choosing among several configurations observed during these iterative methods would be very useful.

The second drawback is that it is very common to try several classifiers or variants of the same classifier for solving a task. It would be very useful to rank them according to their predictive ability on unknown future data.

These two drawbacks are engaged by different methods which are still a subject of research. We are interested here in experimenting a method which covers various Bayesian techniques originally applied to MLPs by MacKay (1992). We would like to see whether it could give in practice an answer to the two drawbacks previously discussed.

The paper is organised as follows. The next section reports some results of MacKay's framework which are useful for this work. In fact, we work here only on a limited part of MacKay's method. In Section 3, Bayesian MLPs will be used on three public classification databases and compared to other methods. Perhaps more important than this comparison, we will see how Bayesian methods can practically rank various MLPs architectures. A discussion will follow in Section 4. Even if definite conclusions cannot be given after running Bayesian MLPs on just three databases, we expect that the tests we have made can be useful to other practitioners or researchers.

2 Bayesian MLPs

We do not intend to introduce here the Bayesian formalism in MLPs. For discussion on Bayes theory the reader should refer, for example, to Bernardo (1995). This Section reports those results of MacKay's framework which are useful for this work. The reader should refer to MacKay (1992) for a full comprehension of the subject. We are going to introduce some notation. A Bayesian scheme for MLPs follows.

We are given a training data set $D = \{(x_1, t_1), \ldots, (x_N, t_N)\}$. We want to link x_i and t_i through the classical relation $t_i = R(x_i) + \epsilon_i$, where ϵ_i is an expression of the various uncertainties, e.g. noise. We want to approximate $R(x)$ by a function $F(x, W)$ where F and $W = (w_1, \ldots, w_k)$ are respectively the non linear transfer function and the weights of a given MLP architecture. The state of each cell y_k which receives information from cells y_l through weights w_l is given by $y_k = f(\sum w_l \cdot y_l)$, where f is usually a sigmoid. In our simulator, SN from Neuristique SA, we have chosen f to be symmetric in the range $[-1.7, +1.7]$. Moreover, we have coded input data x_i in the range $[-1, +1]$. For each (x_i, t_i) from the training set, in the case of a classification task in n classes, t_i is coded as a $n-$dimensional vector in which each component is -1 except the kth which is set to $+1$ when x_i belongs to the kth class.

Bayesian inference uses probabilities for quantifying uncertainties on quantities. Using the well known Bayes formula, we get for a quantity of interest θ: $P(\theta|D, H) = P(D|\theta, H) \cdot P(\theta|H)/P(D|H)$ where H is the knowledge that we have and the assumptions we have made about the given problem before seeing the data. A Bayesian scheme for MLPs adopts two main levels of inference. The first level consists of estimating the weights from the data, conditioned on a parametrized prior distribution on those weights. The second level consists of estimating the parameters of the prior. For MLPs, these new parameters are the standard deviations of the noise and weights distributions, σ and σ_W, that are assumed by MacKay to be Gaussian. Following MacKay, tuning a Bayesian MLP is an iterative process which works as follows. At step t, starting from weights W^t and given parameters σ^t and σ_W^t:

a) Find W^{t+1} which minimize: $\dfrac{1}{\sigma^2} \sum_{i=1}^{N} (F(x_i, W) - t_i)^2 + \dfrac{1}{\sigma_W^2} \sum_{j=1}^{k} w_j^2$ (1)

b) Find the most probable values σ^{t+1} and σ_W^{t+1} using the data.

c) Go back to (a) using now W^{t+1}, σ^{t+1} and σ_W^{t+1}.

Steps (a)–(c) are alternated, as described in Sections 2.1 and 2.2, until the MLP has converged. This is still not the end, because one should find a method for ranking different MLPs configurations at different iteration steps $t, t+1, t+2, \ldots$. Each of these configurations define a new model M^t. The main practical point here is that between two models M^j and M^k, the most adapted to the problem is the one which maximises the conditional probability $P(D|M, H)$, called *evidence*. MacKay gives a formula for computing this expression.

The practical implementation of the above iterative process fulfills the following formulas. The reader could find them in MacKay's work, where formula (1) is written as:

$$\beta E_D + \alpha E_W \qquad \text{being} \tag{2}$$

$$\beta = \frac{1}{\sigma^2} \qquad \alpha = \frac{1}{\sigma_W^2} \qquad E_D = \frac{1}{2}\sum_{i=1}^{N}(F(x_i, W) - t_i)^2 \qquad E_W = \frac{1}{2}\sum_{j=1}^{k} w_j^2$$

2.1 Standard and quick-and-dirty versions

In the above iterative process, step (a) finds weights W^{t+1} which minimize $\beta^t E_D + \alpha^t E_W$, given parameters W^t, σ^t and σ_W^t (or β^t and α^t) and thus local values E_D^t and E_W^t of E_D and E_W. This is done using a conjugate gradient algorithm. MacKay uses the variable metric method found in Press et al. (1992). We have used here the version available in our software SN from Neuristique SA and derived from the work of Moller (1993). Step (b) computes new values of β^{t+1} and α^{t+1} as follows. It first computes γ^{t+1} as:

$$\gamma^{t+1} = k - \sum_{a=1}^{k} \frac{\alpha^t}{\lambda_a + \alpha^t} \qquad \text{and then } \beta^{t+1} \text{ and } \alpha^{t+1} \text{ as:}$$

$$\alpha^{t+1} = \frac{\gamma^{t+1}}{2E_W^{t+1}} \qquad \beta^{t+1} = \frac{N - \gamma^{t+1}}{2E_D^{t+1}} \tag{3}$$

where λ_a is the ath eigenvalue of the Hessian matrix of $\beta^t E_D^{t+1}$ computed using weights W^{t+1}.

To avoid the computation of the Hessian matrix and the search for its eigenvalues in (3), MacKay proposes a cheap (in CPU time) version of the standard method:

$$\alpha^{t+1} = \frac{k}{2E_W^{t+1}} \qquad \text{and} \qquad \beta^{t+1} = \frac{N}{2E_D^{t+1}} \tag{4}$$

In fact, this cheap solution is very interesting, because sometimes it is the only solution allowing the use of Bayesian MLPs for both CPU time and purely numerical reasons.

2.2 Distributed versions of the standard and cheap versions

It can be reasonable to think that the different weights do not have the same role in a MLP. For that reason MacKay has written other versions of formulas (3) and (4). We suppose now that we have different sets of weights: for instance, the weights between each couple of connected layers and, for each layer, the weights between it and the bias unit. Let us write the formulas for such a set c of k^c weights. In the *distributed standard version*, for the given set of k^c weights, the values of $\gamma^{t+1,c}$ and $\alpha^{t+1,c}$ are:

$$\gamma^{t+1,c} = k - \sum_{a=1}^{k^c} \frac{\alpha^{t,c}}{\lambda_a^c + \alpha^{t,c}} \quad \alpha^{t+1,c} = \frac{\gamma^{t+1,c}}{2E_W^{t+1,c}} \quad \text{being} \quad E_W^{t+1,c} = \frac{1}{2}\sum_{j=1}^{k^c} w_j^2 \quad (5)$$

where the λ_a^c are the eigenvalues of the Hessian matrix of $\beta^t E_D^{t+1}$, computed using weights W^{t+1} and associated to the k^c weights considered. But the formula for computing β^{t+1} is still the same and we thus still have the same β^{t+1} for all the weights. Likewise, in the *distributed quick and dirty version* the formula for computing β^{t+1} is not changed and we have

$$\alpha^{t+1,c} = \frac{k^c}{2E_W^{t+1,c}} \tag{6}$$

3 Tests

We are going, on several classification problems, to tune several MLPs architectures using: 1) the "classical" on-line back-propagation algorithm in which weights are updated after each presentation of a learning example; 2) a conjugate gradient back-propagation (Moller, 1993); 3) various versions of MacKay's Bayesian techniques as described in Sections 2.1 and 2.2.

For each net tuned using methods (1) or (2), we choose the "best" weights configuration by estimating its predictive error on test set. We will compute at some steps t in the learning process (3), the MLPs evidence: we expect the "best" net to have the largest evidence value. Nets obtained with methods (1), (2) or (3) are then a-posteriori compared using the test set. The results obtained by the MLPs using either of methods (1), (2) or (3) will be compared to those obtained by other researchers on the same databases.

3.1 Waveform, Cancer and Satellite classification problems

'Waveform' is a three-class problem based on the three waveforms h_1, h_2, h_3 pictured in Figure 1 (Breiman et al., 1984). Let u be a uniform random number and ϵ_i normally distributed with zero mean and variance 1. Vector x in class j will be generated as: $x[i] = u\, h_m[i] + (1 - u)\, h_n[i] + \epsilon_i$ where $m, n \in \{1, 2, 3\}$. Class 1, class 2 and class 3 are respectively combinations of (h_1, h_2), (h_1, h_3) and (h_2, h_3). As in Breiman et al. (1984), the learning set is made respectively of 100, 85 and 115 measurements vectors for classes 1, 2 and 3. Five thousands test vectors have been independently generated with equal proportions for the

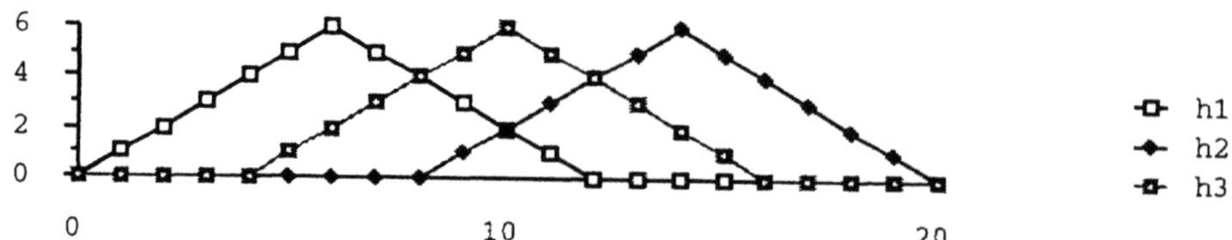

Figure 1: Waveforms.

Theoret. Bayes	DA	LVQ	MLP (1)	MLP (2)	Bayesian MLP
86%	74%	84.6%	81.6%	82%	84.3%

Table 1: Optimal recognition rates on test set in the waveforms problem. Confidence interval = 1.73%.

three classes. An analytical expression can be derived for the Bayes error rate. Using this rule on a test sample of size 5000 gives a recognition rate of 86%.

The 'Cancer' problem refer to medical diagnosis applied to breast cytology. It is a two classes problem: benign or malignant based on cell descriptions. Each example is a real vector of size 9. There are 699 examples in which the 350 first are used as learning set. This database was originally obtained from the University of Wisconsin Hospitals, Madison from Dr. W.H. Wolberg. See Prechelt (1994) for more information.

The 'Satellite' database was originally generated from data purchased from NASA by the Australian Centre for Remote Sensing and used for research at the University of South Wales. It is a 6 classes problem (red soil, cotton crop, vegetation stubble, grey soil, damp grey soil and very damp grey soil). The database is a sub-area of a scene, consisting of 82×100 pixels, each pixel covering an area of 80×80 meters. The information for each pixel consists of its class and the intensities in four spectral bands of the pixel and of the 8 other pixels contained in its 3×3 neighbourhood. Thus, the total information for each pixel is its class and 36 real values. The examples are randomised, so that it is impossible to reconstruct the image of the area. See also Michie et al. (1994) for more information.

3.2 Overview of the results and some remarks on Bayesian MLPs

For each method, we have chosen the "best" case. For each classification task, we give the confidence interval or uncertainty on the performances obtained, as given by the Hoeffding formula. In the tables below, 1NN means the one nearest-neighbour classifier and DA is the Discriminant Analysis method. MLPs (1) are tuned using stochastic back-propagation and MLPs (2) are tuned using conjugate gradient. The architectures and initial weights of the MLPs methods are always the same.

In the waveforms problem (Table 1), the "best" Bayesian MLP is tuned using the distributed standard version (formulas (5)). For the cancer problem (Table 2), the "best" Bayesian MLP is tuned using the standard (non dis-

1NN	LVQ	MLP (1)	MLP (2)	Bayesian MLP
97.8%	97.9%	98.5%	98.8%	98.8%

Table 2: Optimal recognition rates on test set in the cancer problem. Uncertainty = 6.5%.

1NN	DA	LVQ	MLP (1)	MLP (2)	Bayesian MLP
90.6%	72.9%	89.5%	89.1%	89.1%	89.1%

Table 3: Optimal recognition rates on test set in the satellite problem. Uncertainty = 2.74%.

tributed) version (formulas (3)). In the satellite problem (Table 3), the "best" Bayesian MLP is tuned using the distributed cheap version (formula (6)).

The first conclusion is that, in the three classification problems, the different classifiers have similar performances on test set, except Discriminant Analysis on problems 1 & 3 and MLPs (stochastic and conjugate gradient) on problem 1. Moreover, it is easy to get very good results with very simple methods such as 1NN and LVQ. Thus, at first sight, it is not so obvious that using the Bayesian MLPs is very interesting if the only criteria for choosing a classifier is its performances (expected or measured), since tuning a Bayesian MLP is sometimes very complex compared to the nearest-neighbours methods. We could also get the same conclusion when comparing MLPs to other methods such as CART, ID3 or others (Michie et al., 1994).

But, normally, specific methods are needed for tuning the parameters of classifiers (e.g. cross-validation, bootstrap, validation set) and estimating their predictive capabilities. Using all the training set for inferring the parameters and using the test set for ranking the methods or stopping the iterative process as we have done here (or as others have done, Michie et al., 1994) for all the methods except the Bayesian MLPs is not accurate. Thus, in a certain sense, comparing Bayesian MLPs to other methods as we have done it is "biased".

3.2.1 Overfitting

The first observation is that overfitting never occurred in our experiments using Bayesian MLPs while it could occur when using stochastic or conjugate gradient. In a certain sense it is already a success, but there still are a lot of unsolved problems.

3.2.2 When to begin to use Bayesian Learning?

Using Bayesian learning requires that some assumptions are true. The main one is that the posterior on weights must be locally Gaussian and sharp-peaked. A learning procedure always starts using random weights (here taken so as to be in the linear part of the sigmoid). Of course, it is not reasonable to assume the posterior distribution on weights to be sharp-peaked around these random weights. Thus, we must first use a stochastic or conjugate gradient (in our case the conjugate gradient) version to first reach a "good" weight configuration where to begin the learning Bayesian process. If it is not difficult in the two first

Performances	78.2%	84.7%	84%	84.4%	84.3%
Quadratic Error E_D	3762	2843	2826	2830	2803
Log of Evidence	-545.42	-496.45	-492.61	-491.84	-491.67

Table 4: Log of Evidence function of Performances and Quadratic Errors on test set.

classification problems, in the third one, it is not so clear. In the satellite images problem, if the Bayesian process begins "too early" before having reached a "good" weight configuration, the MLP will not overfit but in the same time will never reach an acceptable solution. If the Bayesian learning begins "too late" at a point where (having used the conjugate gradient) the MLP has begin to overfit, the Bayesian MLP will not be able to reach an acceptable solution. In other words, Bayesian learning must start when the conjugate gradient has found a solution not too far from its best possible configuration. Then, it is observed that the Bayesian MLP can improve a little bit the performances of the MLP while not being able to overfit. Then, this suggests to use the Bayesian MLPs as follows: a) use the conjugate gradient algorithm for various iterations steps $t_1, t_2, \ldots$; b) for each t_i start a Bayesian learning procedure; c) at various steps of these Bayesian procedures, compute the MLP evidence; d) at the end of the computation, choose the configuration for which the evidence is maximum. It seemed to us that, at least in our experiments, this methodology worked, but we must be sure that the ability of a Bayesian MLP to have good predictive ability is a monotonic function of the evidence.

3.2.3 Evidence and predictive ability

Since there is, for the moment, no theoretical links between predictive ability (or generalisation accuracy) and evidence, we are here just able to report the experimental results obtained without any possibility of generalisation of it. Table 4, shows for a MLP architecture in the waveforms problems, the values of performances, quadratic errors ED and value of log-evidence at different steps of a Bayesian procedure. We chose the fifth configuration (Log of Evidence = -491.67) as the "best" MLP. We can see that there is a big difference between the first column of results and the others. It suggests that the evidence as a method for ranking the different configurations "works". Otherwise, for the other columns, the values are very close. We have observed the same results on other configurations of the architectures for each classification problem.

3.2.4 Choosing among the Bayesian versions

In Paragraph 3.2, we have seen that various Bayesian procedures have been used for the three classification problems. Why? In fact, it is only due to very practical computation or (unsolved) numerical problems. The waveforms problem was the easiest one and any version worked very well. A very simple network containing less than 100 weights was enough. Even if there were less data, less classes and better performances on test set, the MLPs were more difficult to tune and bigger (150 weights) on the cancer problem. Moreover, we got numerical problems. The routines in Press et al. (1992) for computing the eigenvalues were not able to loop in some cases when we used the standard distributed Bayesian version. The satellite problem was by far the most difficult to process and a network containing hundreds of weights was needed. In that case, it was not CPU-timed reasonable to compute the Hessian matrix. Computing such a matrix can take 24h on a SPARC10 and moreover, as in the case of the cancer problem, many other numerical problems occurred. For that reason, the only possibility was to use the distributed Quick and Dirty Bayesian version.

4 Conclusion

We have tried to experiment Mackay's Bayesian methods for tuning MLPs. We have observed that the performances obtained were equivalent to those obtained with other methods but we pointed out that our methodology for comparing the different classifiers favoured the other methods (see Section 3). Work has to be done to find a good methodology for comparing methods. We also remarked that the Bayesian MLPs never overfit. Another point is that Bayesian methods allow to compare different MLPs (tuned using Bayes). But some numerical points are not clear and the Bayesian method cannot always be applied because of some CPU time or numerical problems.

We must say that we only applied a limited part of the Bayesian formalism described in MacKay's work (in particular we did not use his results on error-bars and his specific work on classifiers). We also did not use the Occam razor aspect which will be the subject of our future work.

References

[1] J. Bernardo and A. Smith. *Bayesian Theory*. John Wiley ed., 1994.

[2] L. Breiman, J.H. Friedman, R.A. Olshen, and Stone C.J. *Classification and Regression Trees*. Wadsworth International Group, 1984.

[3] D.J.C. MacKay. "A Pratical Bayesian Framework for Backprop Networks". *Neural Computation*, 4:448–472, 1992.

[4] D.J.C. MacKay. "Bayesian Interpolation". *Neural Computation*, 4:415–447, 1992.

[5] D. Michie, D.J. Spiegelhalter, and C.C. Taylor. *Machine Learning, Neural and Statictical Classification*. Ellis Horwood, 1994.

[6] M. Moller. "A Scaled Conjugate Gradient Algorithm for Fast Supervised Learning". *Neural Networks*, 6(4), 1993.

[7] L. Prechelt. "Proben1 - A set of Neural Networks Benchmark problems and Benchmarking Rules". Technical report, n. 21/94, Univ. Karlsruher, 1994.

[8] H. Press, S. Teukolsky, W. Vetterling, and B. Flannery. *Numerical Recipes in C*. Cambridge Univ. Press, 1992.

Constructive Fuzzy Neural Networks

F.M. Frattale Mascioli, G. Martinelli, & G.M. Varazi
INFO-COM Dpt., University of Rome "La Sapienza"
Via Eudossiana 18, 00184 Rome, Italy

Abstract

Fuzzy neural networks are a very important and active research topic. In the present paper the ANFIS net, which is a powerful fuzzy neural network, and the Min-Max net, which can be determined under a constructive approach, are merged with the aim of obtaining an original constructive fuzzy paradigm. ANFIS net is very powerful when the IF-THEN rules, characterizing the context to be fuzzy modeled, are well determined. Min-Max net can easily yield an optimized set of these rules. Therefore, the resulting fuzzy paradigm is a very promising tool. To illustrate the behavior of the proposed method, we applied it to the prediction of the chaotic series of Mackey-Glass.

1   Introduction

Fuzzy neural networks are a very promising tool, since they imitate human behavior from different and complementary levels. Their potentiality is well documented in Jang and Sun [1]. The ANFIS net, there discussed, is very powerful and impressive results are quoted with respect to applications, as for instance chaotic series prediction. Further improvements are possible. In fact, the number and locations of the rules characterizing the context to be fuzzy modeled are not simply determined. Their determination can be conveniently guided by the basic Learning Theory by recurring to a constructive approach.

In the present paper, we propose to merge the ANFIS net with another type of fuzzy neural network, the Min-Max net [2], with the aim of obtaining a constructive fuzzy paradigm. The Min-Max net is easily determined under a constructive approach and yields an optimized set of rules. ANFIS net is very powerful when the rules are well determined. Consequently, the resulting fuzzy paradigm is very efficient, as will be illustrated in Sect. 3. In Sect. 2 we briefly describe the two fuzzy nets to be merged and the resulting algorithm.

2   The proposed fuzzy algorithm

In order to understand the proposed method, we shortly describe the two algorithms we intend to merge to obtain the net with their optimal performances. As said, the Min-Max net yields an optimal determination of the rules of the fuzzy model of the context of interest. We consider a constructive version of the clustering type [2]. The ANFIS works very well when the rules are correctly determined.

2.1   Min-Max Net

The net operation is characterized by a suitable mechanism of partitioning the input space. The regions corresponding to the rules are covered by hyperboxes parallel to the coordinate axes; with appropriate membership functions defined for them. The location of each hyperbox is completely defined by two extreme vertices: the 'min' and 'max' vertices. The fuzzy Min-Max algorithm consists in determining the hyperboxes necessary to cover the said regions. It is characterized by a three step process:

1) <u>Expansion:</u> in this step the hyperboxes already constructed are expanded in order to accomodate a new example of the training set. The expansion is limited by a maximum dimension, represented by a parameter θ to be chosen carefully;

2) <u>Overlap test:</u> determination of overlaps among hyperboxes of different rules;

3) <u>Contraction:</u> elimination of overlaps, if they exist.

The complexity of the resulting Min-Max net is directly defined by the final number of hyberboxes. In fact, the net is constituted by an output layer containing a neuron per hyperbox. Each output neuron implements a hyperbox, corresponding to a rule; a parameter γ regulates how fast the membership values decrease outside a hyperbox.

2.2   ANFIS Net

The ANFIS (adaptive neuro-fuzzy inference system) net is based on the fuzzy Sugeno model. It is cheracterized by fuzzy inputs and deterministic outputs. This latter characteristic simplifies the use of the net, since the expensive defuzzyfication is not required. In particular, the attention is limited to Sugeno model of order 1. This model requires to know the regions of the input space where the IF-THEN rules are applied with a proper membership function. We use at this regard the following function without the conjunctive operator:

$$\mu_k(X) = \frac{1}{1 + \frac{\|X - C_k\|^{2b_k}}{a_k^2}} \qquad\qquad k = 1,2,...R \qquad (1)$$

where C_k is the center of region k related to the k-th rule; $X = [x_1, x_2, ..., x_n]^t$ is the input vector; a_k and b_k characterize the shape of the function and how it decreases outside the region. The k-th rule is characterized by:

$$z_k = r_k + \sum_{h=1}^{N} p_{kh} x_h \qquad (2)$$

and the output of the net is:

$$z = \sum_{k=1}^{R} \overline{W}_k z_k \; ; \qquad\qquad \overline{W}_k = \frac{W_k}{\sum_{i=1}^{R} W_i} . \qquad (3)$$

The scheme of the net is shown in Fig. 1. It is characterized by the following sets of parameters:

1) R centers C_k ; a_k, b_k: $\qquad (N+2)R$ parameters;

2) r_k, p_{kh}: $\qquad (N+1)R$ parameters.

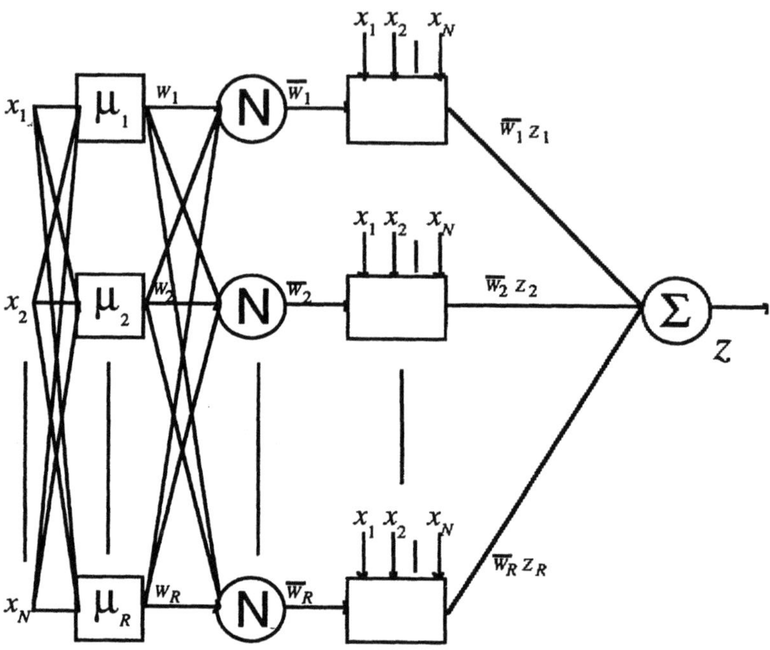

Fig.1: Scheme of the proposed fuzzy neural network.

2.3   Constructive Procedure

On the basis of the performances of the two nets, we propose to determine the rules by the constructive version of the Min-Max algorithm. Then, we determine the two sets of parameters of the ANFIS net by the following iterative procedure, based on a gradient descent technique. The procedure consists of two steps in each iteration:

<u>Step 1:</u> determine set 2) assuming set 1) to be known. Since z is linear in function of the input, this step can be explicitly solved if a quadratic error function is used;

<u>Step 2:</u> determine set 1) assuming set 2) to be known. The inizialization of the procedure is based on the results obtained by applying the Min-Max algorithm.

3   Simulation Results

In order to illustrate the performance of the proposed net, we applied it to the prediction of the chaotic series of Mackey-Glass, which is a classic benchmark. The experiment is carried out as in Jang and Sun [1] to which we refer. In particular, four values of the series are used to predict the successive one. The training set is constituted by 500 samples of the series and the test set by the successive 500. It is important to note that the ANFIS net in the version proposed in Jang and Sun [1] with 104 parameters already overcomes other approaches, as shown in Tab. 1 quoted from Jang and Sun [1].

Method	Number of epochs	Error (NDEI)
ANFIS net	500	0.007
AR model	500	0.19
Cascade-Correlation NN	500	0.06
Back-Prop NN	500	0.02
6th-order Polynomial	500	0.04
Linear Predictive Method	2000	0.55

Tab. 1: Comparison among the considered algorithms.

The procedure requires to choose some parameters. They were fixed after an exhaustive search. The main choices are as follows:
1) Values of θ and γ of Min-Max: they are determined by the constructive approach applied to Min-Max net. In Fig. 2 we show the number of hyperboxes in function of θ.

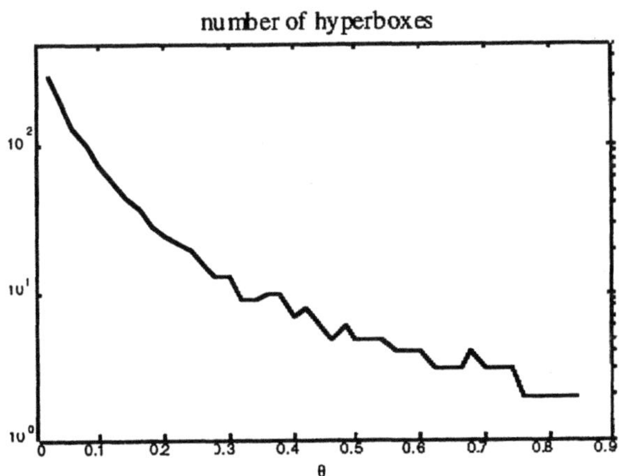

Fig. 2: Number of hyperboxes vs. θ.

2) Initial values of C_k, a_k and b_k of region k: C_k is the center of the hyperbox k. The values a_k and b_k are chosen so that the slope p takes on a suitable value. In Fig. 3, the shape corresponding to different values of p is shown in a mono-dimensional case.

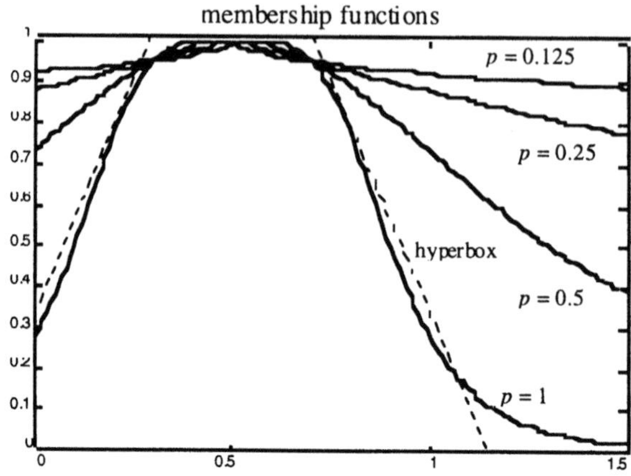

Fig. 3: Membership functions vs. p.

3) The correction rate k of the gradient descent strongly affects the minimization process. Fig. 4 shows the learning curve in function of k. We see that values in the range 0.05÷0.1 are the more appropriate. Moreover, it is convenient to vary its value during learning. We have followed the simple rules:

a) if the function to be minimized is successively reduced four times, we increase k by 10%;

b) if the function decreas and increases two successive times, then k is reduced by 10%.

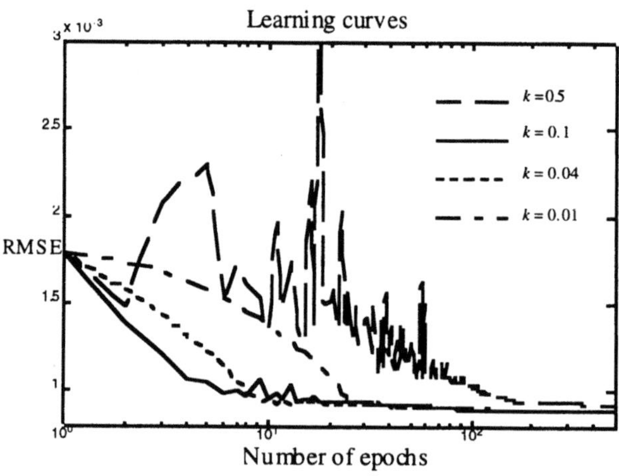

Fig.4: Effect of the initial learning rate.

4) The performance of the net is measured by the NDEI (nondimensional error index), defined as the root mean square error (RMSE) divided by the standard deviation of the target series. We will consider the NDEI both for the training set and the test set.

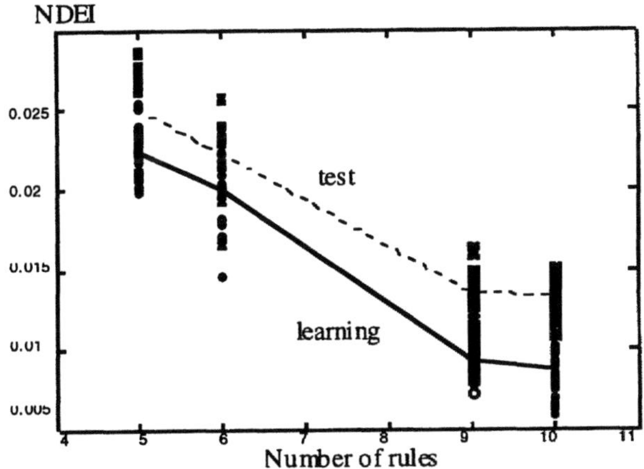

Fig.5: Net performance vs. number of rules.

The results of several experiments are summarized in Fig. 5. The conclusion of our investigation is that the performance of our approach is similar to the original ANFIS net when similar number of parameters are used. However, our approach is much rapid, in the sense that it starts from an error which is much smaller than that obtained without the rules determined by the Min-Max net. The initial advantage is however compensated with the number of iterations. This behavior is clearly shown in Fig. 6. Consequently, our approach is particularly suited to on-line processing.

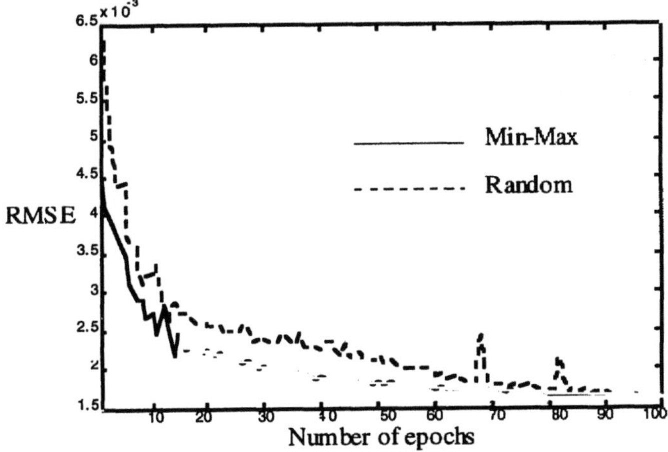

Fig. 6: Random and Min-Max initializations.

4   References

1. Jang J.S.R. and Sun C.T. Neuro-fuzzy modelling and control. Proc. of IEEE 1995. 83: 378-406.

2. Simpson P.K. Fuzzy min-max neural networks - Part 2: clustering. IEEE Trans. on Fuzzy Systems 1993. 1: 32-45.

An Adaptable Boolean Neural Net Trainable to Comment on its own Innerworkings

F. E. Lauria, M. Sette, S. Visco, M. Milo, R. Prevete

Dipartimento Scienze Fisiche, Universita' di Napoli

Mostra d' Oltremare Pad. 19

I-80125 NAPLES, Italy

INFM

e-mail: lauria@na.infn.it

Abstract

To train a Boolean neural network to comment on the progress of the task it is controlling, we have implemented in the network an Hebbian rule with the help of the node assemblies. Actually, our network can comment also on its communications with its environment without any danger to enter in a loop-like endless activity.

1 Introduction

Let us consider a Boolean, ie, a McCulloch and Pitts net, see [1], as formalized by Caianiello, see [2], organized in *assemblies*, see later on, and an Hebbian rule. We discuss how to train the net to comment, in response to our queries, on its control of the task in execution. As is well known, see [3] pag. 87, the problem of the Hebbian rule implementation in a Boolean network has not been solved in its generality. The difficulties depend on the possibility, first, to jumble the previously stored tasks. Second, when the net is controlling a sequence of elementary actions, to trample with the latest on the leading one. And, eventually, to enter in an endless activity whenever the same task has to be executed more than once. As discussed in [4, 5, 6], a net organized in node *assemblies*, corresponding to the *DIDs* in [4], consent us to find a general solution to this kind of problems. Moreover, thanks to them the net is trainable to communicate either its thought when controlling a task or its own thought on the same description, or both.

2 The neural network and the hebbian rule

Our neural network contains N nodes evaluating, synchronously, their input signals with a delay of one tau. The node k is connected to the node h by variously delayed monodirectional connections. The coupling coefficient $C[h, k, l]$, or C, is associated to an *excitatory* connection when positive, to an *inhibitory* when negative. At the time t the node h, as well as its output connection, is in one of two internal states, formally represented by the threshold function $f[h, t]$. To compute the latter, we confront the product of the state vector times the coupling coefficient matrix with the node threshold. If their difference is non negative, then, $f[h,t] = 1$, ie, the node is *excited*. It is zero otherwise. A *firing connection*, an *absolute inhibitory connection*, is a connection associated to a, negative, coupling coeficient greater than the node threshold, in absolute value. We have presented in [7, 8, 9, 10] the low-level language associated to the network, actually a propositional calculus dialect, a proof of its universality, the interpretation of the programs, in terms of the network coupling coefficient values, and an assembler-like algorithm associating the user mnemonics to the node identifyers, with user chosen redundancy to obtain a fault tolerant network. Given $M > 0$, an *Hebbian connection* is a connection with associated a variable, in the time domain, coupling coefficient $C[h, k, l] > 0$ computed every tau by applying the *Hebbian rule*, see [4, 5, 6]:

When $C[h, k, l] < M$ it is increased if $f[h,t] = f[k, t - l] = 1$ it is decreased of a small amount if $f[h,t] = 0$ it is decreased of a greater amount if $f[h,t] = 1 - f[k, t - l] = 1$. Oherwise $C[h, k, l]$ is left unchanged and it is said to be frozen.

Because of the i-assemblies, see below and the discussion in [4], it is useless to extend the rule to the negative Cs. In the following we call *Hebbian* a node receiving at least one Hebbian connection.

3   The node assemblies

For simplicity sake, from now on we consider nets without redundancy: easily, see [10], we can obtain a fault tolerant network from the result here presented. A node assembly, see also [4, 11, 12, 13, 14, 15] and the 1, is a subset of the network nodes with the following properties: first, it contains just one Hebbian node, the *assembly Hebbian node* or the *assembly standard Hebbian node*, receiving also one firing connection, at least. Second, it contains two output nodes, the assembly first and second output, generating either Hebbian connections, if they impinge onto another assembly Hebbian node, or the network output connections, thus impinging onto the *primitives*. By construction, an assembly output nodes are active in the given order. Third, in the assembly A the remaining node activity is a temporary pointer to the, different, assembly containing the Hebbian node activated by the signal last generated in the very same assembly A output nodes. By construction, once the training is terminated and the Cs are frozen, an assembly output node Hebbian connection associated to a frozen C impinges, at the most, on the Hebbian node of just one assembly. As discussed in [4, 5, 14, 15], as the activity in the network output nodes is the command to execute a task, each output assembly activity controls the execution of a task. If the assembly Hebbian node receives Hebbian connections with an associated M value greater or equal, smaller, than its threshold, we call it a v-assembly, an &-assembly. An assembly can be both a v-assembly and an &-assembly if its Hebbian node receives more Hebbian connections, some with M greater or equal and the remaining with M smaller than its Hebbian node threshold. An i-assembly is an assembly with an additional Hebbian node, the i-node, receiving an Hebbian and a firing input connection, at least, and sending an absolute inhibitory connection to the assembly standard Hebbian node. If an i-node receives more Hebbian connections, they can be chosen with M values some equal or greater and some smaller than its threshold. Consequently, an assembly can be trained to compute conjunctions as well as inclusive disjunctions of inhibitory and, separately, of excitatory inputs. That is, the assemblies can be trained to behave as NAND or NOR gates. Assemblies of a special kind, the f-assemblies, can be defined in a net with two input and two output connections. The net, and f-assemblies, inputs can be thought of as the hearing and the seeing sensory organ outputs, respectively . The net, and f-assemblies, outputs as the speaking and the writing motor effector inputs, respectively. An f-assembly is indipendently fired by each of the two input connections and receives inter-assembly connections, with M values greater than their Hebbian node firing threshold. Each f-assembly has a third Hebbian output connection stemming out of one of its nodes, the *third output node* or *third node*. This node is active whenever the assembly is controlling the activity in the subtree having the very same assembly as root. By training, the third output node is inhibited by the activity of the same f-assembly father. That is, any f-assembly is trained to inhibit its son third output node. Thus, once the net has been trained, whenever the third node output connection of an f-assembly is active, no one of its sons can have an active third output node. Eventually, only the assembly at the subtree root has an active third node output connection. We have simulated a six hundred node net, containing sixteen assemblies and up to thirteen primitives: it has been successfully trained to control a robot, first, in drawing some digits and letters as sequences of horizontal and vertical unitary bars and, second, in executing some elementary arithmetic operations on them, see [16, 17].

4   On the arousal of attention

Let us consider the partially ordered subset containing the &-assemblies with the following properties. First, the M and the coupling coefficient values, associated to the net inputs impinging on the subset &-assemblies, are smaller than the node thresholds. Second, the higher elements, with respect to the subset order, send Hebbian connections to the lower elements, or to the assemblies belonging to the subset complement or to both. But they do not send, receive, Hebbian connections to, from, the subset elements having no relation with them. In other words, it is impossible to train the net on loops containing, exclusively, subset elements. As, by definition, there is a direction associated to the Hebbian connections, it is possible to send, in bursts, wawes of activity, the *arousal of attention*, in the opposite direction, without any danger of unwanted reinforcements. The arousal of attention can be realized by means of non-Hebbian inter-assembly connections. Their number should be, at least, one order of magnitude greater and their coupling coefficient values should be, at least, one order of magnitude smaller than an assembly standard Hebbian node firing threshold. That is, the complessive value of the arousal of attention coupling coefficients plus any pair of unreinforced Hebbian connection values should be just above the node threshold. And they should carry batch of signals in the direction opposite to the one common to the Hebbian connections, with a repetition cadence one order of magnitude greater than the inter-assembly connection maximum delay. To avoid interferences with the activity going on

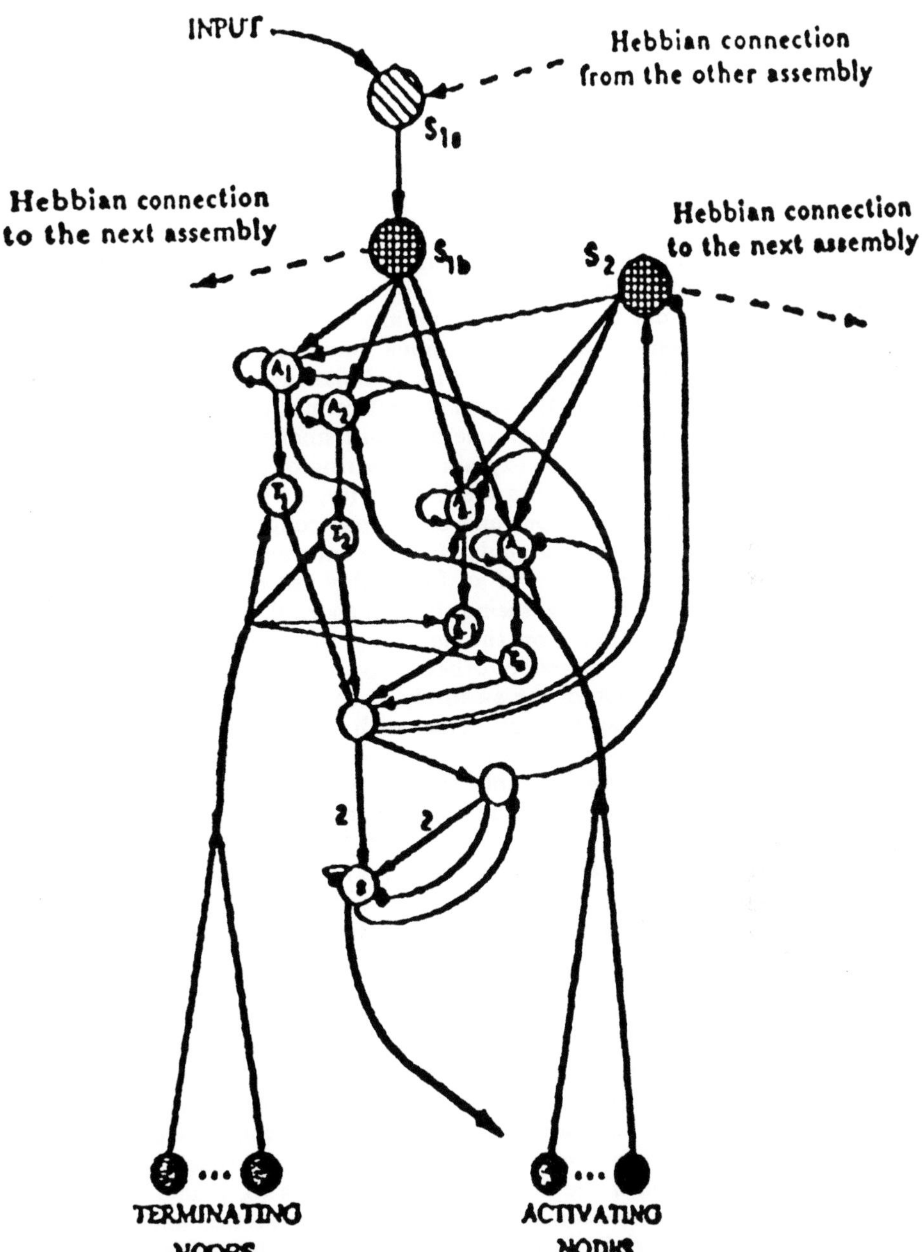

Figure 1: *The neural assembly. From the top: the input to the assembly; it receives also Hebbian connections from the other assemblies. The nodes S_{1b} and S_2 transmit Hebbian connections to the other assemblies. The $A_1, \ldots, A_n$ nodes activate the assembly. The $T_1, \ldots, T_n$ nodes terminate the very same assembly activity.*

elsewhere in the net, the inter-assembly lattice carrying the arousal of attention should make inhibitory contacts on, at least, some of the assemblies belonging to the subset complement.

5   Training the net to comment its own operations

To train the net we need both the f-assemblies and the previously introduced partially ordered subset of &-assemblies, for short the *new assemblies*. That is, some elements of our net can receive an arousal of attention and their connections, with the sensory input connections, have coupling coefficient values greater than one half and smaller than its Hebbian node threshold. The new assembly remaining inputs are the third output connections of the f-assemblies, each with an M value greater than one half and smaller than the new assembly Hebbian node threshold. To explain the wiring of the new assemblies, let us suppose that some f-assemblies are in activity, e. g. in control of a task execution. That is, both an input and an output connection are supposed to be busy. Just one of the new assembly Hebbian connections is the active third output connection stemming out of an f-assembly not engaged with the messages coming out of the hearing, seeing, busy sensor. E. g., the Hebbian signal could be elicited by an input, on the non-busy f-assembly, of the type *what are you doing?*. The new assembly other active input Hebbian connection is the third output connection of the assembly at the root of the tree controlling the writing, speaking, busy effector. Thus with the help of the arousal of attention, of the sub-threshold sensory inputs and of the signals flowing on its Hebbian connections, a new assembly can learn to generate an output of the type *I am executing T*, where T is the name of the task currently controlled by the busy f-assembly. In the next but one paragraph, we discuss how to send a message either on the busy or on the non-busy connection. The wiring here discussed requires as many new assemblies as different T tasks the net has to be trained to nominate, also if they are executed one at the time. Warning: it is better to teach the task and the output message generated by the delayed firing command first, then to teach the auto-comments. Up to now, we have trained the net on a conditional like: A), *if the non-busy and the busy f-assemblies are active on different I/O connections, then a new assembly communicates something, on the non-busy connection, about the busy f-assembly activity.* Now, let 1-assembly and 2-assembly be two new assemblies, ie, two assemblies belonging to the partially ordered subset. From now on, a message flowing on an input connection is said congruent, non-congruent, indifferent, to the carrying connection if the message content points to, points to, does not point to, the output connection and if the carrying connection and the pointed to connection are either the hearing and speaking or the seeing and writing, either the hearing and writing or the seeing and speaking, not mentioned. To a congruent, non-congruent, message is associated a signal on a congruent, non-congruent connection. Indifferent messages or absence of messages on the input connections are associated to the signal absence on both the congruent and the non-congruent connections. Now, we discuss the net training on the two conditionals:

1), *if the non-busy f-assembly receives a congruent message on the hearing,*

*seeing, connection and the busy f-assembly receives the messages on the other
connection, then the 1-assembly, 2-assembly, communicates something, on the
speaking, writing, connection, about the busy f-assembly activity,*
2), *if the non-busy f-assembly receives a non-congruent message on the hearing, seeing, connection and the busy f-assembly receives messages on the other
connection, then the 1-assembly, 2-assembly, communicates something, e. g.
during a pause on the busy connection, about the 2-assembly, 1-assembly, activity.* Obviously, if we send a congruent followed by a non-congruent message to
a non-busy f-assembly, then we shall have messages, on both the non-busy and
the busy output connections, telling us something about the busy f-assembly
and, depending on the connections carrying the input messages, about the 1-
assembly, or the 2-assembly, activity. It is easy to obtain a net trainable on 1),
respectively 2), once we know how to implement A). The former is obtained
by substituting, in the latter, the new assembly with a pair of assembly trees
rooted in the 1-assembly, respectively in the 2-assembly. The 1-assembly and
2-assembly generic connections are the non-Hebbian arousal of attention and
the Hebbian f-assembly third output connections. To learn how to discriminate between hearing and seeing messages, the 1-assemblies, respectively the
2-assemblies, receive the hearing, respectively the seeing, input connection as
specific Hebbian connections. As before, the input connection M values are
greater than one half and smaller than the respective assembly Hebbian node
threshold. To train the net on the conditional discriminating between congruent and non-congruent messages, the first output connection of both the
1-assembly and the 2-assembly must have two sons belonging to the partially
ordered subset and receiving, as generic non-Hebbian connections, the arousal
of attention and the same subthreshold sensory input which is the specific connection of their fathers. Instead, as specific Hebbian input, the 1-assembly,
2-assembly, first son receives a connection, which is active when the hearing,
seeing, connection carries a congruent message. Instead, on the 1-assembly, 2-
assembly, second son impinges a connection, which is active when the hearing,
seeing, connection carries a non- congruent message. The first son has no other
input connections and, in A), its role is played by the new-assembly at the root
of the tree. Instead, the second son receives an Hebbian connection from every
&-assembly of the subset. In this way, the sons of the 1-assembly, 2-assembly,
second son can be trained to generate an output, eventually during the busy
output connection pauses, telling us something on the 2-assembly, 1-assembly,
activity. At the most, the number of the 1-assemblies and, separately, of the
2-assemblies is equal to the number of the f-assemblies which are the root of a
task controlling tree, ie, it is equal to the total number of tasks the net has been
or can be trained on. From a semantic point of view, the environment of the net
containing the 1-assemblies, 2-assemblies, is the net containing the f-assemblies
and the 2-assemblies, 1-assemblies. That is, we can train the system so that a
query triggers a response saying something about the very same system rumi-
nations. We have trained the simulated Boolean net, see 3., to comment, when
so queried, on its execution of the elementary arithmetic operation, see [18]. At
present we are training it to comment on its operation of a production system.

6 Acknowledgements

This research was supported in part through the MPI/INFM 40% fund and the MPI 60% fund.

References

[1] Mc Culloch, W. S. and W. Pitts - *A logical calculus of the ideas immanent in the nervous activity*, Bull. Math. Bioph. 5 (1943) 115-143.

[2] Caianiello, E. R. - *Outline of a theory of thought processes and thinking machines* J. Theor. Biol. 2 (1961) 204-235.

[3] Rumelhart, D. E., J. L. McClelland and the PDP Research Group - "Parallel distributed processing. Vol I Foundations" MIT Press, Cambridge (Mass.) (1986).

[4] Bini Verona, F., P. De Pinto and F. E. Lauria - *Toward a learning boolean net: some cybernetic rules* in R. Trappl ed. "Cybernetic and Systems '92" World Sc. Pub. Co., Singapore (1992) 709-716.

[5] Lauria, F. E. and M. Sette - *A general approach to learning of task sequences* in I. Aleksander & J. Taylor eds. "Artificial neural network, 2" Elsevier Sc. Pub., Amsterdam (1992) 475-478.

[6] Lauria, F. E. - *On some sufficient conditions for the Caianiello-Hebb transform convergence* Rend. Acc. Sc. Fis. e Mat., Napoli, Serie IV- Vol. LXII (1995) (in print).

[7] Abbruzzese, F. and F. E. Lauria - *A Mc Culloch and Pitts network as the actual control system between sensory inputs and motor outputs* Proc. of the IFAC 10th World Congress. Munich 6 (1987) 334-337.

[8] Abbruzzese, F. and F. E. Lauria - *An assembler for a neural network* M. Caudill and C. Butler eds. IEEE First Int. Conf. on Neural Networks. San Diego (CA) Vol. iv (1987) 601-608.

[9] Lauria, F. E. - *A connectionist approach to knowledge acquisition* Cyber. and Syst.: an Int. J. 19 (1988) 35-88.

[10] Lauria, F. E. - *A general purpose neural network* in O. M. Omidvar ed. "Progress in neural networks" Ablex, Norwood (NJ) (1991) 147-173.

[11] Bini Verona, F. and F. E. Lauria - *A connectionist approach to a knowledge based planning system* in E. R. Caianiello ed. "Sixth Italian workshop on parallel architectures and neural networks" World Sc. Pub. Co., Singapore (1993) 301-306.

[12] Romano, A. - *Implementazione di una regola Hebbiana in una rete neurale Booleana* Tesi di laurea in Fisica,Universita' di Napoli a. A. 1991-92.

[13] Bini Verona F., F. E. Lauria, M. Sette, S. Visco - *A cybernetic approach to universal learning* in R. Trappl ed. "Cybernetic and systems '94" World Sc. (Singapore), Vol I (1994) 777-784.

[14] Lauria, F. E. - *On the Hebb rule implementation in a Boolean neural network* Myung-Won Kim & Soo-Young Lee eds. Proc. Int. Conf. on Neural Inf. Proc. ICONIP' 94-Seoul Vol. 2 (1994) 857-864 [invited speech].

[15] Lauria, F. E., M. Sette, S. Visco - *Adaptable Boolean neural networks* (1995) 166 (submitted for publication).

[16] Prevete R. - *Addestramento di un robot controllato da una rete neurale booleana organizzata in assemblee* Tesi di laurea in Fisica,Universita' di Napoli a. A. 1993-94.

[17] Iannelli C. - *Rete neurale booleana organizzata in assemblee come controllo adattativo di un robot* Tesi di laurea in Fisica,Universita' di Napoli a. A. 1993-94.

[18] Pennino G. - *Rete neurale che impara a descrivere il proprio lavoro* Tesi di laurea in Fisica,Universita' di Napoli a. A. 1993-94.

Fast Training of Recurrent Neural Networks by the Recursive Least Squares Method

R. Parisi, E.D.Di Claudio, A. Rapagnetta and G.Orlandi

INFOCOM Dept., University "La Sapienza"

Rome, Italy

Abstract

In this work a novel approach to the training of recurrent neural nets is presented. The algorithm exploits the separability of each neuron into its linear and nonlinear part. Each iteration of the learning consists of two steps: first the descent of the error functional in the space of the linear outputs of the neurons is performed (*descent in the neuron space*); then the weights are updated by solving a linear system with a Recursive Least Squares technique.

The main properties of the new approach are high speed of convergence, favorable numerical conditioning and robustness. The numerical stability is assured by the use of robust LS linear system solvers, operating directly on the data.

1   Introduction

It is well-known from adaptive filter theory that Least Squares (LS) methods generally offer much higher convergence rates w.r.t. gradient-based approaches [1]. In order to apply LS concepts to neural nets it is necessary to construct a linear system; this can be done by exploiting the separability of each neuron into linear and nonlinear part. This idea has been already used in several learning algorithms for feedforward nets [2, 3], leading to fast and robust training algorithms; in this paper we apply this technique to recurrent neural networks. The structure we consider is constituted by a layer of totally recurrent neurons followed by a number of feedforward layers; figure 1 shows the case of two output layers. The final layer should be linear in order not to limit the range of output signals. The linear part of each layer is represented by a weight matrix $\mathbf{W} = \{w_{ij}\}$, where w_{ij} is the weight leading from the i-th input to the j-th neuron. The blocks marked with f are layers of memoryless nonlinearities; in the proposed examples we will refer to sigmoidal-type functions.

In the figure $\mathbf{x}_1(t)$ is the state vector of the system at time t while $\mathbf{y}_3(t)$ is the global output of the network. The whole structure implements the following nonlinear dynamical system:

$$\begin{cases} \mathbf{x}_1(t) & = & \phi\left[\mathbf{x}_1(t-1), \mathbf{u}(t)\right] \\ \mathbf{y}_3(t) & = & \psi\left[\mathbf{x}_1(t)\right] \end{cases} \tag{1}$$

where ϕ and ψ are the functionals modeled by the recurrent and the feedforward sections respectively; $\mathbf{u}(t)$ is a column vector containing one pattern of the

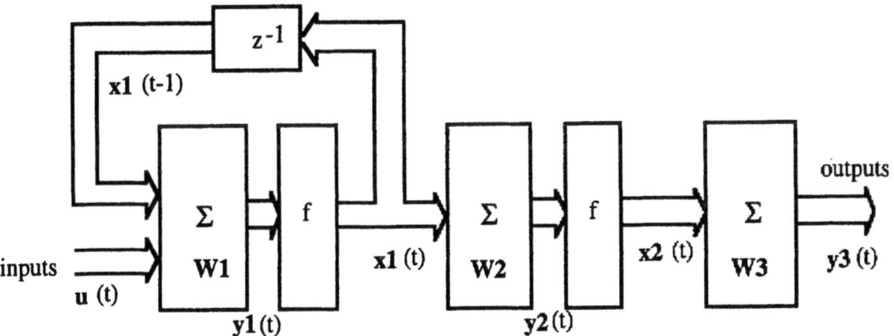

Figure 1: reference network

learning set.

The matrix formulation for the recurrent section of the net is:

$$\begin{cases} \left[\, \mathbf{x}_1^T(t-1) \quad 1 \quad \mathbf{u}^T(t)\,\right]\mathbf{W}_1 &=& \mathbf{y}_1^T(t) \\ \mathbf{x}_1^T(t) &=& f[\mathbf{y}_1^T(t)] \end{cases} \tag{2}$$

where 1 is the bias input.

The final outputs are computed from the following equations:

$$\begin{cases} \left[\, \mathbf{x}_1^T(t) \quad 1\,\right]\mathbf{W}_2 &=& \mathbf{y}_2^T(t) \\ \mathbf{x}_2^T(t) &=& f[\mathbf{y}_2^T(t)] \\ \left[\, \mathbf{x}_2^T(t) \quad 1\,\right]\mathbf{W}_3 &=& \mathbf{y}_3^T(t) \end{cases} \tag{3}$$

2 Description of the algorithm

At each time step, the actual output of the network is compared to the target value and correspondingly the error functional E is built; usually the *Mean Squared Error* (MSE) is considered. The learning process consists in the minimization of E. In the present procedure this minimization is performed at each iteration through two sequential steps:

1. descent of the error in the neuron space;

2. computation of the optimal weights by solution of a proper linear system.

The *neuron space approach* exploits the typical structure of neural networks; it is based on the descent of the error functional E (generally the MSE) in the space of the linear outputs of the neurons (the summation outputs). This is in contrast with the usual *weight space approach*, based on the descent of the error directly in the space of weights.

The neuron space approach has been successfully used for training multilayered structures [3]; its application to recurrent nets is possible using the *time unfolding* technique [4]. Figure 2 shows the net of fig. 1 unfolded in the time

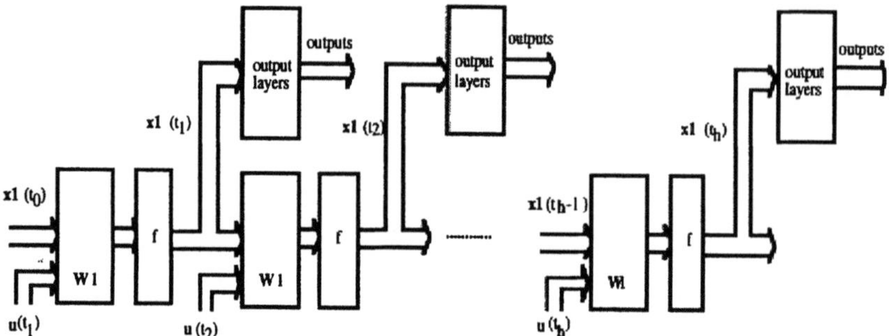

Figure 2: unfolding the network of fig.1

interval $t_1 \div t_h$.

We introduce the following matrices:

1- the matrix $\mathbf{U}$, containing the set of the input patterns between t_1 and t_h:

$$\mathbf{U}(t_1, t_h) = \begin{bmatrix} \mathbf{u}^T(t_1) \\ \mathbf{u}^T(t_2) \\ \ldots \\ \mathbf{u}^T(t_h) \end{bmatrix} \tag{4}$$

2- the matrices $\mathbf{X}_1$ and $\mathbf{Y}_1$, containing respectively the state vectors and the summation outputs for the recurrent section:

$$\mathbf{X}_1(t_0, t_{h-1}) = \begin{bmatrix} \mathbf{x}_1^T(t_0) \\ \mathbf{x}_1^T(t_1) \\ \ldots \\ \mathbf{x}_1^T(t_{h-1}) \end{bmatrix} \tag{5}$$

$$\mathbf{Y}_1(t_1, t_h) = \begin{bmatrix} \mathbf{y}_1^T(t_1) \\ \mathbf{y}_1^T(t_2) \\ \ldots \\ \mathbf{y}_1^T(t_h)) \end{bmatrix} \tag{6}$$

These matrices follow the relationship:

$$[\ \mathbf{X}_1(t_0, t_{h-1}) \quad \mathbf{1} \quad \mathbf{U}(t_1, t_h)\ ]\, \mathbf{W}_1 = \mathbf{Y}_1(t_1, t_h) \tag{7}$$

3- the matrices $\mathbf{X}_{k-1}$ and $\mathbf{Y}_k$, containing the inputs and the summation outputs of the k-th feedforward layer:

$$\mathbf{X}_{k-1}(t_1, t_h) = \begin{bmatrix} \mathbf{x}_{k-1}^T(t_1) \\ \mathbf{x}_{k-1}^T(t_2) \\ \ldots \\ \mathbf{x}_{k-1}^T(t_h) \end{bmatrix} \tag{8}$$

$$\mathbf{Y}_k(t_1, t_h) = \begin{bmatrix} \mathbf{y}_k^T(t_1) \\ \mathbf{y}_k^T(t_2) \\ \ldots \\ \mathbf{y}_k^T(t_h) \end{bmatrix} \tag{9}$$

The following relationship holds:

$$[\ \mathbf{X}_{k-1}(t_1, t_h) \quad 1 \ ]\mathbf{W}_k = \mathbf{Y}_k(t_1, t_h) \tag{10}$$

In the preceding equations 1 is a column vector containing the bias inputs.
The neuron space approach is based on the descent of the error E in the space of the summation outputs. This is expressed by the following general descent formula:

$$\hat{\mathbf{Y}}(t_1, t_h) = \mathbf{Y}(t_1, t_h) + \eta\mathbf{D}(t_1, t_h) \tag{11}$$

In (11) the desired $\mathbf{Y}$ (both for the recurrent and the feedforward sections) is estimated by using a proper direction matrix $\mathbf{D}$; η is the *step-size.*
Different choices of $\mathbf{D}$ are possible; in particular $\mathbf{D}$ can be the opposite of the *gradient matrix* $\nabla_\mathbf{Y}E$, leading in this case to a *gradient descent* in neuron space:

$$\hat{\mathbf{Y}}(t_1, t_h) = \mathbf{Y}(t_1, t_h) - \eta\nabla_\mathbf{Y}E(t_1, t_h) \tag{12}$$

In (12) the error E is evaluated over the entire interval $t_1 \div t_h$. At each step of the learning process the gradient of the error is computed and subtracted to the actual $\mathbf{Y}$, giving the estimate $\hat{\mathbf{Y}}$.
The weights of the network are computed in the following way. The weight matrix $\mathbf{W}_1$ is computed by solving the linear system:

$$[\ \mathbf{X}_1(t_0, t_{h-1}) \quad 1 \quad \mathbf{U}(t_1, t_h) \ ]\mathbf{W}_1^{new} = \hat{\mathbf{Y}}_1(t_1, t_h) \tag{13}$$

The weights of the k-th feedforward layer are computed from the system:

$$[\ \mathbf{X}_{k-1}(t_1, t_h) \quad 1 \ ]\mathbf{W}_k^{new} = \hat{\mathbf{Y}}_k(t_1, t_h) \tag{14}$$

Systems (13) and (14) are solved in the LS sense using a recursive approach based on the *QR decomposition* [1, 5]. It is important to remark that the length of each epoch (e.g. h) can be properly chosen; in particular if h=1 the algorithm is on-line, since each update is performed after the presentation of each learning pattern.
The described approach can be shown to be equivalent to a second-order learning algorithm (*modified Newtons method* [6]) in weight space [7].

3   Experimental results

The new algorithm has been tested in several system identification and time-series prediction problems.
The following example has been drawn from the classical paper by Narendra and Parthasarathy [8]. The problem is the identification of the system described by the equation:

$$y_p(k+1) = \frac{u(k)}{1 + y_p^2(k-2) + y_p^2(k-1)} +$$

$$+\frac{y_p(k)y_p(k-1)y_p(k-2)u(k-1)[y_p(k-2)-1]}{1 + y_p^2(k-2) + y_p^2(k-1)} \tag{15}$$

In the learning phase 100,000 random samples uniformly distributed in the range [-1,+1] have been presented; a network with 12 recurrent neurons, 9 feed-forward neurons and 1 output linear unit has been used. Fig. 3 shows the outputs of the network and of the system to the test input $u(k) = \sin\frac{2\pi k}{25} + \sin\frac{2\pi k}{10}$ after the training. The final MSE/MSS (ratio of MSE to the Mean Squared Signal) is also given as an index of performance; in this case MSE/MSS $= 9.5 \cdot 10^{-3}$. The algorithm has been tested also on time-series prediction problems. As an

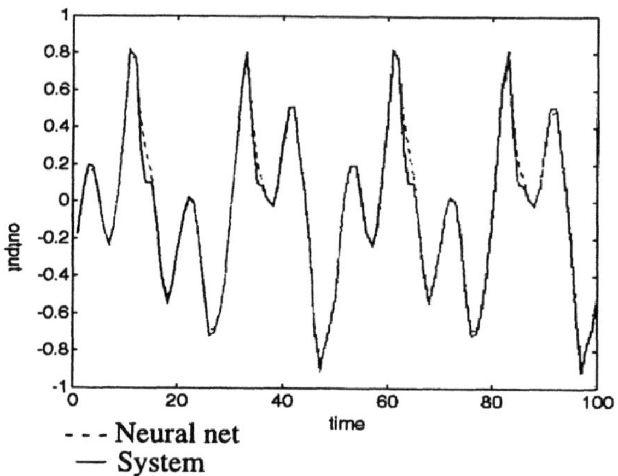

Figure 3: network's and system's response to the test signal for system of eq. (15); MSE/MSS $= 9.5 \cdot 10^{-3}$.

example, we show the results obtained on the sunspot numbers series; this is a quite typical test in prediction theory and modeling. The database is the set of monthly means of sunspot numbers recorded from 1700 to 1984; training was performed on the first 640 samples, while the remaining samples were used for testing. A net with 5 inputs (the five last samples) and 10 hidden neurons was used. Figure 4 shows a typical learning curve and the corresponding error; convergence is reached after about 40 samples.

As a conclusion, in comparison with gradient-based learning algorithms (like [9]), the new algorithm has higher computational cost per iteration (being a second-order method), but requires a reduced number of iterations and a lower CPU time to get the convergence. In addition, numerical stability and robustness are assured by working on raw data and using stable LS solvers like the QR and the Singular Value Decompositions.

Acknowledgment

This work was supported in part by Italian MURST and CNR.

References

[1] S. Haykin, *Adaptive filter theory*, Prentice Hall, 2nd edition, 1991.

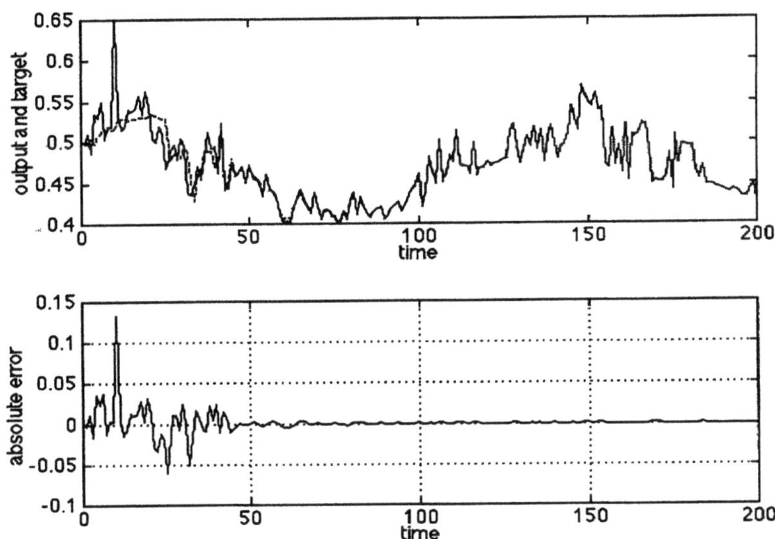

Figure 4: sunspot numbers, learning and error curves (dashed line: net's output, solid line: target)

[2] R.S. Scalero and N. Tepedelenlioglu, "A fast new algorithm for training feedforward neural networks," *IEEE Trans. on Signal Proc.*, vol. 40, no. 1, January 1992.

[3] R. Parisi, E.D. Di Claudio and G. Orlandi, "Fast learning algorithms for feedforward neural networks," *(Review paper) 1995 Italian Workshop on Neural Networks*, Vietri (Italy), May 18-20, 1995.

[4] S. Haykin, *Neural Networks - A Comprehensive Foundation*, IEEE Press, 1994.

[5] G.H.Golub, C.F.Van Loan, *Matrix computations*, The Johns Hopkins University Press, Baltimore, 1989.

[6] D.G. Luenberger, *Linear and Nonlinear Programming*, Addison Wesley, 2nd ed.,1989.

[7] R. Parisi, E.D. Di Claudio, G. Orlandi and B.D. Rao, "A generalized learning paradigm exploiting the structure of feedforward neural networks," to appear on *IEEE Trans. on Neural Networks*.

[8] K.S. Narendra and K. Parthasarathy, "Identification and control of dynamical systems using neural networks," *IEEE Trans. on Neural Networks*, Vol. 1, no. 1, March 1990.

[9] R.J. Williams and J. Peng, "An efficient gradient-based algorithm for on-line training of recurrent network trajectories," *Neural Computation*, no. 2, pp. 490-501, 1990.

AUTHOR INDEX

 MIX
Papier aus verantwortungsvollen Quellen
Paper from responsible sources
FSC® C105338
www.fsc.org

If you have any concerns about our products,
you can contact us on
ProductSafety@springernature.com

In case Publisher is established outside the EU,
the EU authorized representative is:
Springer Nature Customer Service Center GmbH
Europaplatz 3, 69115 Heidelberg, Germany

Printed by Libri Plureos GmbH
in Hamburg, Germany